The Environment in American History

From pre-European contact to the present day, people living in what is now the United States have constantly manipulated their environment. The use of natural resources—animals, plants, minerals, water, and land—has produced both prosperity and destruction, reshaping the land and human responses to it. *The Environment in American History* is a clear and comprehensive account that vividly shows students how the environment played a defining role in the development of American society.

Organized in thirteen chronological chapters, and extensively illustrated, the book covers the following themes:

- Native peoples' manipulation of the environment across various regions
- The role of Old World livestock and diseases in European conquests
- Plantation agriculture and slavery
- Westward expansion and the exploitation of natural resources
- Environmental influences on the Civil War and World War II
- The emergence and development of environmental activism
- Industrialization, and the growth of cities and suburbs
- Ecological restoration and climate change

Each chapter includes a selection of primary documents, and the book is supported by a robust companion website that provides further resources for students and instructors. Drawing on current scholarship, Jeff Crane has created a vibrant and engaging survey that is a key resource for all students of American environmental history.

Jeff Crane is Associate Dean of the College of Humanities, Arts, and Social Sciences at the University of the Incarnate Word in San Antonio, Texas. He is the author of *Finding the River: An Environmental History of the Elwha*, and co-editor of *Natural Protest: Essays on the History of American Environmentalism*.

The Environment in American History

Nature and the Formation of the United States

JEFF CRANE

NEW YORK AND LONDON

First published 2015
by Routledge
711 Third Avenue, New York, NY 10017

And in the UK
by Routledge
2 Park Square, Milton Park, Abingdon, Oxon OX14 4RN

Routledge is an imprint of the Taylor & Francis Group, an informa business

Library of Congress Cataloging-in-Publication Data
The environment in American history : nature and the formation of the
United States / Jeff Crane.
pages cm
Includes index.
1. Human ecology—United States—History. 2. Nature—Effect of human beings on—United States—History. 3. Human beings—Effect of environment on—United States—History. 4. Natural resources—United States—History. 5. United States—Environmental conditions. I. Title.
GF503.C73 2014
304.20973—dc23
2014025235

ISBN: 978-0-415-80871-2 (hbk)
ISBN: 978-0-415-80872-9 (pbk)
ISBN: 978-1-315-81732-3 (ebk)

Typeset in Dante
by Apex CoVantage, LLC

Printed and bound in the United States of America
by Edwards Brothers Malloy

To Jennine, Chloe, and Ella

Contents

Acknowledgments

There are many people that have played some role in the creation and crafting of this book. The external readers will always have my gratitude for their suggestions and critiques. This book is much stronger for their work. Several others provided valuable contributions. Former superstar students Sophia Garee-Boyles Keen and John Rhine shared useful comments, and it was enjoyable to continue my relationship with them via this book. Andrew Orr's willingness to bring his military history expertise to bear in reading my Civil War and World War II chapters was invaluable. Andrew Lopenzina also offered me beneficial guidance. The conservation chapter benefited from the suggestions of Suzanne Orr, and conversations with Adam Sowards and Michael Egan were invaluable in helping me flesh out my ideas.

Routledge editors Mary Altman, Genevieve Aoki, and Kimberly Guinta were wonderful throughout this whole process. They have shown great forbearance as I endeavored to complete this challenging task. I appreciate equally their patience as well as the steady encouragement and nudges to keep me on track. Estella Zamora has provided invaluable assistance with paperwork. Special thanks go to my boss Jack Healy. In my first year in academic administration he has allowed me the time to work on this book, has lent me his ear on numerous occasions as I tried to work my way through some ideas or intellectual thickets, and also provided me with valuable editing. For all that I will be forever grateful. Without his support I doubt I could have finished this book. I am thankful for Ella Duval's steady support and interest in my work. Finally, and always, I have to thank my wonderful wife, Jennine, and daughters, Chloe and Ella. Too many weekends and evenings have been interrupted by work or haunted by thoughts of "the book." I appreciate your continued support and patience. Without you this book would mean nothing.

Preface

Telling the story of a nation's history is a daunting task. When framing that narrative, the historian is forced to consider many factors: economics, politics, immigration, war, culture, race, diplomacy, land use, and so much more. In addition to the factors are the people. *European Americans* is used as a blanket term, but within that you have Irish, Scots-Irish, English, Germans, Eastern Europeans, and many others. African-Americans, Asians, Hispanics, and Native Americans all have their part in the epic tale of America as well. Get that all sorted out, flesh out a rough outline, and then there is the periodization of the past. Which eras deserve the most attention; which less? While moving forward chronologically, what stories or events deserve deeper analysis? Finally, how does a particular historian interpret the process of change and convey that to the audience? In some sense then, writing an environmental history of America should be fairly straightforward, seeming to exclude many of the issues just listed. This turns out to not be true at all. So much of American history can be seen through the prism of environment and ecology that it is tempting to examine every development and event from that perspective. The environmental historian must then be selective in applying this lens. What is integral to understanding the development and formation of the economy and government? How do resource and land-use practices change over time, and what compels those changes? What specifically does examining American history from an environmental perspective offer that is new to our understanding of the nation's history? These are but a few of the compelling questions that the environmental historian must address.

It is often harder to discern whether Americans are more unaware of history or of the environment. Both topics are poorly covered in popular culture and generally are not among the hot topics of discussion in daily life. This has changed somewhat in recent years with increased attention to the subject of climate change. American environmental history might seem to bear a double burden then. Spend some time in the local bookstore in the history section. It is easy to be overwhelmed by the number of books on World War II, the Civil War, the Founding Fathers, the experiences of the military in Iraq and Afghanistan, the Old West, and so on. Rarely is an environmental history book to be found in this section. If the bookstore carries such a book at all, it is more commonly found in cultural studies or in nature/environment, next to books about growing organic tomatoes, raising backyard chickens, and the slow food movement. Keeping all this in mind, it is fair to say that environmental history is one of the most important areas of study today and certainly of more relevance than another book on Stonewall Jackson or about the Special Forces. This is because this field offers a new way of comprehending American history and culture while also increasing environmental knowledge during an era in which understanding the environment and our role within it is absolutely essential.

American historians have told the story of this country in different ways over time, their emphases reflecting prejudices of their era, discoveries of sources, innovative methodologies, and new interpretive approaches. The narrative of American history has traditionally asserted the role of the Anglo-Saxon "race," the Protestant work ethic, capitalism, faith, the frontier, and numerous other characteristics and theories as essential to understanding the rise of America as a powerful democratic, capitalist society in the 19th century and to world dominance in the 20th. Assumed to have played some role but rarely seen as fundamental to American greatness was the great reservoir of natural capital available to Europeans and Americans.

This book does not attempt to retell the political development of the United States, nor does it seek to explain every form of natural resource use and land conservation strategy. Rather, the focus is on how land and resource use influenced American culture and politics as well as attitudes about race, about social mobility, and about America's place in the world, among other things. Moreover, the book explains conflicting ideas about proper use of land and resources and illuminates how those ideas change over time and the consequences of those shifts in perspective.

Indians used and altered the land but did not exploit it in a damaging way. Having lived sustainably for thousands of years, their conquest and displacement meant that Europeans and Americans were able to occupy and exploit hundreds of millions of acres of fertile soils; hundreds of millions of acres of vast, relatively untouched, old growth forests; 15–20 million bison; millions of passenger pigeons; and untold numbers of striped bass, shad, herring, cod, salmon (Atlantic and Pacific), and numerous other species of fish. Millions of beaver and deer were killed for their pelts and skins, and Americans were able to profit from barely touched and often completely unused beds of coal, veins of gold, and deposits of uranium and silver. It is not enough to assert that Americans ingeniously accessed and used these resources. The stark fact is that this vast natural wealth provided a subsidy to American growth and development rarely available to a young, modern nation. Many interpreted this richness as Providence. For example, a 1776 Virginia petition against a dam that was damaging fish runs declared that shad runs were "this great blessing & advantage which kind providence intended bountifully to bestow on all such as should live on or near the [Nottoway] river."[1] This was a common refrain from the Puritans in the 17th century forward to the 20th century, although expressed in more muted terms later in our history. This wealth also made possible the free-wheeling laissez-faire capitalism that is so intertwined with American democracy and instrumental in America's rise to power.

The expansion of the nation and capitalism, creating economic growth and prosperity, also engendered stress and conflict in American life. This book examines the ecological aspect to some of the famous conflicts and wars in the nation's history. As Americans ran up against scarcity caused by aggressive farming, logging, fishing, overhunting, and other reasons, they struggled to craft solutions, and by the late 19th century, it was becoming clear to many leaders that government would have to take on the mantle of conserving resources.

This book takes on the topic of American Environmental History with a few goals. One is to help Americans gain a greater literacy about nature and its role in the development of the nation. For environmental historians, soil fertility, fish runs, railroads, patterns of consumption, and disease hold as much importance to understanding our history as legislative debate might for a political historian or a tank battle for a World War II historian. While the political and military histories are well known and represented, environmental history still has vast inroads to make. Understanding how ecosystems function and how land and resource use affect those processes is certainly something all Americans should strive toward and will inform a more complete and sophisticated understanding of the past.

Another ambition for this book is to demonstrate the ways in which natural resource wealth and vast acreages of fertile soil helped create an ideology of abundance. The Puritans sought a land of milk and honey, and when they and other European observers observed the flocks of passenger pigeons that went on for days and days, the rivers choked with millions of spawning fish, beaches covered with foot long oysters, and the fecundity of the forests and the Indians' farmlands, they knew they had found it. It is not only the idea, pursuit, and celebration of abundance that makes America, it is the reality of this natural capital that is important, too. I am not offering a simplistic interpretation of this nation's history that lays all its accomplishments on the doorstep of beneficent landscapes. Too often, though, the opposite has been true. The wealth of nature has played a bit part in too many of the histories of this nation so focused on the development of our political system, the economy, and even the contemporary work focused on issues of race, gender, and class. Nature provides the very underpinning of the great American moment, and its deterioration and transformation has changed our country over time as well.

Finally, one pressing goal of this book is to tell the story of environmental change, deterioration, and societal responses. While we must be cognizant that there are historians that reject environmental history as merely concerned with ecological degradation, revealing their own unfamiliarity with the field, we must also not back away from a clear-eyed appraisal of the "wastey ways" of Americans. When we look around us at numerous environmental problems today, it is natural to assume the greater competence, thoughtfulness, and moderation of our ancestors while also dreaming of a time when rivers were choked with fish and the sun darkened by flocks of birds. One reason we so glibly make these leaps is that for so long our histories neglected the story of landscape change and decline, leaving us to assume the best. We may be better served in this era with a history that shows us that not only has nature subsidized the country's rise to dominance, but also Europeans and Americans engaged in ecologically destructive actions that required societal and government solutions. Knowing this might better position us to consider current environmental problems and dilemmas.

Note

1 Quoted in Harry Watson, "'The Common Rights of Mankind': Subsistence, Shad, and Commerce in the Early Republican South," in Paul Sutter and Christopher J. Manganiello, editors, *Environmental History and the American South: A Reader* (Athens, GA: The University of Georgia Press, 2009), 145.

Faith in a Generous Land 1

Some time after the animals had been created, along came Coyote. Coyote had all the emotions and problems that humans have today. Everything that the human being was, this is what Coyote was. He was traveling one day when his brother, Fox, stopped him and told Coyote that a great monster was devouring all the animal people. Fox told Coyote that he had to save the animal people. Coyote gave it some thought and decided that maybe he would have to kill this monster. He got five flint knives all sharpened and shaped and put them in his belt. He had a little pouch, and he put soot and pitch and rope in it. Then Coyote went to look for the monster.

He came across the prairie where Grangeville [Idaho] is now, and he hollered over toward the Clearwater Valley. The monster at that time was laying in the valley devouring all the animal people. Coyote hollered, "Monster, here I am! I am right here! Come and get me! You can't eat me like you do all those other people. Come and get me!"

So the monster raised his head over the edge of the canyon and looked out across the prairie toward the place where Coyote was hollering at him.

Coyote said, "Oh, there you are monster. I've come here to see what you are doing to the animal people."

The monster looked over and said, "I am going to eat you, too."

The monster devoured the animal people by sucking them into his stomach with his breath. Coyote knew this, so he tied himself to the mountain on the other side. So when the monster would suck in his breath, Coyote would come to the end of his rope and stop. He teased the monster saying, "You can't get me." And then monster would try again, but Coyote would stop at the end of the rope. The monster did this three times.

Coyote finally decided to go inside the monster so he could rescue the animal people. He reached around behind him and cut the rope and went in the monster's mouth. When he was inside Coyote looked around until he could see the animal people. Some were already devoured, some were half dead, and some were still alive. As Coyote was walking among them, Rattlesnake shook his rattle and struck out at him and Coyote said, "What are you getting mad at me for? I came here to save you, and here you are striking at me." So Coyote stepped on Rattlesnake's head, and that's why today Rattlesnake had a flat head. Coyote went a little further, and Grizzly Bear roared and growled at Coyote. Coyote said, "What are you getting mad at me for? I came here to rescue you." Grizzly Bear growled at him again, so Coyote pushed his nose, and that's why Grizzly Bear has a flat nose. Coyote then told all the animal people that he was going to rescue him and that he was going to kill the monster.

> Coyote built a fire from the pitch in his pouch, and he used the fat from the monster to keep the fire going. The Coyote told the people, "When the monster takes its last breath, you all escape by running out of the holes of the monster. Wait by the holes and when the monster takes its last breath, run out. That will be your last chance to escape." Then Coyote started working. He started to cut the heart away. While he was doing this, sometimes his knife would break, and he would get another knife and keep going. In time he was down to his last knife and the last piece of flesh that was holding the heart. Coyote then told the animals to get ready because the monster was going to die. The animals were waiting by the holes in the monster: by the nose, by the ears, by the mouth, and by that hole underneath the tail. When Coyote made the last cut, the heart came loose. When the monster took his last great breath, all the animals ran out. The last one out was Muskrat. He ran out of the hole where the tail was and the hole closed on his own tail while he was getting out. That's why Muskrat's tail doesn't have any hair.
>
> When the monster was dead, Coyote came out and said, "This place should have some human beings in it. It is such a beautiful place to be. I am also going to create other people, too." Coyote started cutting the monster up; as he did this he would throw pieces of the monster in all directions, and he would create tribes out on the Plains and to the south and east and north and west. So Coyote did all that. He scattered the body parts to the four winds, and that's where the different tribes come from.
>
> Then his brother Fox said to Coyote, "You forgot to put human beings here. You have to create people here, too." Coyote replied, "What I will do is create people this way." Coyote washed his hands in water to get the blood off and scattered the blood droplets on the ground. When those drops of blood hit the earth, human beings sprang up as Nee-mee-poo, the Nez Perce people. That's how we were created: from the blood that hit the earth. Coyote said, "These will be a special kind of people in this valley. They will have strong hearts and strong minds, and they will live well here in this valley." That's how the Nez Perce people came to be. To this day you can still see the heart of the monster where Coyote cut it out at Kamiah, Idaho.[1]

The Nez Perce (Nee-mee-pu in their own language) historically lived on and used a vast range of territory in southeastern Washington, northeastern Oregon, and central Idaho in the area of the Snake, Clearwater, and Grande Ronde rivers. Prior to contact with Europeans they lived primarily on a mix of salmon, bison, camas root, deer, and other resources. Like other native peoples they told and retold an extensive oral culture that communicated ideas of land and resource use, instructions on preserving meat and building tools and weapons, humorous stories, and the dictates and rules of diplomacy and relationships. Their creation stories all illuminated a great deal about their local landscape and their relationship with the land and the animals, birds, and others species that shared that ecosystem with them. Typically in these stories, the humans are but one species among many, maybe distinct, but in no way accorded dominion or superiority over the others. Native culture was different from European culture in that respect; this reflects a faith system built on respect and reverence for the natural world.

Before Europeans arrived in the western hemisphere, a complex and diverse array of Indian cultures used and altered the landscapes of North and South America. The most highly developed cultures, the Incas and Aztecs, were located in South America and Meso-America. While they are not described directly here, Meso-American peoples did influence the rise of agriculture among North American Indians. An analysis of representative groups of the Northeast region, the Pacific Northwest, the American Southwest, and the Southeast, with some consideration of a few Native peoples in other areas, provides an opportunity to better understand the variety of ways that Indians used and adapted to the landscape while even transforming it to suit their own needs.

Native peoples showed great ingenuity and flexibility in their use and management of land and resources across the continent. In the era before contact with Europeans they largely constructed systems of great resilience, emphasizing dependence on multiple food items originating from different sources and ecosystems, including mixed agriculture; hunting of a number of species; fishing and collecting shellfish from beach, river, and ocean; and of course, collection and use of a wide variety of plant species. They did not simply glean from the abundance of the land. Indians helped to increase abundance and diversity through changing ecosystems by use of fire across the continent and because of agriculture and irrigated agriculture in thousands of locations. While Indians faced droughts and sometimes overused local resources, they created a model of sustainable resource use that functioned well for thousands of years with little degradation of the land.

Life in the Bays and Forests of New England

In the American Northeast, Native peoples such as the Pequot, Narragansett, Wampanoags, and others used and changed the environment in myriad ways. In the region of Connecticut and Massachusetts, some tribes gained as much as two-thirds of their diet from agriculture. They grew predominantly corn, beans, and squash, otherwise known as the three sisters for the tradition across North America of planting the crops together. Besides the nutritional benefits of these three crops, plenty of carbohydrates and proteins in addition to important nutrients, these plants complement each other. Corn is a nitrogen thief; this grain will render soil infertile within a few short years. Legumes such as beans host bacteria on their roots that fix nitrogen, introducing it back into the soil in replacement of that consumed by corn. While this does not replace all of the nitrogen taken up by corn, it does extend the usable life of a particular plot. Also, corn stalks provided poles for beans to climb up, and the squash were planted in between corn and beans in order to provide shade from scorching summer sunshine. Squash leaves and vines covered the soil, limiting erosion, and rotting squash leaves release a toxin that slows weed growth. This type of agriculture was practiced as far north as the Kennebec River of Maine and deep into the American Southwest, and it constituted the foundation of most Indian nations' food economy prior to the arrival of Europeans.

Being the staff of life for these Indians, they carefully protected their corn. According to Puritan Roger Williams, crows numbered in the millions and

> are great devourers of the Indian corne as soon as it appears out of the ground . . . against the Birds the *Indians* are very careful, both to set their corne deep enough that it may have a strong root, not so apt to be pluckt up, (yet not too deep, lest they bury it, and it never come up) as also they put up little watch-houses in the middle of their fields, in which they, or their biggest children lodge, and earely in the morning prevent the Birds.[2]

Williams wrote that they avoided killing the crows because of their belief that the Crow had brought the first corn, "an *Indian* Graine of Corne in one Eare, and an *Indian* or *French* Beane in another, from the Great God *Kautàntouwits* field in the Southwest from whence they hold came all their Corne and Beanes."[3] They readily killed large numbers of other birds, but due to the crow's spiritual significance gave it protection. Instead, children threw rocks to try and drive them off without hurting and offending them.

Although they gained the majority of their food from agriculture, these native peoples also followed the seasonal cycle. The rivers of the region hosted large spawning runs of striped bass, alewives, sturgeon, American shad, Atlantic salmon, and other species. Indians traveled to traditional fishing spots on these rivers during the spawn, netting, spearing, and trapping these fish. Profusely laden with clams, with oysters up to a foot long, and lobsters by the thousands available in shallow water during high tide, the beaches also provided access to coastal fishing. Ripening fruit such as strawberries, blackberries, and blueberries also contributed to the diet of Indians of this region as did numerous other indigenous plants. Hunting of deer, turkey, passenger pigeon, rabbits, moose, and others provided needed hides, fat, oils, and protein for their diets. Bones could also be made into tools or sewing needles and internal organs used for storage containers. These foods were so important that the native calendar was organized around the cycle, with months named for when spawning fish arrived or the right time to plant corn. Northeastern native peoples built a broad and complex food economy based on using a variety of foods that supplemented their agricultural production. This provided them with multiple, reliable sources of food so they did not become overly specialized or overly dependent on one particular organism or means of feeding themselves. Such a complex approach made their economy sustainable as well.

The Good Fire

The Puritans and other European settlers justified their dispossessing the Indians of their land by failing to recognize native peoples' changes to and uses of the land while portraying them as savages barely occupying a "howling wilderness." Some historians perpetuate the misrepresentation, portraying most North American Indians as simple hunter-gatherers, which is the exception rather than the rule, and neglecting to discuss agriculture, land use, and ecosystem change. In so doing they sustain the myth of native peoples that justified European and American conquest. Understanding native use of fire is an effective way to reframe our understanding of Indian land use in a way that shows they did alter ecosystems.

Fire was crucial to Northeastern Indians' agricultural strategy. Early colonial visitors and settlers commented on the open, park-like land beneath the canopy, the woodlands largely free of underbrush, with extensive meadows and grasslands. They noted these features without understanding their origins in natives' uses of fire. For the clearing of agricultural fields they girdled and burned trees. Girdling consisted of stripping bark in a circle completely around the tree. This stopped the flow of sap and killed the tree, drying it out and rendering it more flammable. Burning introduced vital nutrients into the soil such as nitrogen and phosphorus and generally extended the life of a particular field from 3–5 years to as long as 8–10 years. When the fields were abandoned, grasses, berries, shrubs, and finally forests would recolonize that ground. This successional stage created habitat and allowed the soil to regain fertility for later use. Fire was also useful for hunting because it cleared out undergrowth and brush, rendering the hunting of animals, particularly large mammals in the fall or winter, much easier. The overall result was a mosaic landscape composed of meadows, farms, and patches of forests composed of different species of trees of varying age and size. The Europeans did not happen onto an ancient forest primeval but rather a dynamic landscape of constant use and change that hosted a variety and abundance of plant and wildlife species.

The sustained use of fire and the creation of a mosaic landscape resulted in large numbers of edge ecosystems throughout the northeastern landscape. These edges between forest and meadow and meadow and farm are rich in both abundance of organisms and in a diversity of flora and fauna. Fire opened forest to sunlight, bringing forth berrying plants, grasses, shrubs, and new types of trees and providing food that both Indians and animals benefited from. In New England deer, elk, wild turkey, bear, and numerous other species thrived on the food produced on these edges. This edge ecosystem did three key things: It produced more edible plants and berries for the Indians, increased the abundance of desirable game species by providing more food for them, and brought these desired species right to where the Indians wanted them, making them that much easier to hunt and trap. What the Europeans saw as natural abundance was in fact partially created by the Indians. Native peoples created a sustainable economy, mixing agriculture with hunting, fishing, and gathering in a way that allowed them to maximize their flexible use of the landscape while even increasing the "natural" carrying capacity of the land. In doing this, the first inhabitants of the continent created a sustainable economy that lasted thousands of years with no significant degradation of the environment.

Fire brought other benefits. Fleas could become a real nuisance in camp sites, making life miserable. Burning forests and grasslands reduced their numbers, at least for a while. Further south, poisonous snakes, fleas, and ticks could be reduced or removed by the use of fire. The practice of burning to clear out nuisance species would continue in some part of the United States into the 1930s. Fire was often used by warring natives to prevent pursuit as well or in some rare cases to drive game away from the area in which an enemy tribe lived and hunted.

In using fire so extensively Indians changed the environment and the composition of forests, creating a "mosaic" of patches of grassland, successional plants, agricultural fields, and different communities of tree varieties. Many of the impressive, large white pines of New England later so coveted by the British for their naval and merchant fleets grew from old burn sites. In New England, trees that sprout from the roots thrived in active burning areas. Oak, hickory, and chestnut constituted a larger part of the forest the Europeans encountered than would have "naturally" occurred without Indian land-use practices. What appeared as simply God's creation to European eyes was actually a consequence of native burning and use.

Native Life in the Southeast and Deep South

Agriculture, like in the Northeast, constituted the most important source of food for Southeastern native peoples. At the same time, they also showed great ingenuity in their comprehensive use of the resources of this landscape. Among Southeastern Indians like the Cherokee, Creeks, Catawbas, Chickasaws, Choctaws, and others, early agriculture focused on native plants such as sunflowers, sumpweed, and knotweed. Native peoples settled along river floodplains where these plants thrived, and they collected, stored, and ate the seeds of these plants. This rudimentary agriculture was upgraded with the arrival of gourds and squash around approximately 1,000 BCE. Not only did these represent a welcome addition to their diet, but also gourds could be dried and used as containers. The squashes originated in Mexico, and corn, which arrived around 200 BCE, also traveled north from Meso-America through trade and adoption by Indians along the way. The third sister, beans, arrived around 1,200 CE. Like Northeastern Indians, these natives employed fire to create and enrich farming soil. They too created a mosaic landscape that increased the flora and fauna abundance of the Southeast.

One native group in the Chesapeake region maintained 3,000 acres of corn, and John Smith commented on extensive cornfields on the Chickahominy River, noting that one powerful woman leader among the Appomattox Indians controlled a field of 100 acres, as did the famous Indian chief Opechancanough (he would later lead the 1622 and 1644 uprisings). He also wrote of

> plenty of corn in the river of Chickahominy, where hundreds of savages in divers [sic] places stood with baskets expecting his coming. And now the winter approaching, the rivers become so covered with swans, geese, ducks, and cranes, that we daily feasted with good bread, Virginia peas, pumpkins, and persimmons, fish, fowl, and diverse sorts of wild beasts as fat as we could eat them.[4]

The Chickahominies grew so much surplus corn that they were able to sign a treaty with the Virginia Colony in 1614 providing 1,000 bushels of corn annually for a handful of iron hatchets. They had clearly crafted an economy that took advantage of natural abundance while producing a variety and profusion of food through agriculture as well.

With the return of spawning fish in March, the men left the villages and migrated to the coastal areas to harvest the abundance from the sea. They used weirs to harvest fish in rivers and tidal zones. These are stone or wooden obstructions that trap the fish. The men then harvested the fish with nets on poles, clubbed them, or speared the fish. In deeper rivers Indians used what is now commonly known as a trot line. Stretching a long line across a river they then attached several shorter strings a few feet apart, hanging with bone hooks baited with shellfish meat and other bait. Like trot-liners today, they checked the lines periodically, gathered their catch, rebaited the hooks, and repeated the process again. Native peoples showed particular ingenuity in one method of fishing. In summer some natives of the region ground up red buckeye nuts and roots into a fine powder, let it ferment in a trough with water, and then spread the toxic brew on a pond. It poisoned the fish, stunning them long enough to float to the surface to be scooped up by waiting fishermen. This caused no permanent damage to the fish or habitat and allowed the Indians to harvest many more fish, more quickly, than by using hook and bait. Southeastern Indians did not make spawning fish the mainstay of their economy; instead, they consumed a large amount of fish during the runs but set very little aside for later use. This prevented dependency on fish and helped conserve that food source. These various methods and locations allowed them to maximize their use of the fishery. Because they were not over reliant on one species or area, if the shad or striped bass run failed, they

Figure 1.1 Belgian Theodor de Bry was a goldsmith and engraver who produced several illustrations of the Chesapeake area based on watercolors by Roanoke colonist John White. These were included with a 1590 publication of *A Briefe and True Report of the New Found Land of Virginia* by Roanoke colonist Thomas Harriott. De Bry depicts natives in stylized poses based on classical statuary and changes their skin tone and face structure to look more European. The value of these images is in their glimpse of the New World by Europeans and how they perceived it. It also serves as a primary source for understanding native culture. Indians in the Southeast used a variety of methods to catch fish in the ocean in addition to those harvested in rivers and lakes. Within this painting of Virginia natives, several methods are shown. Spearfishing from the canoe is the most obvious. In the background of the image several walls can be seen. These were traps behind which fish got caught as the tide left them stranded. Also depicted are Indians tending a weir and trap and many others fishing. De Bry also depicts a seascape of abundance, showing crabs, turtles, fish, skates, and birds. Courtesy of the Jamestown Yorktown Collection, Williamsburg, Virginia.

were not threatened by starvation. Pursuing multiple species of fish also prevented them from overusing any one species, guaranteeing their productivity into the future.

Diversity of diet was the key to survival for Indians, and even with the growth of agriculture in their economy, Southeastern Indians continued harvesting a variety of native plants, including berries, root plants, nuts and fruit, which constituted a major part of Southeastern Indians' food economy. Women maintained the responsibility for the collection of these foods. British naturalist William Bartram described the centrality of hickory nuts in the Creek economy:

> I have seen above an hundred bushels of these nuts belonging to one family. They pound them to pieces, and then cast them into boiling water, which, after passing through fine strainers, preserves the most oily part of the liquid: this they call by a

> name which signifies hiccory [sic] milk; it is as sweet and rich as fresh cream, and is an ingredient in most of their cookery, especially homony [sic] and corn cakes.[5]

These nuts added needed nutrients and fats to the diet and, in addition to enriching their diet, provided another source of food for a time when the deer hunt failed or the fish ran in small numbers.

Some native peoples in this region also made significant changes to the landscape. The Calusas controlled a large part of Florida, including the Caloosahatchee Valley, the area around Lake Okeechobee, the Everglades, and the current Miami region. They hunted deer and harvested shellfish and fish. Fish were harvested from canoes carved from trees and were also caught by trap and weir. Oysters were particularly favored, and in one feast for an early Spanish visitor, they served oysters raw, boiled, and roasted. They suspended nets made from palm fibers in the water with shellfish floats, catching fish like the mullet. Sharks' livers were rubbed on the body to repel mosquitoes, and shark teeth were used for cutting tools, knives, and the tips placed into a board to make a tool for grating roots. An abundance of fruit and nuts, such as saw-palmetto berry, seagrape, pond apple, hog plum, prickly pear cactus fruit, grape, and acorns, helped make for a salutary diet. Apparently subordinate Indians that either enjoyed the protection of the Calusas, or were compelled by them, paid tribute in the form of a popular root used for making bread. This nation also altered the shoreline environment fairly dramatically in the construction of villages facing the water, containing temples and houses. Canals as well as artificial lagoons allowed for water access and transport of people and goods. These canals could be thirty feet wide and six feet deep and typically ran through the middle of towns and connected bodies of water. The need to cut down mangrove and pine trees with shells and rock and then dig extensive amounts of dirt and haul it away required at least thousands of hours of labor. Seawalls and jetties were built, as were defensive walls made from seashells. The home of a town leader could hold as many as 2,000 people, according to Spanish accounts. Like so many of the native peoples in North America, they did not simply live lightly on the land; they also altered it in numerous ways.

The Lower Mississippi River Valley was heavily populated by the Natchez, Tunica, Choctaw, and others. They practiced the typical agricultural model of corn, beans, and squash production, raising their crops in the fertile floodplains of the Mississippi and tributaries like the Yazoo River. They supplemented these crops with game, fish, and plants harvested in alluvial areas and the nearby uplands. In the river bottoms, deer, opossum, raccoon, muskrats, and other species were readily harvested in the fall. Deer hunters covered themselves with the head of a deer and made noise to attract interested bucks during the rut of autumn. Hickory nuts, acorns, and pecans became increasingly important as the season shifted into winter. Tunicas prized the persimmon to such a degree that the entire family participated in its harvest in autumn. These fruit were dried for later consumption and also used to make bread. Oxbows, ponds, and lakes scattered across the flood plains provided for easy fishing with nets, poisonous plants, and trot lines. This type of fishing worked well with agriculture as Indians could take a break from weeding, planting, or harvesting a couple of times a day to check their sites and collect their fish. Similar to other tribes across the continent, natives in this region burned grasses to make travel easier, to lower grass heights so game was more easily observed and hunted, and to bring forth new, green shoots to attract animals.

Their most important game species, deer, moved deep into the swamps in the winter, making the hunt more difficult. Black bears became more highly desired at this point and were harvested for meat and oil rendered from their fat. Mast, the nuts from hardwoods, helped the bears grow fat, and they often hibernated in the hollow trunks of large hardwoods. A French observer described a native bear hunt in the region. The Indians

> gather a heap of dried canes, which they bruise with their feet, that they may burn easier, and one of them mounting upon a tree adjoining to that in which the bear is, sets fire to the reeds, and darts them one after another into the breach; the other hunters having planted themselves in ambuscade upon other trees. The bear is quickly burned out of his habitation, and he no sooner appears on the outside, than they let fly their arrows at him, and often kill him before he gets to the bottom of the tree.[6]

Waterfowl such as goose and ducks gained importance in winter, and vast flocks on Lake Pontchartrain, oxbows, and other wetlands along the Mississippi and its tributaries made harvest relatively easy and dependable.

The Choctaw Indians, located in Alabama, Louisiana, and Mississippi, were master agriculturalists. The Choctaws prioritized agriculture over hunting and gathering, and corn was their most valued crop, followed by beans, squash, pumpkins, and sunflowers. The Choctaw consumed more varieties of corn than we generally see today. There were approximately six different varieties: one was ground into a white flour and another ripened so early that it was possible to reap two crops of that corn in a summer. And while they did clear trees for farming like other Indians, wild fruit trees such as plum and nut trees were left standing by their village for easy harvest. This provided them with a variety of crops, each serving various dietary needs and becoming available at different times of the year.

The Choctaw developed a fairly complex agricultural system starting with their two-field system. Gardens for family use were planted close to homes in the village. Larger fields for the village overall were situated outside the village and worked communally by tribal members. While women were responsible for the majority of agricultural work, men pitched in to sow seed and for the harvest in the fall. The Choctaw girdled trees, killing them and making them easier to burn for clearing fields and introducing nutrients into the soil. In these newly created fields they planted the corn in mounds about three feet apart with the beans planted next to them so the legumes could use the corn stalk as a pole. The Choctaw planted squash and pumpkins between the corn mounds and squash as well as pumpkin and sunflowers on the margins of the plot and field. To the European observers this appeared an untidy mess, not good European agriculture. The fact that the Indians did not cut back the flourishing weeds further darkened their impression of Indian agriculture. What they could not or did not see was that in addition to the benefits provided by planting corn, beans, and squash together, the weeds in fact contained edible plants that contributed to the Indians' diet. The Choctaws hesitated to remove undesired plants for fear of damaging the "weeds" they needed. Europeans were unable to see and understand the complexity, productivity, and sustainability of Choctaw agriculture and this blindness informed their dismissal of Indian farming.

Enhancing the sustainability of their agriculture and increasing their yields was the Choctaw practice of not planting on the sides of hills or using the entire piece of ground. By leaving hillsides undisturbed, and also preserving intact swathes of plant life between plots and fields, they were able to curtail erosion and stem the loss of valuable, fertile soil. In fact, the Choctaw were able to achieve yields of 40–60 bushels of corn, a highly respected harvest in the recent past. The national average of American farmers was less than 40 bushels an acre until 1948 when nitrate fertilizer use became the norm and pushed yields skyward. The amount of corn produced equated to about two-thirds of their caloric needs, but Choctaw did not live on corn, squash, beans, and pumpkin alone. Hunting, fishing, and gathering food also occupied a great deal of their time and diversified and enriched their diet.

Women harvested wild sweet potatoes and Jerusalem artichoke as well as nuts, berries, and fruits like plum and grape. The men focused on deer, and when years were good, meaning no loss of crops to drought or flood, they were careful to harvest the deer after does had time to wean fawns and were able to live on their own. This also coincided with the development of an attractive, full coat on adult deer as fall shifted into winter. While it is not known whether this was an intentional conservation measure, it did help maintain stable deer populations. The deer were pursued primarily for the venison, and many Indians in this region believed that the flesh of this animal gave strength and wisdom to those who ate it. The deerskins were used in multiple ways, such as for clothing, covers of shelters, moccasins, bedding, and so forth. Antler tips were used as arrowheads, and deer tongues were prized for feasts and ritual consumption.

Native peoples of the South and Southeast over thousands of years devised a complex strategy of agricultural production, hunting, and fishing as well as the gathering of wild plants to live successfully in the forests, bays, and river bottoms of this region.

Life on the Edge of the Plains

At the time of European arrival in North America the Pawnees lived on the Platte, Republican, and Loup rivers in the eastern Great Plains. Their mixed economy of agriculture, small game hunting, food gathering, and the pursuit of bison required mobility and a comprehensive use of grasslands and river bottoms. Half the year was spent along the rivers, with the bluffs on the north side providing protection from bitter winter winds, northern blasts that could bring the wind chill down to 70 degrees Fahrenheit below zero; wood for fires and home construction; as well as the soil and water necessary for agriculture. Men and women cleared fields together in the spring, burning weeds and debris to introduce nutrients into the soil. But Pawnee women performed the great share of agricultural work as was true of almost all native peoples except in the Southwest. Given fields of one to three acres, which individual women could use as long as they desired, the Pawnee women planted corn, beans, squash, pumpkins, and sunflowers. Avoiding superfluous labor, they worked along rivers and streams and at the mouths of ravines, places where the soil was already broken. This was also where the best ground was typically found. Using multiple varieties of corn, beans, and squash, they planted beans with the corn but sowed squash separately, organized in plots by breed to prevent cross-pollination. Pawnee women also harvested numerous wild plants; the Indian potato was the most important non-agricultural food source. Women worked hard gathering these in spring, when food supplies were typically at their ebb and still too early to plant summer crops. Exhuming the roots was difficult labor and lasted for two months. Women also gathered ground beans, Jerusalem artichoke, water chinquapin, wild onion, and numerous other plants. These plants added diversity and nutrients to their diet.

While women worked their fields, Pawnee men made their major contribution to sating the stomach through hunting local game. But there was not enough deer, rabbit, or quail to make this the integral part of their diet; more important were the freshwater clams gathered from rivers. Men also hunted bison, but women's work made it possible to abandon the winter villages and journey onto the central plains in pursuit of bison. Provisions of corn, pumpkin, potatoes, and beans were required for the long walk to where the bison gathered in vast herds. Indians hunting this large mammal before the arrival of the horse employed a number of strategies. Buffalo jumps are well known sites where Indians would drive hundreds of bison off of cliffs to their deaths using noise makers, lines of people, and fire. Bison were also hunted on foot by men draping a green bison hide over their body to conceal themselves and their human smell. Encirclements, rock and human lined chutes, and other methods were all part of Indians' arsenal of bison hunting strategies. In this era, for the Pawnee and most Indians living on the edge of the Plains, bison constituted just one species in the complex food economy. Their lack of a domesticated mammal that could be used for hunting kept them from becoming overly dependent on the bison or from overharvesting them.

Their existence was not an isolated one; the Pawnees actively participated in a trade network extending to the Gulf of Mexico with Caddo Indians in southeastern Texas, where the Caddoes lived and farmed. The Apaches were located further west on the plains, and they practiced a greater dependence on bison prior to the arrival of the horse than any other tribe. But man cannot live on bison meat alone, and they engaged in an active trade with Pueblo Indians to their southwest. Historian Elliott West provides a powerful image of an Apache trading mission:

> The Apache caravans must have been an arresting sight. Men and women hefted dried meat and hides to their shoulders and wrestled fifty-pound packs onto their largest dogs. Other animals [dogs] were harnessed to travois. Teams of dogs staggered forward with their loads of jerky and hams, howling from the oozing sores rubbed into their backs. Weeks later caravans returned loaded with corn, other garden foods, and crucial items they could not make for themselves, especially pots and other ceramics.[7]

Life on the Edge of the Sea

Pacific Northwest peoples thrived in a landscape of prodigious natural wealth. From northern California to southeast Alaska, Native American tribes on the coast built villages at the mouths of rivers entering bays and the Pacific Ocean, along estuaries, and on the shorelines of rivers and creeks. In these sites they harvested salmon, trout, halibut, seals, whale, smelt, ducks, and numerous other species. Following the seasonal cycle like other American native peoples, they were able to spend a greater amount of time at their central village because of the abundance of salmon available to them.

On the Pacific coast five species of salmon historically populated bays, rivers, estuaries, and the ocean, providing a consistent supply of nutrient- and calorie-rich food. Ranging from the iconic Chinook salmon, otherwise known as king salmon, a fish that could reach as much as 130 pounds and more than 6 feet in length, to the equally important but somewhat less magnificent silver, sockeye, pink, and chum salmon, the variety of salmon spawning up rivers at different times of the year, at approximately the same time each year, provided steady and dependable nourishment to Indians of this region. They harvested these fish in a number of ways. Sometimes runs were thick enough and so easily accessed that Indians could wade into the stream and simply toss the fish to the bank. More commonly, Indians used spears, nets attached to long poles, fish traps, and complicated weirs 20 to 30 feet in length and several feet high for the collection of the valued fish. Native peoples enacted specific measures to ensure the return of salmon, such as leaving holes at the bottom of weirs, removing traps one day a week, and other steps. They understood that fish had to spawn to reproduce the runs and enacted conservation measures to ensure sustainable salmon populations into the future.

Northwest coastal peoples caught all kinds of fish, not only salmon. Spanish explorers noted one ingenious strategy: The Indians took a harpoon made with a mussel shell point and a hook on the opposite end. "They also took a piece of wood in the shape of a cone, with some thin and flexible strips of bark fastened in the periphery of its base like feathers,

Figure 1.2 Waterfalls and rapids made salmon and steelhead easier to see and catch. Indians built stands and perches at sites like Willamette Falls in Oregon and Celilo Falls on the Columbia River. From these they could spear and dipnet fish. Courtesy of the University of Washington Libraries, Special Collections, Seattle, Washington.

the whole being very much like a shuttlecock." The "shuttlecock" lure was placed on the hook at the end of the long rod and very gently placed on the bottom of the bay or sound next to the head of a fish.

> They then pulled away the hook and the shuttlecock went up to the surface with a rapidity which did not allow the fish to see what it was. Deceived in this manner it followed the other object up to the surface of the water and then the Indian, who had already turned the rod, presented the harpoon, threw it at the fish, usually with such accuracy that they seldom failed to hit it.[8]

Smaller fish that schooled and spawned in great numbers also proved valuable. Herring were usually harvested by means of a rake with teeth made from sharpened elk bones, using it to lift the herring into a canoe. The Klallam Indians of Puget Sound creatively harvested herring eggs, one of their delicacies. After the tide's retreat they laid hemlock twigs on spawning sites on the beach. Later, they collected the egg-coated twigs, allowed them to dry, then shook the dried eggs into baskets. Likewise, Indians' resourcefulness was demonstrated in their pursuit of waterfowl. They erected large poles, typically 40 feet high but

Figure 1.3 Salish Indians in Puget Sound showed great ingenuity in catching flocks of waterfowl. In this image the poles used for nets are clearly evident. Native peoples would spook a flock of birds, while Indians waiting at the base of the poles pulled the nets up the poles to catch the birds. Courtesy of the University of Washington Libraries, Special Collections, Seattle, Washington.

sometimes as high as 80 or 100 feet. Nets lay on the ground between the poles as Indians on each side of the poles waited in hiding. When flocks took flight between the poles, bird hunters would pull on ropes, thus lifting the nets 40 to 80 or even 100 feet, catching the birds in the air like fish trapped in nets. Sometimes nets were left up at night to catch birds in the darkness like weirs or traps in rivers.

Women played a pivotal role in this economy, sometimes assisting in catching halibut, an important food source, or rainbow trout, a small part of the Indian's diet. They also harvested clams, mussels, and other seafood. But the greater portion of women's labor was dedicated to gathering wild plants. With the help of their children and female slaves, women gathered huckleberries, elderberries, blackberries, camas, salal, Oregon grape, strawberries, wild carrots, and the roots and bulbs of various ferns, along with many other plants. This variety of fruits and vegetables provided necessary nutrients and diversified the diet.

While Northwest coastal Indians practiced limited agriculture, it constituted a small part of their overall food economy. They did alter the landscape in numerous, subtle ways. Regular burning in this region led to increased amounts of berrying plants such as blackberries, huckleberries, thimbleberries, and so forth. Indians burned also to sustain camas beds and the growth of ferns as well as to keep grasslands open from trees. This ensured the continued use of the area by deer and elk. Some Indians transferred salmon eggs from one river to another in order to create new or better salmon runs. Also, coastal Indians that traveled to the Celilo Falls area of the Columbia River, a popular area of trade and fish harvest for coastal and interior Northwest Indians, brought camas back to the coastal region from the interior of the state. Camas, a key staple of Columbia Plateau Indians such as the Nez Perce, Palouse, and Coeur D'Alenes, produced tuber roots like small potatoes. Coastal Indians enjoyed them enough to transplant them to the West side of the state, introducing them there.

Even beaches' appearances changed due to Indian practices. Clams were a favored and important part of some coastal tribes' diets and also served as a valued trade item with interior Indians. To increase production of clams some natives built rock walls out into the tidal zone with a retaining wall running parallel to the beach. These structures would collect the sand of longshore currents (the current that moves along the beach line carrying and depositing sediments) and the sediment deposited by waves, which would accumulate behind and inside these rock walls and create a sandy beach for the creation of clam beds. In this particular case, the Indians improved and expanded the clamming beds by modifying beach habitat.

Northwest native peoples did not simply harvest from the abundance surrounding them. They took steps to ensure sustained salmon runs; used fire to change the landscape and guarantee plenty of berries, camas root, deer, and elk; and even modified beach ecosystems to produce more clams.

The Makah, Quileute, and Quinault Indians made their homes on the western side of the Olympic Peninsula of what is now Washington State. Certainly, they used many of the same resources as other coastal peoples, but in one way they were quite distinct; this was in their pursuit of whales. With their villages on the Pacific they could see and pursue these migrating mammals. Humpback and gray whales were the species most actively pursued, and a kill of two or three per village per year was the norm, although there were cases of harvests that dramatically exceeded that.

Hunting whales was complex and dangerous. A canoe of eight men approached the whale at dawn so their shadows would not alert the whale to their presence. As most hunts were conducted at dawn, that meant reaching the other side of the whale and approaching it from the west, the open ocean. A harpoon with a shaft crafted from yew and a tip of mussel shells was used to strike and hold the whale. An adept modification of the harpoon shaft was the splicing of two different pieces of wood together so that if and when the whale thrashed the shaft would break instead of sweeping the men from the canoe. The canoe was backed away quickly while the cord to the harpoon was kept tight. On this cord was strung a series of sealskin floats to keep the whale afloat and in sight. Next, a killing harpoon was used to finish off the whale, and one man quickly jumped into the water and used a rope to tie the whale's mouth shut so it would not swallow water and disappear into the depths.

The highly dangerous hunt was worth it because whale flesh and oil were highly prized. Following skinning and butchering of the catch, the meat and oil was distributed among villagers. A typical whale provided close to 5,000 quarts of oil. Stored in sea lion stomachs, cod stomachs, or even chum salmon stomachs, the oil was an important part of these native peoples' diets. The Northwest coastal peoples over thousands of years crafted a culture and economical system that took advantage of the profusion of nature in numerous ways while living in a manner that did not destroy their needed resources.

Figure 1.4 Asahel Curtis, younger brother to Edward Curtis, embarked on his own remarkable photography career. In this photo taken in 1910, several members of the Makah Nation are shown butchering a whale. Located on Cape Flattery at the far northwestern corner of the continental United States, where Washington State's Olympic Peninsula juts into the Pacific Ocean, the Makah have actively hunted whales for centuries. Members of the tribe worked together to remove the meat and blubber and then distributed it throughout the community to be eaten and used. Courtesy of the Washington State Historical Society.

Farming the Desert

The Southwestern landscape, a mix of high altitude mountains and deserts with most areas averaging five to fifteen inches of rain a year, does not lend itself to a life of easy abundance. In this parched land of limited rainfall Southwestern native peoples created a number of irrigated societies of varying complexity. In fact, Indians of the Southwest region were more reliant on agriculture than other native peoples across North America. Employing numerous irrigation strategies, the Pueblos grew predominantly corn, beans, and squash. More reliant on agriculture than other native peoples in North America, making the desert bloom was no easy task and required creativity and alterations of the land. Precipitation in this region is spotty and unpredictable, so Indians strived to increase their

Figure 1.5 Edward Curtis embarked on a monumental project to photograph 1,500 images of Native Americans from across the nation. The 20-volume series that was eventually completed in 1930, after more than three decades of work, contained 2,200 images of natives from more than 80 distinct nations. This Edward Curtis photograph shows a Pueblo walled garden constructed in a manner that was thousands of years old by the time of this photograph. The stones helped to hold soil and also catch and hold water, allowing it time to subside into the dirt instead of washing away. The stones also helped create microclimates by holding heat during the day and releasing it during the cool desert and mountain evenings to extend the growing day and increase yield. Courtesy of the Charles Deering McCormick Library of Special Collections, Northwestern University Library, Evanston, IL.

odds. While rain clouds might pass a village by, a storm several miles away could bring water to the village via a wash. A straightforward method to use this water was to plant crops at the mouths of washes. Crops planted where the water rushed out had a chance of gaining enough moisture to germinate and produce a crop. Sometimes a stone wall was built to try and hold some water and capture soil carried by the water. Another example of adaptation to environmental limits was the method of planting small gardens inside small plots of 15–25 square feet bordered with rocks

In the Jemez Mountains west of Santa Fe, for example, Pueblo Indians would plant numerous small plots like this across the tops of ridges, hoping that enough of them would catch rain to produce crops. The water trickling down the foot-high walls increased the moisture in the plots, and the radiant heat stored in the rocks and released at night, extended the growing season. Moreover, Pueblo Indians in this region used pumice, a lightweight, porous volcanic rock, as a form of mulch. The pumice absorbed moisture and

Figure 1.6 This Curtis photograph of waffle gardening depicts a system that included the advantages of the small walled garden and maximized space. Pueblos irrigated by carrying water in jugs and by establishing these gardens across the landscape. In this way, they maximized the possibility of obtaining enough precipitation in an arid environment where downpours are often short in duration and cover small areas. Courtesy of the Charles Deering McCormick Library of Special Collections, Northwestern University Library, Evanston, IL.

slowly released it after the rain had evaporated. The rock also absorbed heat, keeping these small garden plots a bit warmer at night, offsetting cool evening temperatures.

The Zuni Pueblos used waffle gardens to similar affect. The small rectangular plots divided by low clay walls helped contain rainfall and could be easily watered with gourds and clay jars. The Hopi showed great ingenuity in finding a way to put the sand dunes to good use. Knowing that water lay underneath the sand, they increased water storage by building fences of shrubs about 20 feet apart. Sand and snow piled up behind them, and the drifts of snow would melt into the sand, extending the planting season and adding to the water stored a few feet below. The Hopi over many generations selected a type of corn with a short germination and that sent its main root as far as eight feet down to reach the water below. Above ground the plant looked nothing like the corn one sees in Nebraska today; it stood about 18 inches high. This provided protection against the loss of energy and water to the constant, dry winds of that region and put more energy into corn production. These variations in irrigated agriculture demonstrate how native peoples of the Southwest adapted to a difficult environment for farming and crafted successful agricultural societies lasting for centuries.

Figure 1.7 Pueblo Indians bred several different types of corn for different dietary uses and to grow well in an arid, desert environment. The type of corn shown in this Curtis photo, grown for centuries and still used today, is quite short compared to the corn most gardeners raise or that we see in Nebraska cornfields. Its shorter height keeps it from losing too much moisture and energy to wind and dry air, while a root up to 8 feet long reaches for water deep below the sand and soil. Courtesy of the Charles Deering McCormick Library of Special Collections, Northwestern University Library, Evanston, IL.

Spanish explorers described the irrigation and land-use methods of Acoma Puebloans located west of present-day Albuquerque. The pueblo sits on a mesa that stands about 600 feet above the valley bottom where the Indians farmed.

> These people have their fields two leagues distant from the pueblo, near a medium-sized river, and irrigate their farms by little streams of water diverted from a marsh near the river. Close to the sown plots we found many Castile rosebushes in bloom; and we also found Castile onions, which grow wild in this land without being planted or cultivated. In the adjacent mountains there are indications of mines and other riches, but we did not go to inspect them because the natives there were numerous and warlike.[9]

These local adaptations are impressive; even more so are the complex irrigation systems created throughout the Southwest, starting more than 3,000 years ago and ending in the early 15th century. The Hohokam of the south-central Arizona region built an intensive system of water use and distribution beginning around 1500 BCE, reaching its peak in the period from 1150 to 1450 CE. They began constructing ditches and canals along the Santa Cruz River and large-scale projects starting in approximately 450 BCE. Corn was their primary food source, and evidence indicates the Hohokam first planted it around 2,000 BCE and began irrigating it in 1500 BCE, using canals to carry water miles from rivers to farms and side channels to spread the water around. Some of these canals were up to 10 feet wide and 15 feet deep. At their peak in the lower Salt River Valley 14 irrigation communities used canals totaling 300 miles in length, supporting agriculture and permanent residence in an approximately 400-square-mile area. The Hohokam raised corn, squash, and beans as well as cotton and probably tobacco. While cotton had been developed in the Old World prior to contact, it is the New World cotton that now dominates world production, accounting for about 95 percent of contemporary production. Strong evidence indicates that the Hohokam domesticated the Agave plant and Cholla cactus, and some communities grew large amounts of Agave. The harvest of fruit from wild-growing saguaro cactus provided an important food source months before farmed crops were ripe. Other Southwestern Indians used irrigation systems like the Hohokam. Canals were sometimes lined with sandstone slabs to reduce water loss, and in some communities canals and ditches extended dozens of miles.

The collapse of the Hohokam society like those of the religious city of Chaco in western New Mexico and Cahokia on the Mississippi River in Illinois gives rise to questions regarding the sustainability of societies so dependent on irrigated agriculture and with high population densities. For a long time it was assumed that salinization of the soil from irrigation was at the root cause of the collapse of the Hohokam in the late 14th century, but the long period of sustained farming in the same area suggests they found a way to prevent or limit that problem. The more likely explanation is the combination of climatic changes leading to more and longer droughts along with numerous floods that inflicted great damage on their canals and dams.

The high population densities of these agricultural communities had a negative impact on game populations such as turkey, deer, rabbit, and lizards, as they were overhunted by Indians desirous and needful of meat-based proteins in their diet. Irrigated agriculture sustained population growth too large for regional game populations, and many species were overhunted. Because of this, native peoples in this region engaged in trade with tribes that were game reliant, essentially swapping carbohydrates for proteins. At the time of Coronado's expedition through this region in 1540–42 the Pueblos had developed a complex and largely sustainable irrigated agricultural society in which a rich material culture flourished.

Faith and Nature

The Indians of North America practiced a variety of polytheistic faiths, seeking harmony and strong relations with an animistic natural world, one in which every living creature and even some inanimate objects, such as lakes and mountains, contained a soul or spirit ally. Indians sought to maintain a healthy and predictable relationship with nature through a variety of means: ritual, worship, prayer, taboo systems, and humility. In so doing, they crafted a method of using and living in the landscape that was for the most part sustainable over thousands of years. To speak of the natural world or Indians' spiritual system is to create a false dichotomy; they were intertwined, one and the same.

There are problems inherent in speaking of Indian faiths in a general way, keeping in mind the thousands of different tribes and nations that inhabited the continent and the array of religious practices. But there are commonalities. The Algonquian term *Manitou* conveys the idea that animals exerted great spiritual force. Every creature from the frog to the moose deserved respect and was dealt with carefully through prayer, ceremonies, and rituals. The power of any particular animal depended on the tribe and the importance of that animal in that culture. The animal's power was based on its importance as a food item; how dangerous it was; whether it was a trickster figure like a crow, raven, or coyote; or on stories and the role the animal played in the Indians' history and development.

One expression of Manitou was the power animals had to withhold themselves, that is, an offended deer spirit might choose to not give itself to Choctaw, Creek, or Cherokee hunters. Similarly, the Moose might disappear for the Mic-Mac, or a salmon run not appear for the Skagit, Tulalip, or Klallam. But their power extended beyond simply withholding themselves. A powerful and offended animal could bring famine, disease, and war upon the Indian community of the offending individual. This culture of respect and deference to a natural world imbued with great spiritual power is a fundamental reason that native peoples were able to live sustainably for so many centuries.

Indians were generally careful, modest, and respectful in their killing and harvest of animals as well as plants, rejecting arrogance or waste. Pacific Northwest Klallam Indians adhered to an oral tradition that warned against denigrating other creatures. The central lesson in most stories is to not mock the fish or play with them as they are waiting to be cleaned. In one story,

> a girl of about ten was swimming in the Dungeness River and made fun of an old salmon. Soon after, she became ill. Her eyes began to look like salmon eyes and her actions were just like the movements of the fish as they swim. Her people asked her if she had played with a salmon. She admitted that she had. The shaman could do nothing for her and she soon died.[10]

In another Klallam story, some bold young men questioned the validity of the first salmon ceremony. They attached parts of their ceremonial headdresses to the tails and fins of the old salmon then released him to see whether the stories were true, whether he would lead another run of salmon. Shocked at this salmon chief's return the following year, they died soon afterwards, thrashing like dying salmon.

Beyond taboos, prayer, and humility, Native Americans conducted numerous ceremonies to demonstrate respect and ensure success in the hunt or harvesting of fish. The First Salmon Ceremony reveals their dependence on salmon and the fish's sacred status. Many of the Northwest coastal Indians believed that they had at one time lived at the bottom of the sea and the salmon were long lost kin. Coastal natives believed that the salmon people were led by a chief from their underwater villages to offer themselves each year. But for the runs to be successful, the chief must return to the village and lead the next group the following year. Therefore, observers were posted to watch for the first salmon in the spawning run, the chief. Catching him and bearing him with respect into the village, the tribal religious leader directed the village in the First Salmon Ceremony. The various members of the tribe would eat the flesh of the salmon, and the bones would be placed back in the river, always handled with reverence and care, so that the salmon might return again.

A ceremony similar to the First Salmon Ceremony in its celebration of an important fish species, but shockingly different in practice, was a religious ceremony of the Calusa of Florida. Mullet represented a prized and important species for the Calusa. The fish migrates from coastal estuaries to deeper waters in the Gulf of Mexico for reproduction. In order to safeguard their return, this tribe selected a victim, usually a captive, non-member of the tribe, beheaded the victim, and then presented the head to an effigy of the mullet, which then consumed the sacrificed person's eyes. The significance of this ritual is revealed by the Calusa belief that the immortal soul remaining after death was contained within the eyes. This individual's soul was consumed to ensure the mullets' return. Deep faith and religious practice was essential to maintaining a balance of nature and the appropriate relationship to key species to sustain harvests and Indian society.

Powhatan hunters in the Chesapeake region gave thanks after a successful deer hunt by making offerings of blood, suet, and tobacco to the deer spirit while gathered around what appeared to be stone altars. Treatment of bones with reverence was integral to the cosmology of Algonquian Indians as others. To treat bones as refuse or to allow dogs to gnaw on them was deeply disrespectful and could create anger. They were careful to burn the bones in fire to avoid giving offense.

Hunting required careful preparation and reverence for the target of harpoons, arrows, and spears. Even joining the ranks of Northwest coastal Indian whalers was a daunting process, requiring a special invitation by the chiefs and arduous preparation. One whale hunter recalled the requirement to bathe regularly for four years prior to becoming a whaler, having to leave the village for periods of four days to bathe in ten different rivers in that period. This was done while fasting. The aspirant would return to and remain in his home for ten days, and then venture forth again and repeat the process of fasting and cleansing. Makah whale-hunters shrouded their preparations for the hunt in great secrecy, spending the night in bathing and prayer, then sprinkling ash from the fire on their hair when entering their home to conceal their activities from their family. Social anthropologist Ann Tweedie writes,

> the hunter himself never ate meat and avoided intercourse with women during hunting season in case they were ceremonially unclean. For the same reason, women were never allowed to touch whaling gear. During the hunt, the hunter's wife was required to lie still so the whale would be docile and was prevented from combing her hair, because it was thought that breaking any strands would cause the lanyard to snap.[11]

For tribes like the Pawnee, the hunt likewise required prayer, careful preparation, and caution to understand the auspices for the hunt and avoid giving offense to the Bison spirit ally. Special chiefs of the hunt were appointed and represented the Sky-gods, holding absolute power over the preparations and the conduct of the hunt itself. Many Plains Indians danced in order to bring the bison closer to camp while also preparing with prayers, sweats, and the burning of sweetgrass to drive away evil spirits. It is important to remember that these ablutions and ceremonial preparations are ways not only to ensure the animal gives itself in the hunt but also to avoid giving offense and suffering repercussions in the future.

Although primarily an agricultural people, the Pueblos of Acoma also needed meat and prepared carefully for the hunting of deer, rabbit, antelope, and others in a deeply spiritual manner. Four days of song, prayer, and smokes were led by a hunt chief, and, according to historian Ramón Gutiérrez:

> During this time the eldest male of each household brought his lineage "offspring" animal fetish to the kiva and placed it next to the hunt chief's "mother" fetish on the society's altar. There the hunt chief empowered the fetishes with animal spirits for a successful hunt by bathing them in nourishing blood and feeding them small bits of the animal they were going to hunt. These fetishes contained the living heart and breath of the animals they depicted.[12]

To purify himself, a hunter avoided sex for four days before and four days after the hunt and tried to remove any smell or thought of women from his body and heart because of the belief that spirit allies were offended by the smell of women and the sexual dissolution of men. Upon killing a deer, the animal was accepted into the home by the women offering cornmeal while extending the invitation, and family members attempting to gain the essence of the deer's beauty and strength by rubbing their hands on the deer and then on their own faces. Before the meat was divided, the hunter used juniper smoke to purify himself to prevent haunting by the deer spirit. The Acoman Pueblos did not simply eat deer and use parts of its body for tools and containers, they derived physical and spiritual strength from their harmonious relationship with this animal and the rest of the natural world.

Native Peoples and Violence

To argue that native peoples lived sustainably prior to the first European footfall in the New World is not to make the naïve argument that they lived peacefully in a pastoral idyll. Indians did go to war over control of resources as well as for other reasons, such as to acquire slaves, to gain vengeance, and control trade. Overhunting an area of deer, bison, or other species would compel Indians to move into other areas such as borderlands between tribes or even onto sovereign territory of another tribe. This typically led to conflict with the stronger natives forcing the losers to find new territory. Northwestern Indians occasionally raided to steal food from villages, and in the Southwest extended droughts in the historical past led to the abandonment of cities and villages and occasionally brutal, violent attacks in which whole communities were slaughtered. Violence of this sort indicates not only an effort to gain access to water or crops but also likely a failure of their belief system because of their desperation and the violence that ensues in such situations.

Similarly, there has been much debate over Indian-caused extinctions. The tendency to view Indians as environmentalists reflects an anachronistic perspective and the effort by some Americans to create a usable past, depicting a version of Indian life that supposedly provides an alternative to the problems and complexities of the modern era. While the vast majority of Native peoples in North America lived sustainably within their environment for long periods of time, it is important to not confuse this with environmentalism. The question is, how much damage did the Indians inflict on the environment? First, it is important to delineate between pre-contact and post-contact native culture. Post-contact Indians inflicted serious damage on their ecosystems in many cases, but discussing that without considering the impacts of disease and conquest, the allure and destabilization of European trade goods, and the transformation of Indian attitudes regarding nature is to construe the Indian as static and is simply bad history. What damage did pre-contact Natives perpetuate?

The most popular argument in this regard is the overkill hypothesis proposed by anthropologist Paul Martin. He argues that newly arrived Clovis hunters over-hunted a number of large species, or megafauna, into extinction approximately 11,000 years ago at the end of the Pleistocene, not long after their believed arrival via the Bering land bridge and migration south. The short-faced bear, a species that approached 1,500 pounds; the giant beaver (which was 6 to 8 feet in length—imagine the dam!); the saber-toothed cat; and the long-horned or giant bison were some of the roughly 35 species Martin believes Indians wiped out. He compares the wanton killing of Clovis hunters to the German strategy at the opening of World War II—the fast striking, wide-ranging front of blitzkrieg warfare. His thesis proved very popular with many, particularly scholars or pundits seeking to counter the idea of an environmental or sustainable Indian culture. The continuing fascination with the overkill hypothesis reveals the inherent tension in the depiction of the Indian as victim or aggressor.

Some problems with the overkill hypothesis emerged with careful study over the decades since the provocative idea was offered. Very few kill sites have been found, only 50 large ones in fact. Also, the human population that existed at that time, even if committed to nothing but the destruction of megafauna, requiring the abandonment of carcasses across the landscape, could not have killed the number of species that existed at that time. Even the premise that all of the Indians pursued the large mammals is deeply flawed—many Indians practiced a variety of food hunting and gathering strategies that did not include extraordinarily large animals. Additionally, not all Indians were Clovis hunters, and, therefore, not all Indians were sufficiently armed for the destruction of giant sloths, short-faced bears, and the very large beaver.

The other key problem with the overkill hypothesis is its neglect of the role of climatic factors at the end of the Pleistocene when temperatures rose as much as 13 degrees Fahrenheit. Increased aridity, changes in plant growth patterns, and the growing scarcity of the large animals' favorite foods as well as colder winters produced a cumulative effect on the now extinct species. One last factor to consider is the extinction of numerous other small species, such as waterfowl, flamingos, vultures, and cowbirds during the same era. A more realistic assessment of the extinction of the 35 megafauna, along with the other species that died off, includes human hunting pressure as one factor among many driving these animals over the brink.

Native Americans used and altered the land and resources upon which they were dependent in a number of ways. Because of limited populations, a deep reverence for a spiritually imbued natural world, and the constant use of complex, varied food economies, they were generally able to live sustainably for long stretches of time without causing severe ecosystem degradation. Fishing the rivers of the Northwest, harvesting whales, driving bison off of cliffs, and growing corn, beans, and squash in the forests of New England, along the lower Mississippi River, and in the deserts of what is now southern Arizona, they had no inkling of what was coming. European explorers, fishermen, traders, and colonists would trigger a tsunami of ecological and cultural change that Indians were ill-prepared for and would have to rapidly adapt to in order to survive.

Document 1.1 A Key into the Language of America, *Roger Williams*, 1643

Roger Williams was a Puritan theologian, critic of the Puritan community, and dissenter to policies of the Massachusetts Bay Colony. He arrived in 1631 and was exiled in 1636 when he established New Providence. His study of native culture provides a wealth of material that historians have used extensively to better understand Native American land use, farming, and religious beliefs.

1. What are the types and variety of species hunted and harvested by Indians of New England? From this document, what do you conclude about their food economy and use of the landscape?
2. What strategies do Indians employ in their pursuit and production of food? How versatile and diverse are these efforts?

Divers part of the Countrey abound with this Fish [bass, lamprey, herring, cod, sturgeon mentioned by the author]; yet the Natives for the goodnesse and greatnesse of it, much prize it and will neither furnish the *English* with so many, nor so cheape, that any great trade is like to be made of it, until the *English* themselves are fit to follow the fishing.

The Natives venture one or two in a Canow, and with an harping Iron, or such like Instrument sticke this fish, and so hale it into their Canow; sometimes they take them by their nets, which they make strong of Hemp. [Their Nets] Which they will set thwart some little River or Cove wherein they kil Basse (at the fall of the water) with their arrows, or sharp sticks, especially if headed with Iron, gotten from the English . . . Of this fish [probably bluegill or other perch] there is abundance which the Natives drie in the Sunne and smoake; and some English begin to salt, both ways they keepe all the yeare; and it is hoped it may be as well accepted as Cod at a Market, and better, if once knowne.

. . . This is sweet kind of shellfish [clams], which all *Indians* generally over the Countrey, Winter and Summer delight in; and at low water the women dig for them: this fish and the naturall liquor of it, they boile, and it makes their broth and their *Nasaũmp* (which is a kind of thickned broth) and their bread seasonable and savory, instead of Salt . . . This the English call Hens, a little thick shel-fiish, which the Indians wade deepe and dive for, and after they have eaten the meat there (in those which are good) they breake out of the shell, about halfe an inch of a blacke part of it, of which they make their *Suckaũhock*, or black money, which is to the pretious . . . The Periwinkle Of which they make their Wòmpans, or white money, of halfe the value of their Suckáwhock, or blacke money . . .

The Natives take exceeding great paines in their fishing, especially in watching their seasons by night; so that frequently they lay their naked bodies many a cold night on the cold shoare about a fire of two or three sticks, and oft in the night search their Nets; and sometimes goe in and stay longer in frozen water.

Document 1.2 New Relation of Gaspesia, with the Customs and Religion of the Gaspesian Indians, *Chrestien Leclercq,* 1691

Chrestien Leclercq was a Jesuit Missionary to the Mic-Macs of the Gaspé Peninsula of Canada in the 1670s and '80s. He published two lengthy works on his observations, experiences, and the colonial and religious mission in French Canada that are now considered valuable sources for both colonial and Native American history.

1. Describe the variety and complexity of hunting strategies of the Mic-Mac. What does this reveal of their knowledge of nature and Indian appreciation of these species?
2. What taboos did the Indians practice in regard to handling of the beaver? What does this reveal about their relationship with this species?

The most ingenious method which our Gaspesians have for taking the Moose is this. The hunters, knowing the place on the river where it is accustomed to resort when in heat, embark at night in a canoe, and, approaching the meadow where it has its retreat, browses, and usually sleeps, one of them imitates the cry of the female, while the other at the same time takes up water in a bark dish, and lets it fall drop by drop, as if it were the female relieving herself of her water. The male approaches, and the Indians who are on watch kill him with shots from their guns. The same cunning and dexterity they also use with respect to the female, by counterfeiting the cry of the male.

The hunting of the beaver is as easy in summer as it is laborious in winter, although it is equally pleasing and entertaining in both of these two seasons, because of the pleasure it is to see this animal's natural industry, which transcends the imagination of those who have never seen the surprising evidences thereof. Consequently the Indians say that the Beavers have sense, and form a separate nation; and they say they would cease to make war upon these animals if they would speak, howsoever little, in order that they might learn whether the Beavers are among their friends or their enemies.

The Beaver is of the bigness of a water-spaniel. Its fur is chestnut, black, and rarely, white, but always very soft and suitable for the making of hats. It is the great trade of New France. The Gaspesians say that the Beaver is the beloved of the French and of the other Europeans, who seek it greedily; and I have been unable to keep from laughing on overhearing an Indian . . . "In truth, my brother, the Beaver does everything to perfection. He makes for us kettles, axes, swords, knives, and gives us drink and food without the trouble of cultivating the ground."

. . . They build causeways and dams of a breadth of two or three feet, a height of twelve or fifteen feet, and a length of twenty or thirty; these are so inconvenient and difficult to break that this is in fact the hardest task in the hunting of the Beaver, which, by means of these dams, makes from a little stream a pond so considerable that they flood very often a large extent of country. They even obstruct the rivers so much that it is often necessary to get into the water in order lift canoes over the dams . . .

The Beaver does not feed in the water, as some have imagined. It takes its food on land, eating certain barks of trees, which it cuts into fragments and transports to its house for use as provision during the winter. Its flesh is delicate, and very much like that of mutton. The kidneys are sought by apothecaries, and are used with effect in easing women in childbirth, and in mitigating hysterics.

Whenever the Beaver is hunted, whether this be in winter or in summer, it is always needful to break and tear down the house, all the approaches to which our Indians note exactly, in order, with great assurance of success, to besiege and attack this animal which is entrenched in his little fort.

In Spring and Summer they are taken in traps; when one of these is sprung a large piece of wood falls across their backs and kills them. But there is nothing so interesting as the hunting in the winter, which is, nevertheless, very wearisome and laborious. For the following is necessary; one must break the ice in more than forty or fifty places: men must cut the dams: must shatter the houses: and must cause the waters to run off, in order to see and more easily discover the Beavers. These animals make sport of the hunter, scorn him, and very often escape his pursuit by slipping from their pond through a secret outlet, which they have the instinct to leave in their dam in communication with another neighboring pond . . .

The bones of the beaver are not given to the dogs, since these would lose, according to the opinion of the Indians, the senses needed for the hunting of the beaver. No more are they thrown into the rivers, because the Indians fear lest the spirit of the bones of this animal would promptly carry the news to the other beavers, which would desert the country in order to escape the same misfortune.

They never burned, further, the bones of the fawn of the moose, nor the carcass of martens; and they also take much precaution against giving the same to the dogs; for they would not be able any longer to capture any of these animals in hunting if the spirits of the martens and of the fawns of the moose were to inform their own kind of the bad treatment they had received . . .

. . . The women and girls, when they suffer the inconveniences usual to their sex, are accounted unclean. At that time they are not permitted to eat with the others, but they must have their separate kettle, and live by themselves. The girls are not allowed, during that time, to eat any beaver, and those who eat of it are reputed bad; for the Indians are convinced, they say, that the beaver, which has sense, would no longer allow itself to be taken by the Indians if had been eaten by their unclean daughters. Widows never eat of that which has been killed by the young men; it is necessary that a married man, an old man, or a prominent person of the nation shall be the one who hunts or fishes for their support. So scrupulously do they observe this superstitious custom that they still at this day relate with admiration how a Gaspesian widow allowed herself to die of hunger rather than eat moose or beaver which was left in her wigwam even in abundance, because it was killed by young men . . .

Document 1.3 Benavides' Memorial of 1630, *Alonso de Benavides*

Alonso de Benavides was a Franciscan missionary who immigrated to New Spain in 1598 and joined the conversion efforts in New Mexico in 1626. He established the mission at the Santa Clara Pueblo in 1627 and ten new missions in the region, running the New Mexico missions from 1626–1629. His memorials provide a great deal of useful material for historians of that region.

1. What does Alonso de Benavides say about the use of the American bison by the Apache? How abundant is this resource according to the author?
2. What economic opportunities does the author see for Europeans in regards to the bison? Does he view bison through the lens of European economic values?

All this nation and province sustains itself on cows which they call cows of Cibola. [They are] like our cattle in greatness, but very different in the form, because it is very short in [the] legs, as if hipped and very high in hump and chest, [with] horns very small and sharp, straight upward; very great manes on the forelock which obstructs their vision, and very curly, and the same on the chins and the knees. And all [are] of a dark-brown color, or black, and [it is] a marvel [when] one is seen with any white spot. There meat is more savory and healthful than that of our cows, and the tallow much better. They do not bellow like our bulls, but grunt like hogs. They are not long-tailed, but [the tail is] small and with little wool on it. The hair is not like that of our cattle, but curly like very fine fleece. Of it are made very good rugs, and of the new ones . . . very fine hats . . . Of the skins of the heifers, clothing is lined, as if they were [skins] of martens. I have told so at length of these cattle, because they are in such great number, and so widespread that we have not found [the] end of them. And we have information that they run from the Seat of the South [the Pacific] to the Sea of the North [the Atlantic], and so many that they gorge the fields. These cattle alone were not enough to make a prince very powerful, if they could be, or might offer, a plan whereby they might be brought out to other lands. Troops there are of more than forty thousand bulls to [all] appearances; without there being among them one single cow; because they always go separate until rutting time. They are not cattle that let themselves be rounded-up, though as a means they take among them our tame cattle. And so, at the time of calving, the Spaniards go to catch the little heifers and bring them up with she-goats. As these cattle are so many, and shed or change their hair every year, that wool remains in the fields, and the airs keep drifting it up to trees, or into sundry ravines, and in such quantity that it could make many rich—and it all is lost.

By these cattle, then, all these Vaquero Apaches sustain themselves; for the which they go craftily to their watering-places, and hide themselves in the trails, painted with red-lead and stained with the mud of the same earth; and stretched in the deep trails which the cattle have made, when the [cattle] pass they employ the arrows which they carry. And as [these] are dull cattle, though very savage and swift, when they feel themselves wounded they let themselves fall after a few paces. And afterward the [Indians] skin them and carry off the hide, the tounges, and tenderloins, and the sinews to sew and to make strings for their bows. The hides they tan in two ways; some leave the hair on them, and they remain like a plush velvet, and serve as bed and as cloak in the summer. Others they tan without the hair, and thin them down, of which they make their tents and other things after their usage. And with these hides they trade through all the land and gain their living. And it is the general dress as well among Indians as Spaniards, who use it as well for dress as for service as bags, tents, cuirasses [armor], shoes, and everything else that is needed. And although each year so many cattle are killed, they not only do not diminish but are each day more, for they gorge the plains and appear interminable. These Indians, then, go forth through the neighboring provinces to trade and traffic with these hides. At which point I cannot refrain from telling one thing, somewhat incredible, howsoever ridiculous. And it is that when these Indians go to trade and traffic, the entire rancherias go, with their wives and children, who live in tents made of these skins of buffalo, very thin and tanned; and the tents they carry loaded on pack-trains of dogs, harnessed up with their little pack-saddles; and the dogs are medium sized. And they are accustomed to take five hundred dogs in one pack train, one in front of the other; and the people carry their merchandise loaded, which they barter for cotton cloth and for other things which they lack.

Notes

1 Dan Landeen and Allen Pinkham, *Salmon and His People: Fish and Fishing in Nez Perce Culture* (Lewiston, ID: Confluence Press, 1999), 51, 52.

2 Roger Williams, *A Key Into the Language of America* (Ann Arbor: The University of Michigan Press, 1971), 89.

3 Ibid., 89, 90.

4 John M. Thompson, editor, *The Journals of Captain John Smith* (Washington, DC: National Geographic Adventure Classics, 2007), 15.

5 William Bartram, *Travels Through North & South Carolina, Georgia, East & West Florida* (New York: Penguin Books, 1988), 57.

6 Daniel H. Usner, Jr., *Indians, Settlers, & Slaves in a Frontier Exchange Economy: The Lower Mississippi Valley Before 1783* (Chapel Hill: The University of North Carolina Press, 1992), 152, 153.

7 Elliott West, *The Contested Plains: Indians, Goldseekers, and the Rush to Colorado* (Lawrence: University Press of Kansas, 1998), 43.
8 John E. Roberts, *A Discovery Journal: Vancouver's First Survey Season, 1792* (Victoria, BC: Monk Office Supply Ltd, 2001), 158.
9 George P. Hammond and Agapito Rey, *Don Juan de Oñate: Colonizer of New Mexico, 1595–1628* (Albuquerque: University of New Mexico Press, 1953).
10 Erna Gunther, "Klallam Ethnography," *University of Washington Publications in Anthropology*, vol. 1 (Seattle: University of Washington Press, 1927), 203.
11 Ann M. Tweedie, *Drawing Back Culture: The Makah Struggle for Repatriation* (Seattle: University of Washington Press, 2002), 31, 32.
12 Ramón Gutiérrez, *When Jesus Came, the Corn Mothers Went Away: Marriage, Sexuality, and Power in New Mexico, 1500–1846* (Stanford, CA: Stanford University Press, 1991), 31.

Further Reading

Bartram, William. *Travels: Through North & South Carolina, Georgia, East & West Florida* (1791). New York: Penguin Books, 1988.
Cronon, William. *Changes in the Land: Indians, Colonists, and the Ecology of New England*. New York: Hill and Wang, 1983.
de Benavides, Alonso, translated by Peter P. Forrestal. *Benavides' Memorial of 1630*. Washington, DC: Academy of American Franciscan History, 1954.
Kirk, Ruth. *Tradition and Change on the Northwest Coast: The Makah, Nuu-chah-nulth, Southern Kwakiutl and Nuxalk*. Seattle: University of Washington Press, 1986.
Krech III, Shepard. *The Ecological Indian: Myth and History*. New York: W.W. Norton & Company, 1999.
Leclercq, Chrestien, translated by William F. Ganong. *New Relation of Gaspesia, with the Customs and Religion of the Gaspesian Indians* (1691). Toronto, ON: Champlain Society, 1910.
Merrell, James H. *The Indians' New World: Catawbas and Their Neighbors From European Contact Through the Era of Removal*. New York: W.W. Norton & Company, 1989.
Silver, Timothy. *A New Face on the Countryside: Indians, Colonists, and Slaves in South Atlantic Forests, 1500–1800*. New York: Cambridge University Press, 1990.
Suttles, Wayne. *Coast Salish Essays*. Seattle: University of Washington Press, 1987.
Thompson, John M., ed. *The Journals of Captain John Smith: A Jamestown Biography*. Washington, DC: National Geographic Adventure Classics, 2007.
Usner, Daniel H., Jr. *Indians, Settlers, & Slaves in a Frontier Exchange Economy: The Lower Mississippi Valley Before 1783*. Chapel Hill: The University of North Carolina Press, 1992.
Weber, David J. *The Spanish Frontier in North America*. New Haven, CT: Yale University Press, 1992.
White, Richard. *The Roots of Dependency: Subsistence, Environment, and Social Change among the Choctaws, Pawnees, and Navajos*. Lincoln: University of Nebraska Press, 1983.

Timeline

Christopher Columbus's Arrival in Hispaniola	1492
Hernán Cortéz's Conquest of the Aztecs	1521
Hernando De Soto's Exploration of the Southeast	1539–1542
Francisco Vasquéz de Coronado's Search for Gold in the Southwest	1540–1542
Juan de Oñate's Settlement in Central New Mexico	1598
The Establishment of the Jamestown Colony	1607
Introduction of Tobacco in the Jamestown Colony	1612
Widespread Epidemics in the Northeast	1616
Beginning of Puritan Settlement in New England	1620
Powhatan Confederacy Uprising	1622
The Creation of the Massachusetts Bay Colony	1628
The Pequot War	1637
Powhatan Confederacy Uprising	1644
Metacom's War	1675–1676
Yamasee War	1715
Natchez Uprising	1729

Pathogens and Plows in the Land of Plenty

2

In 1648 Massachusetts Bay Colony leaders and ministers organized a church meeting in Cambridge to create a constitution of the Congregational Churches and make this the formal church of the colony. During the meeting a snake appeared and created great nervousness. One man held it down while another killed it with a pitchfork. There was no doubt in the leaders' minds that the snake was the devil trying to interrupt the synod and destroy the Christian faith of the colony. But their faith was strong enough to vanquish this evil and formalize the Congregational Church. In an incident nine years earlier a boat containing Indians and English sailors passed a rock where a large snake was sunning itself. As the sailors began shooting at it the Indians remonstrated, asking them to stop because if they failed to kill it they would all be in danger. While sailors interpreted this as danger to the boat, the Indians certainly referred to Manitou and the ability of the snake spirit ally to punish them.

Europeans' and Indians' religious views and attitudes about nature are revealed in these events. Whereas these might seem like incidental moments, the responses revealed attitudes about nature, animals, and spirituality that were implicit in the competing cultures and determined the ways that Indians and Europeans used nature. When English, French, and Spanish viewed North America, as they glimpsed the waves breaking on the beaches of Cape Cod or slogged, parched, across the Jornada del Muerto (route of the dead man) in what is now New Mexico, what did they expect to see? How did they interpret what was before them? Their perspective on the land was determined by their experiences and needs in England, France, or Spain; promotional writing about the New World; and by the pursuit of wealth in their colonies. The markers of success included the ability to enslave Indians and later Africans for mining, sugar harvest and processing, and tobacco agriculture, among other things, as well as freeholders gaining access to their own piece of land and achieving economic success and social mobility.

The Spanish expedition led by Francisco Vásquez de Coronado into the American Southwest, across New Mexico, through the Llano Estecado (staked plains) of Texas and all the way into Kansas was guided by stark expectations. They wanted gold. The Spaniards were convinced that cities of gold awaited them on the plains and were disappointed in the dusty pueblos and villages they found time and again. Finally, at the end of their journey in Kansas the exhausted men who hoped to return to Mexico City weighed down with gold gazed upon Querechos raising corn, beans, squash, melons, and sunflowers. As historian Elliott West writes, "It's always disappointing to ask for gold and be handed melons."[1] To Coronado this was a wasteland to be abandoned as quickly as possible. While they returned home with empty packs, they left behind organisms that initiated a wave of ecological change that would devastate Southwestern Indians even before the Spanish undertook serious colonization in that region.

Our history has taught us to nod knowingly when told that Coronado sought gold in New Mexico, Texas, and Kansas, mostly because the Spanish had already been so successful in that quest among the Aztecs of Meso-America and the Incas of Peru. But the first permanent European settlers in Virginia also sought gold. That group of Englishmen making their way up the James River of Virginia in 1607 and establishing the commercial colony of Jamestown sought neither freedom nor to create a new society. They were explicit in their expectations. Gold, rubies, furs, and Indian slaves were their goal; they hoped to emulate the Spanish model of colonization in Mexico and Peru.

The expectations of wealth and the reports of explorers and traders taught the colonists to see or expect more natural richness than actually existed. Early explorers, colonists, and promoters used exuberant descriptions and held high expectations of the new land. Their view of the beaches, rivers, and forests was driven by the needs of the Old World. Vast forests of white pines standing more than 200 feet in height meant masts for sailing ships. Massive schools of cod off the shore were food not only for colonists but for sale in the markets of Europe. And so much firewood! The immeasurable forests would allow the poorest colonist to consume as much firewood as gentry or royalty in Europe. Social mobility via fire on the hearth was easily achieved. The consumption of firewood by colonists was so extreme many Indians assumed that Europeans were there to harvest it for export to Europe. What was seen, desired, and described in breathless, hyperbolic prose was informed by both what was lacking at home and what sold well in the marketplace. They saw commodities of value and the bases of wealth and power.

The rhetoric of copiousness was so common that British writer and explorer Christopher Levett parodied the descriptions in a 1628 book.

> He would not tell you that you may smell the corn fields before you see the land; neither must men think that corn grow naturally, (or on trees) nor will the deer come when they are called, or stand still and look on a man until he shoots him, not knowing a man from a beast; nor the fish leap into the kettle, nor on the dry land, neither are they so plentiful, that you may dip them up in baskets, nor take cod in nets to make a voyage, which is not truer than that the fowls will present themselves to you with spits through them.[2]

But flocks of passenger pigeons flying overhead for days on end, rivers so filled with fish that one could not throw a rock without striking a fish, and beaches where lobsters could be gathered at low tide speak to the very real abundance of the "green breast of the new world."

The wealth of America's nature drew explorers, settlers, religious leaders, and capitalists to the shores of the Atlantic, Gulf Coast, and the Southwest. In their quest to expand new societies and extract resources for the betterment of themselves, their faith, nation, and God, they unleashed organisms on New World ecosystems that inflicted great damage on native peoples and ecosystems, paving the way for European settlement and conquest. As Europeans established themselves, adapted to the landscape, and expanded, Indians fought to survive and adapt to this world transforming so quickly around them.

Pathogens and Conquest

The most devastating weapon brought by the Europeans was one that neither they nor the Indians could see or understand: pathogens. Small pox, measles, influenza, cholera, chicken pox, and others were all Old World diseases against which the Indians had no immune defense and were highly vulnerable to. Puritans moving into the interior of New England in the 1630s commented on the empty villages and the huts full of corn, as well as the untended fields of beans, squash, and corn, and they interpreted this empty land and bounty as proof of God's ordination of their mission, as his preparing a way for them in the wilderness. John Winthrop captured the sentiment well in 1634 when he wrote, "for the natives, they are neere all dead of small Poxe, so as the Lord hath cleared out title to what we possess."[3] This virgin soil epidemic played the central role in the conquest of the Indians, their dispossession of the land, and the transformation of the landscape. Natural wealth stored up over thousands of years in North America joined with Old World pathogens to ease the process of European conquest and colonization.

The earliest epidemics occurred with the Spanish explorations of the interior from 1513 through the 1540s in the Southeast and Southwest. Evidence indicates that epidemics of

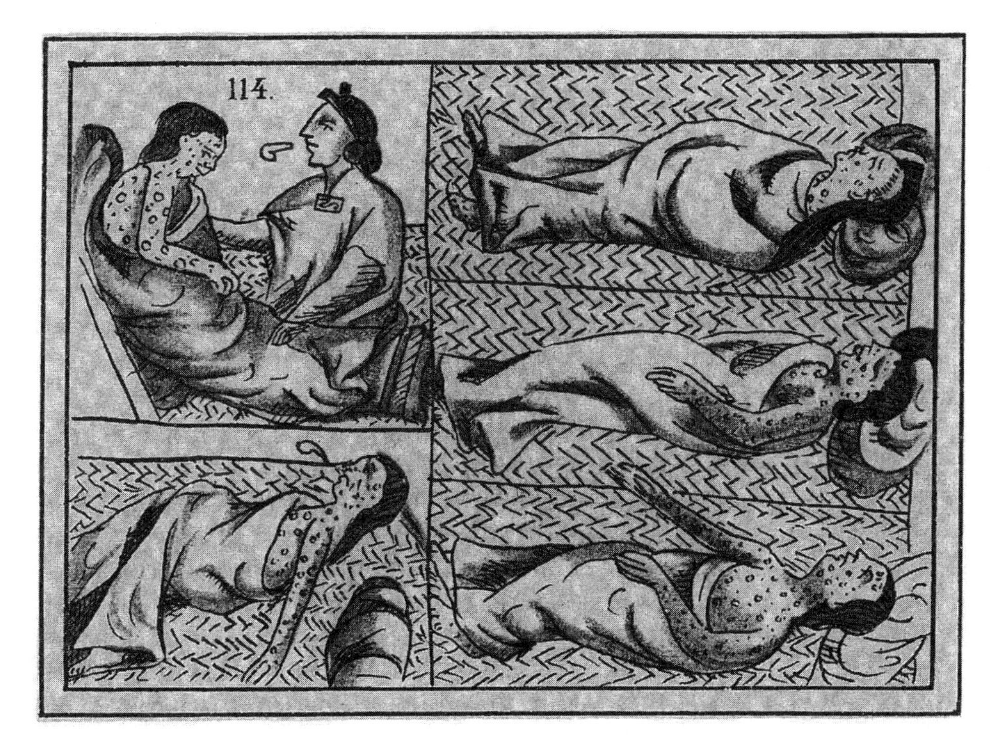

Figure 2.1 This image of ill Aztecs from the *Florentine Codex* is the first depiction of small-pox in the New World. This disease was particularly devastating for the Aztecs as it struck during Spanish commander Hernán de Cortéz's attack on Tenochtitlan. If the leadership and army of the great Aztec city had not been struck by this disease, they might have withstood this initial assault. After six months of brutal fighting the city fell to Spain. Pathogens were integral to European conquest across South and North America, and small pox was the most devastating of the Old World diseases. From the *Florentine Codex*.

the 1520s and 1530s devastated native peoples far to the north as the result of these Spanish incursions and tribal trade and travel. Smallpox was the deadliest of the diseases and aided Spanish conquest by laying low many of the Aztecs defending Tenochtitlan against Cortez's Spanish soldiers and Indian allies.

Historian Alfred Crosby describes small pox as

> a disease with seven-league boots. Its effects are terrifying: the fever and pain; the swift appearance of pustules that sometimes destroy the skin and transform the victims into a gory horror; the astounding death rates, up to one-fourth, one-half, or more with the worst strains. The healthy flee, leaving the ill behind to face certain death, and often taking the disease along with them.[4]

Small pox in the New World often killed at a rate of 30 percent or higher in the first wave with repeated waves of disease possibly exterminating up to 90 percent of a tribe. Survivors built a resistance to the next round of disease, but mothers did not pass antibodies on to their children; following epidemics continued to exact an onerous toll. One theory for the higher death among Native Americans is that, in addition to the lack of acquired immunities, this homogeneous population's general immune system was similar throughout the New World, whereas that of the Europeans and Africans was much more diverse. Within that Old World population there existed a wide range of immune systems with varying capacities to handle smallpox, leading to an overall lower death rate. Even though smallpox was the deadliest, measles, influenza, and other diseases also felled Indians by the hundreds of thousands.

The Europeans' Greatest Ally

Variola major wreaked havoc on human populations throughout the New World and was greatly feared in the Old. Part of the pox family that includes chicken pox and cow pox, small pox is large for a virus and has a single linear double-stranded genome of 186 genes. It assumes multiple infectious forms and enters the cell in a number of ways. It is unique from most viruses in that its long DNA strand allows it to replicate inside the cytoplasm of the cell rather than using the nucleus to reproduce. Airborne transmission of the virus is the primary method of contagion; there is no animal or insect by which the disease is spread. Contact with fluids of the infected person's body, clothing, bedding, and even dry, separated scabs can also spread the variola virus. The incubation period is usually 12 to 14 days but can have a wider range of 7 to 17 days.

Carried into the lungs, the virus moves into the cells of the lung's air sacs. Once inside they begin rapidly replicating inside the cytoplasm. Consuming the cell's nutrients and energy, the virus replicates until the cell bursts, releasing a horde of viruses to enter tissue fluid and set up shop in the lymph nodes. While the lymph nodes normally spot and direct attacks against infections, the variola virus uses the nodes as a key site of viral reproduction thus circumventing the body's defenses. As the viruses replicate, they flood the body with the multitudes issued forth from the lymph nodes via the blood stream.

During this long incubation period the host shows no symptoms, a factor that worsened contagion as many who were sick appeared healthy and were given refuge. At the tenth to twelfth day, the body and immune system is completely overwhelmed, with the fever peaking to 104 degrees Fahrenheit and the victim wracked by brutal headaches and severe back pain in many cases. Two or three days after the onslaught of initial symptoms, blisters develop in the throat and mouth and are quickly followed by the now revealing and horrifying evidence of smallpox, the spread of flat, red spots on the skin. These rise as the cells underneath swell up with replicating variola. These blisters sometimes burst, covering skin in pus and leading to an even more terrible condition called semi-confluent or confluent scabbing when the whole surface of the skin scabs over. Most victims died from a circulatory system collapse and organ failure due to the damage by the virus. Some died of secondary bacterial infections, particularly through the open blisters, while others died of cardiac failure due to the massive loss of blood through open wounds. Evidence suggests that in the three centuries leading up to the 18th century, the disease actually grew more virulent with ever-increasing mortality rates over that period. In the 17th and 18th centuries, one of three people suffered from small pox with one out of twelve killed over that period. In one 25-year period in the 18th century an estimated 15 million people succumbed to this disease, and the estimates of death to small pox in the 20th century range from 300 million to half a billion.

Survivors of the variola major virus gained comfort from the knowledge of their protection from future outbreaks. (The lucky ones were those who caught variola minor—it generally resulted in a 1 percent death rate but still conferred immunity to future outbreaks of the minor and major version of the pox.) Others knew of their experience through the visible evidence: bodies covered with deep scar pits, with some victims suffering as many as 500 scars on their face. Blinding was also an occasional outcome; the disease caused approximately one-third of the cases of blindness in Europe in the 18th century.[5]

God's Invisible Hand

New England peoples were struck by contagion in 1616 before the arrival of the Pilgrims in 1620, leaving the voyagers to their land of milk and honey and to comment on the empty villages and abandoned cornfields. As much as half to three-quarters of the native population may have died even before the pilgrims set foot ashore. Those epidemics continued even as the English were trying to avoid starving and established a beachhead in New England. A European observer in the Boston Bay area in 1622 described the Indians having

> died on heapes [sic], as they lay in their houses; and the living, that were able to shift for themselves, would runne [sic] away and let they dy [sic], and let there [sic] Carkases [sic] ly above the ground without burial . . . And the bones and skulls upon the severall places of their habitations made such a spectacle after my coming into those partes, that, as I travailed in the Forrest near the Massachusetts, it seemed to me a new found Golgotha.[6]

The death toll from the epidemics was worsened by the inability of villages stricken by sickness to plant, tend, or harvest crops. Puritan and long-time Plymouth Plantation governor William Bradford described Indians' helplessness. The small pox victims

> fell down so generally of this disease as they were in the end not able to help one another, no not to make a fire nor fetch a little water to drink, nor to bury the dead.

> But would strive as long as they could, and when they could procure no other means to make fire, they would burn the wooden trays and dishes they ate their meat from, and their very bows and arrows.[7]

Similarly, an epidemic might occur at an important part of the seasonal round, such as the fish harvest or deer season. In such cases, deaths from disease would be followed by deaths from starvation or Indians succumbing to other sicknesses because of their weakened physical state and despair. Historian Timothy Silver writes,

> confronted with the toxic red rash of measles and the draining pustules, callous scabs, and horrible scars of smallpox, many Indians elected to take their own lives. While among the Cherokees, James Adair saw Indians cutting their own throats, shooting themselves, or jumping into raging fires to escape the indignity and humiliation of disfigurement.[8]

They did not simply give up but found that their medicine was unequal to the diseases, as was European medicine. Indian use of prayers, sweat lodges, cold water treatments, and cleaning of sores spread the contagion or worsened secondary illnesses. Medicine men sometimes attempted to suck the poison from small pox pustules, thereby carrying the disease to others while growing sick and dying in the process of moving from sickbed to sickbed. Also, families often participated in caring for the sick as part of Native ceremonial treatment. They picked up the disease, and it spread like a fire.

In the American Southwest the Coronado expedition of 1540–1542 unleashed disease through New Mexico, Texas, Oklahoma, and Kansas so that before the next major southwestern expedition or, as the Spanish called it, Entrada, led by Juan de Oňate in 1598, Indian populations throughout the region were already severely reduced. European conquest was eased and, in many cases, made possible by the fact that Indians were already greatly weakened before contact in most cases and certainly at the point that they launched attacks to drive Europeans from their land.

A passage from John Winthrop's journal in 1633 serves as a reminder that colonists were not completely spared the devastation of epidemics:

> John Sagamore died of the smallpox, and almost all his people (above 30 buried by Mr. Maverick of Winesmentt in one day). The towns in the Bay took away many of the children, but most of them died soon after. James Sagamore of Saugus died also, and most of his folk.[9]

While the Puritans did suffer, their population increased through both immigration and natural reproduction, while the Indians suffered multiple waves of epidemics and their numbers plummeted. The eliding of the Indian presence through disease helped the Europeans gain control of land and expand. The system of land ownership created by Europeans also assisted them in gaining control of land from the Indians that survived.

"Proper" Land Use and Ownership

Europeans brought to the New World a system of land use and ownership foreign to Indians. That does not mean, however, that they had no conception of sovereignty, territory, or rights to parts of the landscape. In fact, Indian tribes and nations knew their territory and boundaries and fought to protect those or to expand onto new land when resources ran low or when suffering from drought. Moreover, within Indian communities there was knowledge of land planted and used communally versus plots of soil that were used by particular individuals and families. Also, harvesting sites for fish or other species were generally divided by individuals within a tribe by rank. The use of land and resources wasn't as breezily egalitarian as some have assumed. However, the European model of land ownership and usufruct rights was completely strange to them and proved integral to dispossessing surviving Indians of their land.

The alienation of land simply refers to the relatively new legal and economic system of selling land and transferring ownership. In the 16th and 17th centuries this was still a relatively new process in European culture, following the devastation of the Black Plague epidemics of the 14th century and accompanying the rise of a yeoman farmer class in Europe. An Old World pathogen, which wiped out approximately half the population of Europe in the 14th century, led to dramatic reimagining of land ownership principles. Ironically, Old World pathogens unleashed in the New World would make it easier to apply these legal conceptions of land acquisition and ownership.

The Puritans and other English settlers viewed dominion over the land and its conversion to useful production as central to their mission in the New World. Genesis 1:28 provided God's order to be fruitful, multiply, and subdue the earth. They took on this task with great energy. Regardless of the evidence of Indian land use, such as vast fields of corn, squash, and beans as well as fire employed to clear out underbrush and improve hunting, the English viewed natives as heathens, living in sloth on an abundant landscape. One English observer in Virginia noted, "Savages have no particular propertie in any part or parcel of that Country, but only a general recidencie there, as wild beasts have in the forest."[10] Since the Indians did not fence their lands or build European style structures upon it, it was "clear" to the British that the land was not properly subdued and used; it was considered empty land and available for use by English settlers. Although Indians counted on rotation of land, so were actually using more land than Europeans comprehended, the forests and fields that were not farmed or did not contain domesticated animals were simply available for the taking. Depending on relationships and attitudes of colonists, some did pay for land and help provide fencing for Indians against European livestock, but as relationships worsened and land became scarce, it was easier to marginalize native peoples and simply take their land by occupation or by force.

Environmental history can at times be too materialist, too focused on ecology, economics, the law, and other factors in history. Any student of this era would be well-served to remember the zealous sense of religious mission that animated Puritan and other deeply religious settlers. Puritans viewed North America as a land of milk and honey and their voyage and settlement there as a new version of the Israelites among the Canaanites. Driven by the idea of creating a "City on a Hill," a beacon of light, with the goal of reforming the world in God's name by creating a theocracy in New England. The Puritans of course viewed all of their actions as reflecting the will of God. This land of abundance was their due reward, provided by a demanding but generous Lord. The wealth of this Canaan would support the creation of this new Christian society, but to accomplish this Indians would need to be set aside, displaced from the land. Viewing them as savage and as minions of Satan certainly made their dispossession much easier.

Animals, Faith, and Land Use

While deeply Christian, Europeans still held on to much of their folklore about animals as portents of change or omens of bad luck or death. Animals were not simply commodities to them but had distinct personalities, and their behavior and actions were closely watched and interpreted. But the signals and actions of animals fit within their Christian faith and could be explained as reflecting God's will and design. God and humans held dominion over animals wild and domesticated as well, so Europeans did not lend these creatures power to bring them ill will or bless them with abundance and fortune—that was solely the province of God.

This perspective stood in strong contrast to native people's belief in Manitou and the power of animals themselves to afflict the people with disease or famine or, alternatively, reward them. Where native peoples and Europeans may have found common ground on their relationships with animals, instead for Europeans the native belief in Manitou and animals' power proved their savagery and inferiority. Adding to the difference was the way that native beliefs seemed to situate them outside God's love. Within their Christian framework the English did believe that certain animals could be witches' familiars and that native medicine men used black magic in league with the devil. In that context, native spirituality and ascribing great power to animal spirit allies proved that Indians were under the sway of Satan. Rather than finding common ground in their views of animals, Europeans used Indians' attitudes toward and reverence for animals as a marker of separation and further proof that native peoples were savages that did not own title to the land.

Huge feral pigs, cattle larger and stranger looking than deer, and other domesticated animals confounded native peoples. The Aztecs and others referred to the horse as a large deer, having never seen such an animal before. Indian warfare was completely unequipped for fighting cavalry tactics, and the first Meso-American Indians to manage to kill a horse in Cortez's cavalry spent the night in celebration and sent pieces around to other Indians to consume and to acknowledge the great victory. However, it would not take native peoples long to adapt to and use the horse as well as any European. Feral cattle and pigs often preceded the settlement of Europeans; an Indian that had never seen a Spaniard, Frenchman, or Englishman might stumble upon such an animal and struggle to fit it into their existing cosmology. Naturally, they accorded cows, horses, pigs, and sheep strong spiritual power. They did this not only because of their size and strangeness but because of their association with Europeans and what was clearly a very powerful god. Likewise, they did not see animals as property; they became such only when killed. Native peoples were compelled to change their attitudes dramatically to be able to protect their crops from these European livestock and to begin incorporating them into their own cultures and economies.

Europeans had a long tradition of animal ownership and husbandry. They lived and organized their lives around raising, feeding, herding, and harvesting cattle, pigs, sheep, chickens, and other species. From their perspective, the fact that Indians had not domesticated animals weighed against them as proof they did not truly use the land. It wasn't simply animal use that was important; it was the ordering of the landscape and time around agriculture and labor, creating an agrarian pastoral that resembled England. Indians' failure to do this also justified occupation of their lands.

An Army of Livestock

Indians may not have domesticated animals, but the arrival of Europeans and their animal companions had immediate impacts on them. The lack of fences hurt Indians in ways beyond the debate over legal use and ownership of the land. Unfenced land could not be protected from the wanderings of livestock, and neither could crops. Settlers across the New World in most cases allowed pigs, sheep, and cows to run loose in the new, rich environment. It was easier to let the animals fend for themselves and harvest them as opportunity arose than to build fences, grow hay, build barns, and so forth. Pigs, in particular, were adept at defending themselves and quickly went feral. In fact, feral pigs tracing their ancestry to this early period of American history cause significant environmental damage in numerous regions across the continent today. The conquest of Native Americans by Europeans was not only the result of superior warfare technology but was enabled by the impact of pathogens, organisms like cattle and pigs, land-use laws and practices, and the capitalist economic system.

Disease epidemics followed by free-ranging livestock consuming both domestic and wild plants devastated native peoples. Indians struck by disease were often unable to tend their fields or harvest crops, so when they did recover, they did not have enough food to make it through the winter and suffered and starved. Starvation and malnutrition also rendered them more susceptible to disease and worsened the impact of epidemics. Free-ranging cattle and pigs worsened damage by consuming Indians' crops, stripping them of food resources. Spanish priests in California complained about this problem in 1782. They described how Spanish livestock run loose and mix with animals owned by Indians and the missions and

> have caused unceasing damage to the crops put in by the Indians. Now all this took place before they were in formal possession even of the land on this side of the river—and the boundary line comes right up to the ground which the Indians have placed under cultivation—it is evident that the damages—and nobody can deny it—will be far greater. The consequence will be, perhaps, that the Indians will have to stop their field work, so as not to labor in vain; and they will have to rely for their food on the herbs and acorns they pick in the woods—just as they used to do before we came. This source of food supply, we might add, is now scarcer than it used to be, owing to the cattle.[11]

Even as they stripped agricultural fields bare, these animals also consumed grass and brush, undermining the game species Indians relied upon as well as causing great ecological damage. Sheep in Latin America and the American Southwest stripped grass to bare soil, again removing food that native games species needed to live. Historian Elinor Melville famously referred to this as a "plague of sheep."[12] The ungulates would explode in population as they consumed all native grass. With massive die-offs the land had an opportunity to recover, but native plants were replaced by European weeds and other invasive plants. This accidental transformation of ecosystems accompanied the more deliberate European efforts—both undermined the Indians' ability to survive until able to gain immunities and adapt to the new organisms and changed landscape.

Swine made themselves a nuisance for settlers as well as Indians, causing such damage that a series of laws were passed in New England in the 1630s–1650s to limit the destruction. In established communities, Europeans began fencing the pigs in, feeding them a mix of fish, corn, and food scraps. But they also let them run loose still in areas further from town, like islands in Narragansett Bay. As Cronon writes,

> Along the coast, the animals wreaked havoc with oyster banks and other Indian-shellfish-gathering sites, but caused little trouble to the English. Roger Williams described how "the English swine dig and root these Clams wheresoever they come, and watch the low water (as the Indian women do)." In one important sense, then, English pigs came into direct competition with Indians for food: according to Williams, "Of all English Cattell, the Swine (as also because of their filthy dispositions) are most hateful to all Natives, and they call them filthy cut throats."[13]

Also, the process of setting pigs loose on the front edge of European settlement was repeated time and again over the next two and half centuries in places like the forests of East Texas, the hill country of northern Mississippi, and the camas prairies in northern Idaho. Indian farms or managed landscapes as well as wild plants provided a natural subsidy for the rapid growth of feral livestock and early economic growth in settlement communities.

From the Three Sisters to Grain, Livestock, and Tobacco

Along the Atlantic seaboard settlers moved west in the beginning of a process of expansion, conquest, and settlement that would continue for the next three centuries. In the initial stage of westward expansion the first lands they occupied and used were abandoned Indian fields. One colonist wrote, "Wherever we meet an Indian old field or place where they have lived, we are sure of the best ground."[14] The first wave of settlers benefited from the land having already been cleared and fertilized by Indians. The next, less fortunate wave typically cleared the successional stages of berry bushes, brush, and small trees covering abandoned fields. Later arrivals were compelled to clear forests in the old-fashioned way. European colonists used similar methods to Indians to remove trees. When converting full forest to agriculture, they were compelled, like the Indians before them, to plant around the roots of trees that had been girdled and killed. As they established themselves and acquired more labor and wealth to purchase domestic animals and tools, they attacked the stumps and root systems to put the entire land to work. The Atlantic economy is integral to this process because as farmers were able to realize profits from their various crops and livestock sold in urban areas on the coast, in the Caribbean, and even Europe, those profits were funneled into more land, indentured servants, slaves, and hired wage labor in the North. Also, they were able to purchase more tools and beasts of burden. The rewards of the marketplace increased the pace and intensity of landscape change all along the eastern seaboard.

In New England colonists farmed for both subsistence and with an eye to market demands. From Jamestown and the Pilgrims on, settlers typically had investors that required payment in natural and agricultural resources, so participation in the Atlantic economy was required. Corn was valued because it could be both sold and used to fatten domesticated food animals and for its toughness against extreme weather. Wheat was preferred for consumption and also because it was an excellent commodity for trade and sale. It, however, required extensive labor and was vulnerable to high winds, hailstorms, and rain occurring during the harvest, leading to sprouting from the grain heads or rot. Kitchen gardens produced a wide variety of foods that could be cooked, traded, and stored for several months. Turnips, radishes, cucumbers, peas, beets, strawberries, lettuce, and other foods fed the colonists and also became desired objects of trade for the Indians.

Cattle accompanied or led colonists every step of the way. They provided dairy products in the forms of milk, butter, and cream as well as meat and hides for numerous uses. Possibly more important was their use in trade to coastal markets and distant economies. The sugar plantations in the Caribbean consumed their landscape in great sugar cane fields and were compelled to import food, particularly the meat of cattle and pigs. Hence the eyes of the colonists turned toward distant markets, whether urban areas on the Atlantic coast or the bustling plantation economies of the Caribbean. Cattle and pigs were one of the best ways to gain precious currency, and Indian farms and the wild landscape served as a natural subsidy to the colonial economy and undergirded economic growth in this period. Moving the animals to market necessitated that roads be cleared; most of them were easily more than a hundred feet wide. These bovine and porcine thoroughfares not only resulted in increased ecological damage with the removal of trees and erosion from the roads but also strengthened the ties between urban centers and rural producers, driving the process of producing goods for the market while purchasing European manufactured goods and, therefore, extending the reach of the market economy. Commodities such as cattle, swine, logs, fur, and agricultural products were more aggressively harvested and produced for use in the expanding market economy. The roads also made migration and settlement in new (Indian) lands easier, playing an important role in both geographic and economic expansion.

Cattle also require a great deal more land than what is needed for subsistence agriculture, generally at least two to ten times as much depending on the quality of grasses. This ongoing need for land, in conjunction with rapidly growing herds, helped drive expansion of colonial communities further into Indian lands. The impacts of the market were noted early by William Bradford, the five-time governor of Plymouth Plantation, who felt that the destabilizing force of cattle expansion undermined the strength of the religious project of the Puritans. Communities were disrupted by people being forced to leave in search of more grass for their cattle.

Allowing cattle to forage freely hurt colonists in other ways. Farmers were unable to collect and use bovine manure because of the lack of stables, corrals, and barnyards. This has been common practice in Europe, but feral husbandry predicated on an abundance of land and feed for swine and cattle made manure collection to improve soil practically impossible and consequently hurt farming. The loss of this brown gold led farmed soil to become infertile more quickly, and this drove even more expansion on to new, fertile lands. In some cases, instead of abandoning a hard-earned farm, colonists instead harvested large numbers of spawning fish like alewives and menhaden and buried them in the soil as fertilizer. The earliest documented example in New England was in 1634, and the practice continued into the late 18th century even as fish runs were collapsing across the region. Natural abundance provided economic growth to the colonies but at the cost of declining soil and community instability. The wealth of nature led to a careless husbandry that caused the decline of agricultural soils and necessitated further expansion for better land.

Removing Predators

The first English settlers in New England knew next to nothing about hunting. In Europe hunting was reserved for the gentry and nobility, a sport of recreation and privilege. As a consequence English colonists viewed hunting as recreation, not work. This gave more credence to their view that Indians lived lightly on the land and the men were lazy and women essentially oppressed, performing all the agricultural labor. While early colonists were incompetent hunters due to their lack of experience, they quickly adapted to the new opportunities and needs and learned to hunt game for food, becoming particularly adept at hunting waterfowl with shotguns. They also hunted to remove predators such as bears, wolves, cougars, and bobcats. Using dogs, tripguns, pitfalls, and offering bounties, they quickly reduced the predator population. The assumption of dominion over nature and the obsession with a well-ordered pastoral landscape increased their aggressiveness, and in some cases Europeans demonstrated great cruelty.

Because livestock ran feral in many cases and, when tended, required more protection than could be reasonably provided, they were highly vulnerable to wolves in particular. As Barry Lopez has demonstrated in *Of Wolves and Men,* there is a long history of fear and hatred of wolves that shows that their destruction is usually not only for economic reasons.[15] The fear of this animal also hid the complex behaviors and social order of this fascinating predator until research conducted in the 20th century. Certainly the English feared the wolves howling in the swamps and forests and hated them for that fear as well as for the economic competition they presented. Sheep, pigs, cattle, and horses were easy prey for these animals that hunted in packs, and the English used a number of tactics to try to exterminate the predator. One common tactic was to bury large fish hooks inside a ball of fat and leave it in the open. Wolves who gobbled up the bait would die later of painful internal hemorrhaging. Guns attached to trip wires were an easy way to kill wolves but also killed livestock, and in one community a small child. Regulations on these limited their use. Pitfalls were another tactic, but the most common was a bounty system throughout the colonies paying English and occasionally Indian hunters for wolf heads. In New England wolf heads were nailed to meetinghouse walls in a demonstration of dominion. In the ongoing effort to reduce the lupine presence, colonists were encouraged to buy mastiffs to hunt wolves. One community actually paid more for wolves killed by dogs than by poison in an effort to incentivize dog ownership and hunting.

British traveler and writer John Josselyn described torturing wolves, revealing hatred, fear, and the hubris of dominion over the natural world.[16] In one incident in Maine he and his hunting partners ran down a wolf with a mastiff, which grabbed the animal by the throat and held it down for the men to bind the animal to carry it to their house for the night. That evening Josselyn and his companions set the wolf loose with the dogs in the room, hoping to see a terrific fight. The dogs sat quietly, and the wolf slouched without making a sound, steadily watching the door. Bored, they carried it outside and smashed in the animal's brain. In 1664 his mastiff caught another wolf. Carrying it home, Josselyn and his fellow hunters staked it out in the yard and let small dogs attack it. The wolf was crippled by a broken leg, and again the bored men caved in this animal's brain as well.

Cruelty directed at wolves and other predators would remain common into the 20th century. In the colonial period the Europeans were efficient in their destruction of predators, and by the early 18th century these animals were drastically reduced in number, effectively exterminated in areas of colonial settlement and land use.

Tobacco Land

The Virginia Colony barely survived its first three years after the original landing in 1607, losing half the population to death while many of the others wisely retreated to England. They would not have survived without the thousands of bushels of corn provided willingly by Indians or stolen by the starving colonists. It didn't help that the rats they accidentally introduced to the New World then wrought havoc on their food supplies. The colony did shift to production of a New World crop, tobacco, quickly after introduction of a popular new breed by John Rolfe in 1612 and owe its success to that crop. This mainstay of the colony drove expansion, economic growth, and ecological change in that region for the next two centuries.

The tobacco economy was practically metastatic in its growth. The first shipment to England was in 1617; by 1624 the colony was exporting 200,000 pounds, and by 1638 the number of pounds of leaf exported exceeded three million. Early tobacco farmers, eager to reap the high prices available in England in the 1620s, used Indian fields, then aggressively cleared forest. The nitrogen and other nutrient supplements provided by burning was not enough to effectively extend land use as tobacco sucked the fertility from the soil. Planters abandoned land typically after three to five years to open up new forests for tobacco. Believing that hardwood forests, particularly along river bottoms, indicated the most fertile soil along with sufficient rain for agriculture, they converted these forests to farmland, reducing the hardwood forests and also removing mast, an important food source for species such as deer and bear, among others. This rapid expansion undercut the ecology of the region while leading to a revolt that shook the Virginia colonial government.

Indian Resistance

Native peoples did fight back and try to hold onto their lands in numerous instances. Some of the bloodiest conflicts in American history in terms of proportion of the population killed and wounded are these battles and wars occurring during the colonial era. Virginia

Indians attempted to drive the colonists into the sea in 1622 and again in 1644. They managed to kill a third of the English population in the 1622 war. The Virginia Company went bankrupt, and the colony reverted to the rule of the British government. Colonists in retaliation and in pursuit of land attacked regional Indians on a regular basis until the second uprising in 1644. This led to a treaty to limit growth onto Indians lands and exacted a beaver pelt tribute to be paid to the colony.

One of the most horrific conflicts in this period was the Pequot War of 1637. Furious at their loss of land in the lower Connecticut River Valley the Indians began striking at homes and small towns then disappearing again back into forests and swamps. The Pequots were winning the war until the Narragansetts allied with the Puritans and led the English force to Mystic Fort. The Pequot warriors had hidden their children, wives, and elders there, and the Puritans surrounded the stockade, set it on fire, and gunned down the women, children, and older people as they attempted to flee. The Narragansetts were horrified at the fury of European warfare, and William Bradford wrote of the battle,

> it was a fearful sight to see them thus frying in the fire and the streams of blood quenching the same, and horrible was the stink and scent thereof; but the victory seemed a sweet sacrifice, and they gave the praise thereof to God, who has wrought

Figure 2.2 This wooden engraving of the massacre of Pequot Indians at Mystic Fort was made in 1638 by two militia captains from the battle. Library of Congress.

> so wonderfully for them, thus to their enemies in their hands and given them so speedy a victory over so proud and insulting an enemy.[17]

The few Pequots that were captured were sold into slavery in the sugar plantations of the Caribbean.

It was the Narragansett's turn next in 1675 and 1676 when they rose up in alliance with the Wampanoags and other Indians under the leadership of Metacom in what is known as Metacom's War. The Indians were winning this war in the early stages as well. More than 50 of 90 Puritan towns were attacked and a dozen completely destroyed. The tide turned when the Iroquois allied with the British. With Metacom dead and his head nailed to a wall like the wolves' heads, the colony was a smoking battleground of thousands dead on both sides. It would take another 40 years for the English to recover the ground lost in this conflict. Because of this war the British crown took control of the colony away from the Puritans.

There were numerous other conflicts along the eastern seaboard throughout this period—some all-out wars like the Yamasee War of 1715 and some local, small skirmishes. The fact that many of these Indians nations were fighting at 10 to 30 percent of their original pre-contact strength speaks to the integral role of pathogens and livestock in conquest. Deaths from disease and famine made Indians weaker and less able to defend their lands and culture. Indian attacks on the colonies also gave further credence to European notions of Indian savagery and demonic nature. For example, Metacom was referred to as an agent of Satan and a "hellhound, fiend, serpent." Viewing the Indians as savages and as opponents directed by the Devil to oppose Christianity provided Europeans further justification for ongoing expansion onto Indian lands and conquest.

Land Hunger and Revolt

Instability created by the tobacco economy led to a violent uprising of yeoman farmers and servants in the Virginia Colony in 1676. Bacon's Rebellion was primarily a conflict born in the competition for land to produce tobacco. Because this crop quickly diminished soil quality, tobacco growers were always on the move, seeking new land. When Indians of the region were defeated in a brutal war in 1622 and ceded large quantities of their territory for farming, prices climbed ever skyward in Europe. This labor intensive crop also required vast quantities of human toil; each acre of tobacco required 900 man-hours of labor. This was more than the landowners could or would provide, so indentured servitude emerged as the dominant labor system of the colony. Servants from England provided strenuous labor for four to seven years with the hope of gaining 100 acres of land to farm for themselves. Of course they had to survive to earn the fruits of their labor. Unfortunately, 50 percent of indentured servants in the Virginia colony died before gaining their freedom. Indentured servants had few rights, were regularly beaten, sometimes physically mutilated as punishment, and even gambled away as property. Those who managed to survive were determined to become successful planters and advance economically and socially. This required ever more land.

Expansion bumped up against Indian borders, increasing the value and cost of available land. In addition to population pressures, treaties between the colonial government and local Indians after a brutal war from 1644–1646, removed land from English use and settlement. The treaties protected Indian lands in order to prevent further conflict. Within 20 years, overproduction of tobacco, with a corresponding glut and drop in price, increased taxes, and limited representation in the colonial legislature for lower class settlers, led to a colony brimming with turmoil and tension. At the same time Indian populations continued to decline as a result of repeated epidemics. By 1670 the English population in the colony had climbed to 40,000, whereas the Indian population had plummeted to a few thousand with some tribes completely extinguished.

The rebellion began with conflicts between land-hungry colonists and Indians. A group of Doeg Indians, angry at not being paid for goods by a trader, attempted to steal his swine as compensation. The trader's overseer killed one of the Doeg, and the Indians responded by killing him. Violence erupted. Several neighboring planters killed members of the Doeg and some friendly Susquehannock Indians not involved in the conflicts. When the governor of the Virginia Colony, Sir William Berkeley, failed to provide reparations to the Susquehannocks, they struck and events spun out of control. Nathaniel Bacon, a planter, the governor's cousin, and a member of his colonial council, insisted upon a stronger response to the Indians, while Berkeley preferred a defensive policy that would leave the treaty relationships between the colony and the Indians intact. Bacon and other freeholders, as well as servants aspiring to be farmers, believed this policy was designed to retain the control of land and government in the hands of planter elites. They were particularly furious at what they perceived to be overly generous treatment of Indians, keeping lower-class farmers from gaining promised and earned land on which to grow tobacco.

Led by Bacon, a large irregular force of white and black indentured servants as well as lower-class white and black farmers moved against the Indians to seize their land. When Berkeley expelled Bacon from the council and briefly arrested him, events escalated into an open revolt against the colonial government. Bacon issued a manifesto declaring that the wealthy were "parasites" on the lower classes and demanding that Indians be removed or killed to open their land to tobacco farming. Bacon's army soon occupied and then burned Jamestown, the colonial capitol, to the ground. On October 26, 1676, the erstwhile revolutionary died of dysentery, and Berkeley moved against the rebellion, executing 23 rebels and quelling the insurrection.

The rebels achieved their primary goal of weakening Indians and taking their land. A few limited reforms such as greater representation in the colonial legislature and lower taxes helped assuage popular dissent. Nonetheless, the tidewater aristocracy continued to dominate the economy and government. The most important consequence of Bacon's Rebellion was the increased reliance on the institution of African slavery as the dominant labor system in the tobacco economy. At the impetus of the planter elites, the colony switched from a society with slaves to a slave society. They believed that African slaves were easier to control than a united lower class of indentured and free whites and blacks. This conflict rooted in access to land to grow tobacco was so damaging and threatening to colonial government that it gave birth to slave society in the tobacco fields of Virginia, the birthplace of slave society in America. Not only did fertile and abundant land provide the foundation for the burgeoning tobacco economy and the wealth produced therewith, but it also gave rise to slavery in America. There are few better examples of the impact of environment on culture than this.

Furs, Fish, and Agriculture: French Colonization

French Colonialism in the Northeast, the interior of Canada, and on the St. Lawrence waterway primarily focused on resource harvest in the late 16th and early 17th centuries. By 1610 approximately 1,000 French ships a year were gathering cod off the rugged North Atlantic coastline and shipping the fish back to Europe. French businessmen traded with Hurons, Algonquians, and Iroquois for the highly prized beaver pelts. After several failed attempts at colony building in the North, Samuel de Champlain, with the support of businessmen, the French monarchy, and the cooperation of the Roman Catholic Church, managed to establish colonial communities in the Quebec area. Like the Spanish and English, the French colonists planted European organisms immediately, particularly wheat, rye, and other grains, as well as kitchen garden foods such as peas, cabbage, turnips, radishes, and others that stored well in cellars. Also, like the English, they quickly learned how to raise corn from Indians.

The long winters required the production of more forage for livestock as well as limiting the growing season. Strongly encouraged by Champlain, French colonists more readily adapted to local resources than did their European neighbors to the south. In addition to adopting and using corn extensively, they also learned from Indians how to harvest and cook spawning eels and shad. The colonists became particularly fond of shad, adding some bacon for flavor. French settlers also hunted more than their English and Spanish counterparts did at this point in colonial development, pursuing deer, ducks, geese, and other species. Native plants such as blueberries, currants, gooseberries, and strawberries were incorporated by the French into their diets. Sometimes starvation was pressing so hard at the door that they scavenged and ate a variety of wild plants to survive. Regardless of these efforts, the colonies could not sustain themselves adequately for decades, requiring a steady stream of rations from France to fend off hunger. Fortunately for them, the Indians were happy to bring them the beaver pelts they needed to send home to pay for more European food.

Because France did not suffer from an excess of population like England, immigration was weak and population growth slow. By 1660 there only 3,000 residents in the St. Lawrence River Valley compared to more than 25,000 in the Massachusetts Bay Colony and almost 40,000 in the Virginia colony. The number of French in that colony only grew to 75,000 by 1760. In French Canada colonization remained focused on trading networks and posts, Franco-Indian alliances, and the export of pelts. In French Louisiana colonial settlement and growth was likewise slow and unsteady. This claimed area was expansive, covering an approximate 20-state area along the Mississippi River watershed. Much of the territory claimed and governed by Louisiana was unsettled or lightly populated by French citizens. The first post was established at Biloxi, in what is now Mississippi, in 1699 to block British access to the Mississippi River. It was abandoned for one in Mobile, and that was soon shuttered as well. This colony was run by a commercial operation from 1712–1731, and its original goal was to find and extract gold and silver. It failed in this effort and turned to an increased focus on agriculture and trade with the Indians for deerskins. Settlements associated with posts were established in northern Louisiana, Mississippi, and Alabama, but by 1717 there were still only 550 colonists in the area.

Colonial leadership began focusing its settlements along the Mississippi River, and population increased somewhat with the importation of indentured servants and criminals

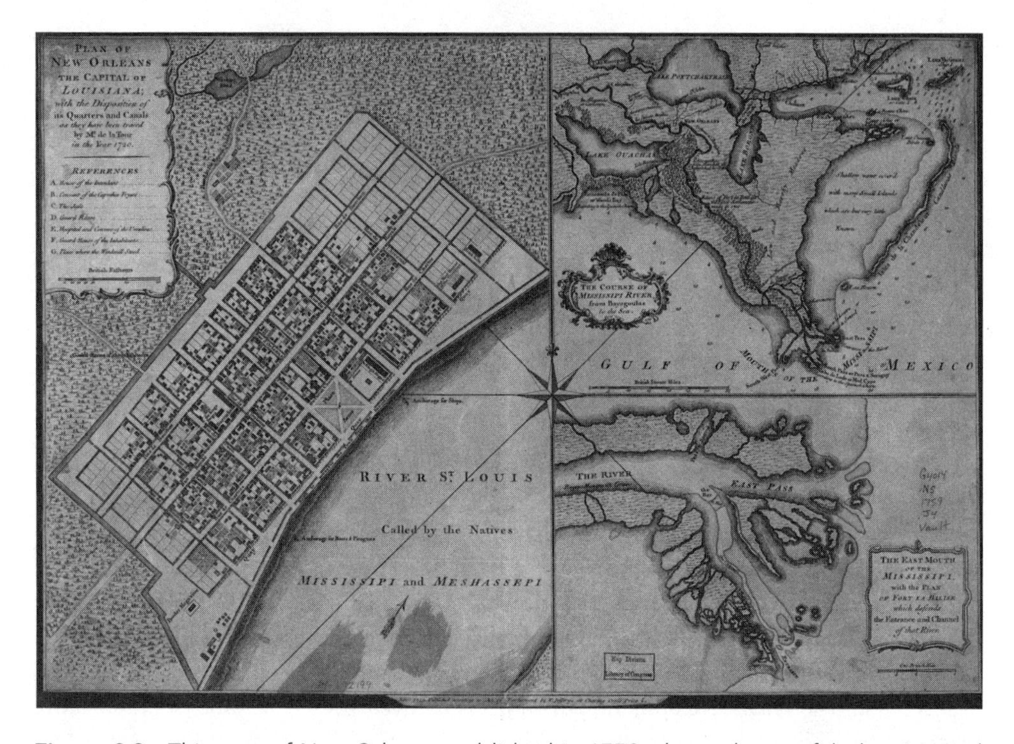

Figure 2.3 This map of New Orleans, published in 1759, shows the careful planning and intent to create a well-ordered European city in a land of bayous and swamps. The French struggled to get their farms established. Floods, humidity, and insects made cultivation difficult. Early levee construction on the river led to an increase in spring floods, and by 1722 New Orleans was already building its first levees for protection. Library of Congress.

banished from France. Like the English and Spanish, they moved onto and farmed Indian lands. The pathogens and livestock introduced by the French reduced Indian populations and destabilized the region. French efforts to establish an export economy of sugar cane and cotton failed, but they were able to export some tobacco and indigo. The Natchez uprising of 1729 stalled the colony's growth when Indians burned the town to the ground and killed 230 Frenchmen while sparing African slaves and most French women. As a result of this massacre and ongoing tensions, the colony returned to French government control in 1731. France's inability to generate increased settlement in its colonies in Canada and in the Louisiana Colony limited economic growth and is one of the main reasons the French lost the French and Indian War in 1763 and ceded this land to the British and Spain.

Creating a Spanish Land

The Spanish system of colonization was better organized and more centrally controlled than that of the British or French in North America. In pursuit of land, resources, and souls, the Spanish integrated a strong commitment to conversion of Indians to Christianity (desired or not) into their plans and practices of expansion. The Encomienda system gave

Spaniards control of Indian labor located on the land they were awarded by the Spanish crown. It was replaced by the Repartimiento system in 1542; this allowed Spaniards to compel part-time Indian labor. Native work was essential because they sought to create Spanish towns and Spanish landscapes as they moved onto Indian lands. The Spanish entered the New World determined to transform the landscape and people in order to create an extended Spanish Catholic society.

The missions established by the Roman Catholic Church throughout New Spain anchored colonial communities, converted Indians to Christianity, enforced that conversion, and played an integral role in Spanish colonialism. What is less understood is their substantive role in the spread and establishment of Old World organisms and European land-use practices. Spanish priests and brothers believed that Indians must learn to dress, talk, eat, and work like a European, in fact, *become* European, for the conversion and salvation of their souls. Also, in the case of the Southwestern missions, the market demand of northern Mexican silver mines for food and bison hides provided an economic incentive to train Indians to produce European agriculture.

Figure 2.4 The church was at the very center of the colonial process for the Spanish as cathedrals and chapels occupied a central physical location in the town plazas of new Spanish towns and Indian towns taken over by Spanish. Indian labor was used to construct the religious sites and to support clerics. The chapel in Taos Pueblo houses an altar that captures the process of colonialism and native adaptation quite effectively. It shows Mary surrounded by corn stalks and squash. This particular church was constructed in 1850, but it reveals the blending of native faith and Catholic Christianity that occurred in many Pueblo communities and persists to this day. Library of Congress.

Tending the fields of the missions served the Spanish priests and brothers' goals of conversion, and kitchen gardens introduced a number of Old World foods that enriched and expanded the diet of Southwestern Indians. The kitchen gardens introduced melons, cantaloupes, peaches, vine grapes, cucumbers, beets, asparagus, lettuce, onions, eggplants, and other foods to Indians. This is one of the benefits of the Columbian Exchange—the enrichment and improvement of diets on both sides of the Atlantic. For Southwestern Indians such as the Pueblo already practicing irrigated agriculture, adopting these organisms did not represent a major transformation in their economy but did necessitate a mastery of more agricultural knowledge, for planting, tending, and harvesting more species successfully as well as greater skill with a wide variety of European agricultural implements. Of course, this was exactly as the missionaries would have it as it served their goal of turning Indians into Spaniards.

As the missions produced more for southern markets (or in Florida for Havana), they expanded their agricultural economy to include exportable amounts of wheat and other grains. With the establishment of new missions and colonial communities, the first order of

Figure 2.5 The Acoman Pueblos lived 400 feet above the valley floor on top of a mesa for at least four centuries before the entrada of Juan de Oñate in 1598. They avoided swearing allegiance to the Spanish as other Pueblos did, and when the nephew of Oñate and other Spanish demanded food in an aggressive fashion in January 1599, the Acomans responded with force, killing the nephew and 12 others. The Spanish governor of New Mexico struck in typical European fashion, amassing 70 soldiers that scaled the walls of the mesa, hauling a cannon up with them. For three days they razed the village, killing 800 Acomans, enslaving all men older than age 12, and amputating one foot from several of the fighting age men. Young children were seized and given to clerics to raise; many of them became servants in households. While Spanish colonialism was not as immediately ecologically destructive as the English model in North America, the Spaniards did require submission of native peoples. Library of Congress.

business was the creation of the Spanish irrigation system. Pueblo Indians were put to work building an *acequia madre* (mother ditch) and then a series of lateral ditches for moving the water to the fields. When irrigation already existed, then it was modified and added to in order to build the Spanish system. Dams and aqueducts were used to help store inconsistent rainfall and snowmelt and move water over rugged terrain to thirsty fields. As the Franciscans taught them European agriculture, the Pueblos moved from digging sticks to metal hoes, spades, axes, and the use of iron-tipped plows drawn by beasts of burden. As they adopted European organisms, they also used improved agricultural tools and worked the land more intensively.

The increasing use of wheat and the baking of bread led to the adaptation of still commonly used *hornos,* adobe outdoor ovens. The Hopi began planting peach trees in their sand dunes, an agriculture practice still in use today, and the Navajos planted their famous peach orchards of Canyon de Chelly.[18] European livestock also made their way into Pueblo agriculture and diets. Similar to the introduction of Old World plants, the Spanish at the missions also demonstrated the uses of livestock to Indians and thereby promoted their adaptation by the Natives, although they picked and chose between species based on their own needs and priorities.

The chicken was popular as an easy, cheap source of protein in Mexico but never became popular with the Indians of the Southwest in this time period. Similarly, the pig was not adopted because of its vociferous appetite and ability to destroy unfenced Indian fields. Two species that did become popular with the Pueblos were the horse and sheep. They were trained by the missionaries to use horses for agricultural work and travel, and they gained possession of these animals by either theft or from the wild horse herds that began to grow from those that escaped missions. Not only did the Pueblos incorporate the horse into their economy, but also their growing horse herds made them a target for tribes such as the Navajo, Apache, and, eventually, the Comanche.

The Churro sheep is probably the species most commonly associated with Pueblo culture. While its wool is not as fine as the Merino's, the mutton is superior, and the coarse wool of the Churro better lends itself to hand weaving and spinning. It also effectively absorbs the colors the Pueblos were (and are) fond of using. Like the cattle on the Atlantic seaboard of North America, sheep initiated and dramatically intensified ecological change by both overgrazing grass and other native plants, leading to ecological problems like erosion, and by introducing European weeds and grasses through their droppings. Over time, their importance to the Pueblo and Navajo economies rose to such a degree that by the 20th century the Southwest was a land of sheep. Spanish colonialism was not as immediately disruptive and rapidly expansionist as English settlement on the east coast, but it still led to dramatic transformations to the land and cultures of native peoples.

Early Colonialism and Ecological Decline

One of the costs of subduing the land and putting it to economic use was the deterioration of streams and rivers throughout the colonies. Clearing forests caused substantial damage. The removal of forest cover resulted in high rates of erosion, sloughing soil and gravel into streams and decreasing water quality significantly. The deposition of dirt sediment on gravel spawning beds of fish ranging from bass to Atlantic salmon choked fish eggs in their nests. Fish beds are built by fish out of gravels or sands to allow the flow of water through and provide aeration of the eggs and fish as they hatch. Soils suffocated eggs and fish inside the nests. Increased stream turbidity,[19] because of heavier sediments, damaged fisheries in at least a couple of ways. Dirt coating fish gills can cause suffocation and increases disease. Turbidity also raises stream temperatures. The dark soil particles absorb sunlight and heat, release the warmth into the water, and warm the water past the temperature where some species can survive.

Farming, ranching, and logging meant that forests were sheared from the land in numerous colonies. This resulted in several negative consequences. Forests regulate water in the trees and root systems. Removal of plant cover increased water runoff. With less water stored in the soils, streams disappeared, springs dried up, and river flow fluctuated more widely between flooding and low-water events. There were even local seasonal changes as a result of forest removal. With trees gone, snow melted earlier and faster in the spring, causing drier conditions. Forest destruction not only negatively affected river ecosystems but also caused problems for human use of the streams for transport, fishing, sawmills, and other uses.

Another important function of forests is to cool streams, and the removal of trees and their canopy contributed to rises in water temperatures. Log transport contributed even more to the collapse of river ecosystems. The rolling and dragging of logs to the stream banks for floating downstream destroyed undergrowth, created furrows that hastened erosion, and also compacted soil, reducing soil quality and plant growth while increasing water runoff. Along the fall line[20] from Maine to Georgia, rivers were used to transport manufactured goods, agricultural commodities and resources, and lumber. Floated in rafts of up to hundreds and even thousands of logs at one time, their movement downstream added even further insult to ecosystem injury. Bouncing and jostling along the rivers, logs drug and pushed furrows into bottoms, stirring up gravel, sand, and soils, shredding the riverbed, and destroying fish nests. Also, the trees shed bark, leaves, and pine needles. This organic material piled up on river bottoms, choking the life beneath them, and when they decomposed, sucked oxygen out of the water, reducing overall water quality and, in worst-case scenarios, suffocated fish, amphibians, and river insects. This practice persisted for another three centuries before being outlawed in the mid-20th century. The wealth gained from lumber as well as the "cheap" cost of transport, which did not include the costs of decimated fisheries, subsidized the growth of the lumber economy and played an integral role in the development of the Atlantic Economy with the ships built from this resource. This all came at an ecological expense that remained outside the financial cost of doing business.

Free-ranging livestock continued to inflict their share of damage. Beaches, for example, absorbed a heavy toll. One area near Saybrook, Connecticut, lost up to 3 feet of soil to wind erosion on a stretch grazed too heavily by cattle. Cronon writes about extensive damage from cattle on Cape Cod:

> Cattle had so seriously damaged the beach grass there by the 1730s that sand was blowing into dunes and encroaching upon the town's meadows. The Massachusetts General Court therefore passed an act in 1739 forbidding all grazing of cattle in the affected areas. The law was repeated many times, and, each April, inhabitants were required as one of their services to the town to plant beach grass on all land whose sandy soil had become exposed.[21]

Cattle consumed food needed by wildlife, driving down populations of deer, elk, rabbit, turkey, and others, and also contributed to erosion by muddying streams and overgrazing areas. Similar processes occurred in the Southwest and California as sheep overgrazed the land, causing erosion and undercutting game populations and native diets. Additionally, as the livestock moved through the forests and grasslands grazing and defecating, the seeds of European domesticated plants and weeds began replacing native flora.

As the British, French, and Spanish established their beachheads in North America and began expansion into the interior, they quickly began putting the land into production, extracting resources for consumption in Europe as well as making the land produce more crops than it ever had under Indian agriculture and management. They did not, however, simply transplant European agricultural practices. Rather, the wealth of the land and its abundance fostered a recklessness that in many cases meant the abandonment of or at least relaxed practice of careful land husbandry practices necessary in Europe. In short, expansion was not merely the result of large immigrating and reproducing populations and the use of livestock for a market economy; it was also the consequence of failing to protect and nourish the land for sustained use. Colonial European agriculture was unsustainable from the very beginning.

Pathogens, livestock, European land-use practices, and legal principles, as well as the expanding capitalist economy, all combined to substantively weaken tribes from the MicMac of Maine and Canada to the Acomans of New Mexico. But native peoples would not simply go quietly into the night. They would attempt to take advantage of many of these items in order to protect, transform, and enlarge their own cultures.

Document 2.1 Of Plymouth Plantation, *William Bradford,* 1651

William Bradford was a member of the Puritan Separatists that founded Plymouth Plantation in 1620. He rose to the position of governor of the colony and served that office 30 times on a recurring annual basis. As governor he was responsible for the finances of the colony, management of laws and the courts, relationships with foreigners and native peoples, and numerous other tasks. Bradford began working on his history of the colony *Of Plymouth Plantation* in 1630 and completed it in 1651. It is one of the key sources of this historical era.

1. According to Bradford what business problems hindered the creation of this Puritan colony in 1620?
2. What are the economic expectations of the member of this colony? How does this influence the way they use the land and see the resources and opportunities available to them?

. . . About this time, also they had heard, both by Mr. Weston and others, that sundry Honourable Lords had obtained a large grant from the King for the more northerly parts of that country, derived out of the Virginia patent and wholly secluded from their Government, and to be called by another name, viz., New England. Unto which Mr. Weston and the chief of them began to incline it was best for them to go; as for other reasons, so chiefly for the hope of present profit to be made by the fishing that was found in that country.

But as in all businesses the acting part is most difficult, especially where the work of many agents must concur, so was it found in this. For some of those that should have gone in England fell off and would not go; other merchants and friends that had offered to adventure their moneys withdrew and pretended many excuses; some disliking they went not to Guiana; others again would adventure nothing except they went to Virginia. Some again (and those that were most relied on) fell in utter dislike with Virginia and would do nothing if they went thither. In the midst of these distractions, they of Leyden who had put off their estates and laid out their moneys were brought into a great strait, fearing what issue these things would come to. But at length the generality was swayed to this latter opinion.

But now another difficulty arose, for Mr. Weston and some other that were for this course, either for their better advantage or rather for the drawing on of others, as they pretended, would have some of those conditions altered that were first agreed on at Leyden. To which the two agents sent from Leyden (or at least one of them who is most charged with it) did consent, seeing else that all was like to be dashed and the opportunity lost, and that they which had put off their estates and paid in their moneys were in hazard to be undone. They presumed to conclude with the merchants on those terms, in some things contrary to their order and commission and without giving them notice of the same; yea, it was concealed lest it should make any further delay. Which was the cause afterward of much trouble and contention.

It will be meet I here insert these conditions, which are as followeth:

Anno: 1620. July 1

1. The Adventurers and Planters do agree, that every person that goeth being aged 16 years and upward, be rated at 10e and 10 pounds to be accounted a single share.
2. The he that goeth in person, and furnisheth himself out with £10 either in money or other provisions, be accounted as having £20 in stock, and in the division shall receive a double share.
3. The persons transported and the Adventurers shall continue their joint stock and partnership together, the space of seven years, (except some unexpected impediment do cause the whole company to agree otherwise) during which time all profits and benefits that are got by trade, traffic, trucking, working, fishing, or any other means of an person or persons, remain still in the common stock until the division.
4. That at their coming here, they choose out such a number of fit persons as may furnish their ships and boats for fishing upon the sea, employing the rest in their several faculties upon the land, as building houses, tilling and planting the ground, and making such commodities as shall be most useful for the colony.
5. That at the end of the seven years, the capital and profits, viz. the houses, lands, goods and chattels, be equally divided betwixt the Adventurers and Planters; which done, every man shall be free from other of them of any debt or detriment concerning this adventure.
6. Whosoever cometh to the colony hereafter or putteth any into the stock, shall at the end of the seven years be allowed proportionably to the time of his so doing.

7. He that shall carry his wife and children, or servants, shall be allowed for every person now aged 16 years and upward, a single share in the division; or, if he provide them necessaries, a double share; or, if they be between 10 year old and 16, then two of them to be reckoned for a person both in transportation and division.
8. That such children as now go, and are under the age of 10 years, have no other share in the division but 50 acres of unmanured land.
9. That such persons as die before the seven years be expired, their executors to have their part or share at the division, proportionably to the time of their life in the colony.
10. That all such persons as are of this colony are to have their meat, drink, apparel, and all provisions out of the common stock and goods of the said colony.

Document 2.2 The Journals of Captain John Smith, *John Smith,* 1612

A member of the Virginia Colony, John Smith commanded from 1608 to 1609. His journals provide a great deal of information on Indian land use, nature, and the travails of the colony in its early years. Made famous by American history and Disney animation, probably best known for the highly questioned rescue by Chief Powhatan's daughter Pocahontas, Smith was an effective leader of the Jamestown Colony until he departed for England in 1609. The colony almost failed after his departure and attempts by the Powhatan Confederacy to starve them out. In fact Jamestown was abandoned in 1610, but a supply fleet from England arrived in the James River and the colony reestablished.

1. In this section from Smith's journal, how are the desires of the colony and its residents revealed in his narrative of resources and the testimony of Indians?
2. What does Smith have to say about the abundance of nature in this region?

Having gone so high as we could with the boat, we met divers savages in canoes, well loaded with the flesh of bears, deer and other beasts, whereof we had part. Here we found mighty rocks growing in some places above the ground as high at the shrubby trees, and divers other solid quarries of divers tinctures. And divers places where the waters had fallen from the high mountains they had left a tinctured spangled scurf [*deposit*], that made many bare places seem as gilded. Digging the ground above in the highest cliffs of rocks, we saw it was clay sand so mingled with yellow spangles as if it had been half pin-dust [*dust from grinding brass pins*].

In our return, inquiring still for this *matchqueon* [*Indian word, possibly meaning "something pretty"*], the king of Patawomeck gave us guides to conduct us up a little river called Quiyough [*Aquia Creek*], up which we rowed so high as we could. Leaving the boat with six shot and divers savages, he marched seven or eight mile before they came to the mine, leading his hostages in a small chain they were to have for their pains, being proud so richly to be adorned.

The mine is a great rocky mountain like antimony, wherein they dug a great hole with shells and hatchets. And hard by it runs a fair brook of crystal-like water, where they wash away the dross and keep the remainder, which they put in little bags and sell it all over the country to paint their bodies, faces, or idols, which makes them look like blackamoors dusted over with silver. With so much as we could carry we returned to our boat, kindly requiting this kind king and all his kind people.

The cause of this discovery [*expedition*] was to search this mine, of which Newport did assure us that those small bags (we had given him) in England he had tried to hold [*prove*] half silver. But all we got proved of no value.

Also [*we came*] to search what furs, the best whereof is at Kuskarawaok, where is made so much *rawranoke* (or white beads) that occasion as much dissension among the savages as gold and silver amongst Christians; and what other minerals, rivers, rocks, nations, woods, fishing, fruits, victual, and what other commodities the land afforded; and whether the bay were endless or how far it extended.

Of mines we were all ignorant. But a few beavers, otters, bears, martins and minks we found, and in divers places that abundance of fish, lying so thick with their heads above the water as for want of nets (our barge driving amongst them) we attempted to catch them with a frying pan. But we found it a bad instrument to catch fish with. Neither better fish, more plenty, nor more variety for small fish, had any of us ever seen in any place so swimming in the water. But they are not to be caught with frying pans!

Some small cod also we did see swim close by the shore by Smith's Isles, and some as high as Rickard's Cliffs. And some we have found dead upon the shore.

. . . our boat by reason of the ebb, chancing to ground upon a many shoals lying in the entrances, we spied many fishes lurking in the reeds. Our captain, sporting himself by nailing them to the ground with his sword, set us all a-fishing in that manner. Thus we took more in one hour than we could eat in a day.

But it chanced our captain, taking a fish from his sword (not knowing her condition), being much of the fashion of a thornback, but a long tail like a riding rod-whereon the midst is a most poisoned sting of two or three inches long—bearded like a saw on each side, which she struck into the wrist of his arm near an inch and a half. No blood nor wound was seen, but a little blue spot. But the torment was instantly so extreme that in four hours had so swollen his hand, arm, and shoulder we all with much sorrow concluded [*expected*] his funeral, and prepared his grave in an island by, as himself directed. Yet it pleased God by a precious oil Doctor Russell at the first applied to it when he sounded it with probe, ere night his tormenting pain was so well assuaged that he ate of the fish to his supper, which gave no less joy and content to us than ease to himself, for which we called the island Stingray Isle after the name of the fish.

Document 2.3 "Reasons and Considerations Touching the Lawfulness of Removing out of England into the Parts of America," *Robert Cushman,* 1622

The author of this piece identified himself as R.C. and is believed to have been Robert Cushman, a member of the Plymouth Colony and passenger on the Mayflower who returned to England in 1621 to promote the colony and handle its business.

1. How does the author use the Bible to justify the seizure of land from the Indians?
2. How does the colonists' role as subjects of the British King affect their rights to the land and responsibility for occupying it?

Letting pass the ancient discoveries, contracts, and agreements which our Englishmen have long since made in those parts, together with the acknowledgement of the histories and chronicles of other nations who profess the land of America from the Cape of Florida unto the Bay Canado (which is, south and north, three hundred leagues and upwards; and east and west, further than yet hath been discovered) is proper to the king of England, yet letting that pass—lest I be thought to meddle further than it concerns me or further then I have discerning—I will mention such things as are within my reach, knowledge, sight, and practice, since I have travailed in these affairs.

And first, seeing we daily pray for the conversion of the heathens, we must consider whether there be not some ordinary means and course for us to take to convert them, or whether prayer for them be only referred to God's extraordinary work from heaven. Now it seemeth unto me that we ought also to endeavor and use the means to convert them; or they come to use. To us they cannot come, our land is full; to them we may go, [since] their land is empty.

This then is sufficient reason to prove our going thither to live lawful; their land is spacious and void, and they are few and do but run over the grass, as do also the foxes and wild beasts. They are not industrious, neither have [they] art, science, skill or faculty to use either the land or the commodities of it; but all spoils, rots, and is marred for want of manuring, gathering, ordering, etc. As the ancient patriarchs therefore removed from straiter places into more roomy [places], where the land lay idle and wasted and none used it, though there dwelt inhabitants by them (as in Gen. 13: 6, 11, 12, and 34: 21, and 41: 20), so is it lawful now to take a land which none useth and make use of it.

And as it is common land or unused and undressed country, so we have it by common consent, composition, and agreement, which agreement is double: First, the imperial governor, Massasoit, whose circuits in likelihood are larger than England and Scotland, hath acknowledged the king, majesty of England, to be his master and commander, and that once in my hearing, yea, and in writing under his hand to Captain Standish—both he and many other kings under him, [such] as Pamet, Nauset, Cummaquid, Narrowbiggonset, Namaschet, etc., with diverse others that dwell about the bays of Patuxet and Massachusetts. Neither hath this has been accomplished by threats and blows or shaking of sword and sound of trumpet; for as our faculty that way is small and our strength less, so are warring with them is after another manner, namely, by friendly usage, love, peace, honest and just carriages, good counsel, etc.—that so we and they may not only live in peace in that land, and they yield subjection to an earthly prince, but that as voluntaries they may be persuaded at length to embrace the prince of peace, Christ Jesus, and rest in peace with him forever.

Secondly, this composition is also more particular and applicatory, as touching ourselves there inhabiting; [for] the emperor [of the Indians] by a joint consent hath promised and appointed us to live at peace where we will in all his dominions, taking what place we will and as much land as well will, and bringing in as many people as we will, and that for these two causes. First, because we are the servants of James, King of England, whose the land (as he confesseth) is; second, because he hath found us just, honest, kind, and peaceable, and so loves our company. Yea, and that in these things there is no dissimulation on his part, nor fear of breach (except our security engender in them some unthought of treachery, or our incivilities provoke them to anger) is most plain in other relations, which show that the things they did were more out of love than out of fear.

It being then, first, a vast and empty chaos, secondly, acknowledged the right of our sovereign king, thirdly, by a peaceable composition in part possessed of diverse of his loving subjects, I see not who can doubt or call in question the lawfulness of inhabiting or dwelling there. But [it is clear] that it may be as lawful for such as are not tied upon some special occasion to there as well as here; yea, and as the enterprise is weighty and difficult, so the honor is more worthy, to plant a rude wilderness, to enlarge the honor and fame of our dread sovereign, but chiefly to display the efficacy of power of the gospel both in zealous preaching, [and in] professing. . . .

Notes

1 Elliott West, *The Contested Plains: Indians, Goldseekers, and the Rush to Colorado* (Lawrence: University Press of Kansas, 1998), 43.
2 Quoted in William Cronon, *Changes in the Land: Indians, Colonists, and the Ecology of New England* (New York: Hill & Wang, 1982), 34.
3 Quoted in Alfred W. Crosby, *Ecological Imperialism: The Biological Expansion of Europe, 900–1900* (New York: Cambridge University Press, 2004), 208.
4 Ibid., 201.
5 An international effort to eradicate small pox realized success in 1979 when the World Health Organization declared the virus officially destroyed. A small stock of the variola virus remained in laboratories under the control of the United States and the Soviet Union and was scheduled for eradication in 1993. This would have been the first time in world history that a species was intentionally made extinct. However, concerns stated by the American and Russian government over the potential use of variola as a biological weapon of terror led to the decision to hold onto those small samples in case future vaccinations were needed.
6 Quoted in Alfred W. Crosby, *The Columbian Exchange: Biological and Cultural Consequences of 1492* (Westport, CT: Greenwood Press, 1972), 42.
7 Crosby, *Ecological Imperialism*, 202.
8 Timothy Silver, *A New Face on the Countryside: Indians, Colonists, and Slaves in South Atlantic Forests, 1500–1800* (New York: Cambridge University Press, 1990), 81.
9 Winthrop quoted in Crosby, 60.
10 Quoted in Virginia DeJohn Anderson, *Creatures of Empire: How Domestic Animals Transformed Early America* (New York: Oxford University Press, 2004), 79.
11 Cronon, *Changes in the Land*, 100.
12 Elinor G. K. Melville, *A Plague of Sheep: Environmental Consequences of the Conquest of Mexico* (New York: Cambridge University Press, 1997).
13 Cronon, 136, 137.
14 Quoted in Hu Maxwell, "The Use and Abuse of the Forests by the Virginia Indians," *The William and Mary Quarterly* 19:2 (October, 1919): 81.
15 Barry Lopez, *Of Wolves and Men* (New York: Scribner Publishing, 1979).
16 John Josselyn, *New England's Rareties Discovered* (London: G. Widdows, 1672; rpt., Boston, MA: Massachusetts Historical Society Picture Books, 1972).
17 William Bradford, *Of Plymouth Plantation, 1620–1647* (New Brunswick, NJ: Rutgers University Press, 1952), 296.

18 These were later destroyed by U.S. troops under Kit Carson's command when the Navajo were forcibly removed from their traditional lands to the Bosque Redondo reservation in Central New Mexico.
19 Turbidity refers to the increased amounts of suspended sediments in water.
20 The fall line refers to the slope descending from the crest of the Appalachian Mountains to the Atlantic Ocean. The Appalachians run from northern Georgia to Maine, and one of the great environmental advantages the English colonies and later America had was these hundreds of rivers with steady flow to the ocean. They were good for transport and trade and would later provide dependable energy for water wheel–powered mills and foundries.
21 Cronon, *Changes in the Land,* 149.

Further Reading

Coleman, Jon T. *Vicious: Wolves and Men in America.* New Haven, CT: Yale University Press, 2004.
Cronon, William. *Changes in the Land: Indians, Colonists, and the Ecology of New England.* New York: Hill and Wang, 1983.
Crosby, Alfred. *The Columbian Exchange: Biological and Cultural Consequences of 1492.* Westport, CT: Greenwood Press, 1972.
———. *Ecological Imperialism: The Biological Expansion of Europe, 900–1900.* New York: Cambridge University Press, 2004.
DeJohn Anderson, Virginia. *Creatures of Empire: How Domestic Animals Transformed Early America.* New York: Oxford University Press, 2004.
Dunmire, William. *Gardens of New Spain: How Mediterranean Plants and Foods Changed America.* Austin: University of Texas Press, 2004.
Dunn, Richard S. and Laetitia Yeandle, eds. *The Journal of John Winthrop, 1630–1649.* Cambridge, MA: The Belknap Press of Harvard University Press, 1996.
Fenn, Elizabeth A. *Pox Americana: The Great Smallpox Epidemic of 1775–82.* New York: Hill and Wang, 2001.
Fischer, David Hackett. *Champlain's Dream: The European Founding of North America.* New York: Simon & Schuster, 2008.
Meinig, D.W. *The Shaping of America: A Geographical Perspective on 500 Years of History, Volume I, Atlantic America, 1492–1800.* New Haven, CT: Yale University Press, 1986.
Merrell, James H. *The Indians' New World: Catawbas and Their Neighbors from European Contact Through the Era of Removal.* New York: W.W. Norton & Company, 1989.
Silver, Timothy. *A New Face on the Countryside: Indians, Colonists, and Slaves in South Atlantic Forests, 1500–1800.* New York: Cambridge University Press, 1990.
Usner, Daniel H., Jr., *Indians, Settlers, & Slaves in a Frontier Exchange Economy: The Lower Mississippi Valley Before 1783.* Chapel Hill: The University of North Carolina Press, 1992.
Weber, David J. *The Spanish Frontier in North America.* New Haven, CT: Yale University Press, 1992.
West, Elliott. *The Contested Plains: Indians, Goldseekers, and the Rush to Colorado.* Lawrence: University Press of Kansas, 1998.
Williams, Gareth. *Angel of Death: The Story of Smallpox.* New York: Palgrave Macmillan, 2010.
White, Richard. *The Roots of Dependency: Subsistence, Environment, and Social Change among the Choctaws, Pawnees, and Navajos.* Lincoln: University of Nebraska Press, 1983.

Timeline

Beaver Wars	1630–early 1700s
Bacon's Rebellion	1676
Pueblo Revolt	1680
First Recorded Comanche Attack on New Mexican Pueblos	1706
Comanches Begin Raiding on Apaches	1720s
French and Indian War	1754–1763
Declaration of Independence	July 1776
Captain Cook's Crew Sells Sea Otters in China	1779
Publication of Captain Cook's Logs	1784
Comanche Raids into Northern Mexico	1820s–1840s

A Great Fur and Hide Marketplace

3

In 1680 the Spanish Governor of New Mexico and hundreds of Spanish settlers huddled inside the Palace of the Governors as furious Indians circled, yelled, and chanted the Roman Catholic liturgy outside. Determined to destroy the Spanish presence in the region, the Pueblo Indians had risen up under the leadership of Taos Pueblo religious leader Po' pay. Shocked by their losses to disease in the 17th century and impressed by both Spanish military might and the ability of European Christians to survive numerous epidemics, many Pueblos had converted to Christianity. Working in the Pueblos and for numerous Spanish missions, they learned to farm like Europeans, growing numerous Old World crops such as wheat, peaches, oats, and grapes and raising animals such as sheep and cattle. The Spanish priests worked them hard, too, producing agricultural goods for sale in the mining districts of northern Mexico. As the century proceeded and Indians continued to die of smallpox, influenza, and other invasive pathogens while diligently working the fields and building churches and missions, their resentment grew, and they listened closely to Pueblo religious leaders calling for a return to native faiths, for rejection of Christianity and European ways.

While Christianity did nothing to stop the scourge of disease, the Pueblos' increased production of agricultural goods of great value and the presence of European trade goods increased raiding on their villages by the Utes, Apaches, and Navajos. Proving further that the Spanish god was not protecting them, a severe drought struck the Southwest in the mid-1670s. Increased raiding by desperate nomadic Indians and starvation among the Pueblos created a crisis in which the traditional native religious leaders gained greater prominence and support. The Spanish priests responded to Indians leaving Christianity or mixing traditional native beliefs with Christianity by destroying Indians' religious items while whipping and even executing Pueblo shamans. One of those whipped was Po' pay, and from the Taos Pueblo he organized the Pueblo Revolt of 1680. When these Indians rose up to remove the Spanish and return to their traditional ways, they killed 22 of 33 missionaries while also slaughtering 400 of the 2,500 Spanish living in the region. Furious Indians also damaged or destroyed every church in the region. The governor was able to secure the release of the terrified Spaniards inside the Palace of the Governors, and they fled south to El Paso.

Having dislodged the Spanish, native faith leaders then launched a revitalization movement, divorcing Indians married as Christians, unbaptizing Christian Indians, and destroying Christian artifacts. They also demanded the rejection of European trade goods and organisms. This proved much more difficult than killing and driving the Spanish out of New Mexico. Peaches, sheep, wool cloth, cattle, and metal tools had improved the Pueblos' quality of life and they would not give them up. The Spanish launched their Reconquista

12 years later, and many of the Pueblos accepted Spanish resettlement because they needed an alliance and protection against raiding Indians. This event shows the transformative power of European trade goods and organisms as well as the impacts of land-use changes and climate in historical events. However, the greatest consequence resulted from untended haciendas and towns where horses were stolen or escaped to form the basis of the great wild horse herds of the West. The populations flourished and expanded, forming the basis of one of the great cultural transformations of American history and the rise of the Comanche, Sioux, Cheyenne, and other tribes with powerful consequences for American history.

The arrival of Europeans in the New World set loose a series of changes that are still difficult to fully understand and measure. The organisms they brought both intentionally and incidentally, along with their highly desired trade goods, helped drive the transformation of Indian cultures and resulted in dramatic ecological change. American Natives learned to commodify different animals to participate in the market economy for any number of reasons. A successful trapper or hunter could more quickly improve his political and economic status via acquisition of European trade goods and distributing them as gifts. Also, as older native leaders, war chiefs, medicine men, and others died off from disease and war, much knowledge was lost in these oral cultures, and it was simpler to replace the complex knowledge and process involved in constructing a bow and arrow, for example, with the purchase of a firearm, powder, and balls. This created dependency for native peoples and led to an increase in the amount of European goods they consumed and the animals they killed for those items. As some Indian nations obtained weapons in order to enhance their own military prowess and political strength, other Indians were then compelled to also harvest pelts and hides for weapons in order to protect themselves. Whatever the reasons, as complex and powerful as they were, the result was chaos for native peoples and degraded landscapes across the continent.

Emptying the Beaver Lodges

It is hard to understand the ability of one species to so completely capture the imagination and desires of both Europeans and Indians with such dramatic unforeseen and severe political, economic, and ecological repercussions across the North American continent. The beaver is a species that is monogamous and mates for life. Beaver kits will remain with the parents to approximately two years of age before venturing forth to find their own stream, build a dam, mate, and start a family. Their average adult weight is 40–50 pounds, but some individuals have reached a whopping 110 pounds. Because of four, curved, ever-growing incisor teeth, beavers must constantly chew to prevent the teeth from growing too long and even entering their brains. Beavers eat the twigs and bark of deciduous trees, leaves and roots of aquatic plants, and some grasses, berries, and flowering plants. Beavers also build dams, and their ponds are critical components in a complex ecosystem with diverse flora and fauna. With their sharp incisors and powerful jaws, beavers can cut down trees up to 3 feet in diameter. Dragging logs and branches to the dam site, they interlock them, filling in the gaps with mud, grass, and rocks, creating a relatively non-porous structure. Dams built in fast moving water are anchored with strong limbs driven into the stream bed and supported with rocks. Ponds and lakes swelling up behind dams that in some cases reached hundreds of yards in width, provide habitat for trout and perch, frogs, salamanders, turtles, snakes, loons, ducks, grebes, geese, and herons as well as ideal browsing sites for moose. When abandoned ponds fill with soil from erosion and deposition, they eventually transition into marshes and over more time into meadows, hosting another set of species such as deer, elk, bear, coyotes, raccoons, various songbirds, hawks, and so forth. Those ponds, wetlands, and meadows were important patches of a mosaic ecosystem adding complexity and abundance to the landscape. Also, beaver dams help to regulate and control the flow of water, slowing down flood waters and storing precious water during drought events. In ecological terms, the beaver is much more important than the value of its pelt. Across North America they added a level of biological complexity and abundance that is simply incomprehensible.

But that pelt! With up to 23,000 hairs per square centimeter it is very dense, and the further north or the colder the weather in which the beaver was trapped, the more luxurious the pelt. Colored typically brown or red, sometimes black and, very rarely, white, the beaver became the object of fervent desire in Europe in the late 16th century. Castoreum had long been popular. This syrupy fluid obtained from beavers' castor sacs was used for any number of medicinal purposes, including treating insomnia, epilepsy, poor vision, and other ailments. However it was passion for the pelt that drove the beaver's near demise. According to historian Eric Jay Dolin,

> at its most elemental the felt-making process involves taking an animal's shorn fur, agitating, rolling, and compressing the fibers in the presence of heat, moisture, and sometimes grease, to the point where the fibers intertwine so tightly that they form a strong fabric. Although virtually any fur can be used in this process, beaver is "the raw material *par excellence* for felt" because the hairs of the beaver's wooly undercoat are barbed, which makes them ideal for interlocking with one another, creating an exceptionally dense, pliable, and waterproof felt that maintains its shape even in the toughest conditions.[1]

The high quality of this hat made it fashionable across Europe. It is not enough however, to note the dictates and whims of fashion. Status was displayed by the size of the beaver hat worn. Not utility then, or simply fashion, drove the near extinction of the North American beaver. Rather, it was the ongoing and never-ending quest for status and its conspicuous display that proved so transformative and damaging to the beaver, Iroquois, Hurons, Susquehannas, Europeans, Americans, and the ecological health of North America.

The desires of Europeans raised the status of the beaver in Indian's eyes. The mammal had no particular economic or dietary significance for native peoples until the market demands of Europe and traders taught them to commodify and overharvest this species for tribal and individual gain. Stunned at first at Europeans' willingness to part with large amounts of tools, pots, and cloth for the fur of an animal that was common and easily caught, native peoples from the Kennebec River Valley of Maine, the St. Lawrence Waterway in Canada, and into the South began harvesting them to gain access to alluring and useful European goods. Axes, hatchets, brass pots and pans, iron tools, colored cloth, beads, steel sewing needles, and many other objects could be easily obtained with pelts. Europeans originally traded for the furs for individual profit or to subsidize colonial settlements.

English fishing crews began taking on furs as they traveled the Atlantic coast, and the settlers at Plymouth Plantation were dependent for a while on the income generated by these pelts. Over time a whole economic and trade system would be built around harvest of this pelt.

The conflicts for furs and the arming of Natives began very early in American colonial history, enough so that the Plymouth Colony was bedeviled by local Indians provided with weapons by English trading ships seeking the highly desired beaver pelts. Governor William Bradford wrote,

> Oh! That princes and parliaments would take some timely order to prevent this mischief, and at length suppress it, by some exemplary punishment upon some of these gain thirsty murderers (for they deserve no better title) before their colonies in these parts be over thrown by these barbarous savages, thus armed with their own weapons, by these evil instruments, and traitors to their neighbors and country.[2]

The arrival of Europeans changed Indian warfare almost immediately. Northeastern tribes traditionally fought en masse, large groups facing off with clubs, tomahawks, and bows and arrows while wearing wooden armor. The Europeans' firearm in the early 17th century, the arquebus, was not the battlefield threat we might assume. With its slow muzzle velocity, Indians easily adjusted by simply dropping to the ground upon hearing the sound of gunpowder lighting and firing. What changed native peoples' methods of war initially was the use of brass and iron arrowheads cut from European pots, pans, and tools. These enhanced arrowheads rendered arrows much deadlier as they now easily penetrated the wooden armor that had stopped traditional flint arrowheads. The increase in battle deaths, anathema to Indians, compelled them to switch from their traditional fighting style to hit and run tactics, small band fighting, and skirmishes rather than battles en masse. In any exploration of environmental history it is requisite to also examine the impact of resource use and commodities on culture. Indians switching from traditional flint arrowheads to enhanced points manufactured from European trade goods transformed Indian warfare to what has been considered "traditional" Indian tactics. The use of hit and run tactics and small band fighting in the early 17th century arose from economic and environmental change initiated by Europeans.

The Iroquois adjusted to the new opportunities and challenges and moved aggressively to gain greater control of the beaver trade over enemies like the Algonquians, Hurons, Susquehannas, and others in what has been deemed the Beaver Wars in the middle 17th century. Using the weapons gained through the aggressive harvest of beaver pelts, the Iroquois expanded their effective range, conducting extensive raids in order to control the beaver trade by stealing supplies and beaver pelts, to gain access to areas that had not been over trapped, and to capture people to repopulate their own confederacy. Beaver and pathogens created the conditions giving rise to the Beaver Wars. The virgin soil epidemics came late to the Iroquois, brutally hammering them in the 1630s and 1640s with lethal outbreaks continuing through the rest of the century. The epidemics in the 1630s and '40s reduced the Iroquois population by half with some tribes likely experiencing death rates approaching 75 percent. When this powerful confederacy began regular raiding in the 1630s and 1640s, the fighting most closely resembled the traditional mourning wars. In this type of warfare

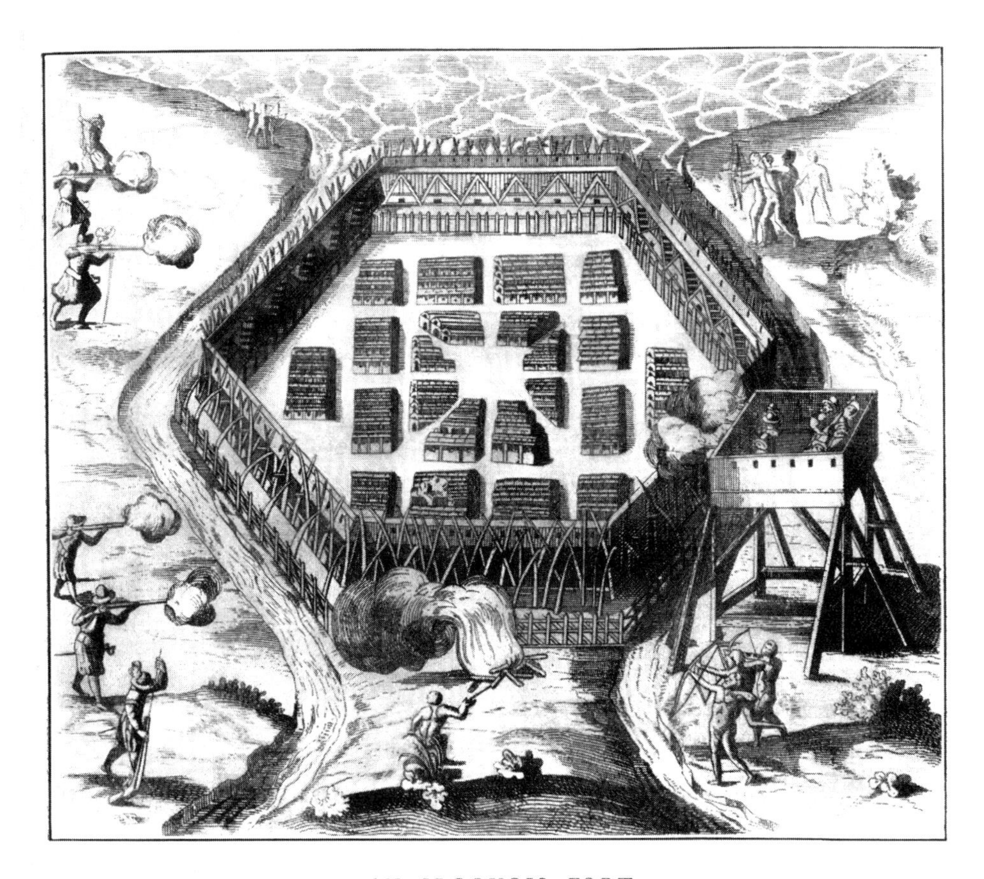

AN IROQUOIS FORT

(Believed to have stood on the shore of Onondaga Lake. Besieged by Champlain in 1615.)

Figure 3.1 Samuel de Champlain first allied with Hurons and Algonquians in 1609 and used his musket to devastating effect, killing two Iroquois warriors with one shot. In 1615, he, 10 Frenchmen, and 300 Hurons attacked this Onondaga Fort in current-day New York State but were defeated and driven off. Destabilizing warfare was one reason the Iroquois moved so aggressively to control the beaver trade and expand economically, politically, and militarily. From *The Voyages of Samuel de Champlain*.

Indians conducted military campaigns to avenge the loss of loved ones or leaders and to capture Indians from other nations to replace those lost in their own community. The fighting coarsened as economic interests intertwined with the desire for vengeance and captives. In a raid on a Huron village in 1642 the Iroquois killed almost all the inhabitants and stole away with a large quantity of supplies and furs. From 1643 on the Iroquois began using blockades and ambushes designed to stifle other Indians' trade with Europeans and capture supplies and furs. A Jesuit missionary of that era wrote,

> So far as I can divine, it is the design of the Iroquois to capture all the Hurons, if it is possible; to put the chiefs and a great part of the nation to death, and with the rest to form one nation and one country.[3]

Historian Daniel Richter describes the warfare and expansion of the Iroquois in the 1650s and 1660s:

> For most of the next decade, Iroquois warriors pursued the Huron refugees relentlessly and overran in turn each of the Iroquoian peoples that sheltered bands of them. The Petuns were dispersed by 1650, the Neutrals by 1651, and the Eries by 1657 . . . Iroquois justified their wars in the name of long-standing feuds and in traditional mourning-war terms; then they ruthlessly attacked successive villages, forcing those they captured to carry plundered furs and trade goods to Iroquoia. Those who eluded attack sought refuge and military allies among neighboring peoples.[4]

They continued to wage war against other Indian nations, attacking Abenakis in New England, ranging south against Indians on the upper Mississippi, and conducting raiding excursions in the Virginia area.

These wars brought devastation and heartbreak on the heels of the first waves of contagion from European diseases, even while those outbreaks continued. Some Indian nations enhanced their power and status but only with a corresponding diminishment and destruction of many other Indian peoples. The beaver population was annihilated, the species almost extirpated. In striking this devil's bargain, all Natives, weak or strong, found themselves increasingly dependent on Europeans and their merchandise.

Deerskins for Europe

In addition to the demographic collapses caused by small pox, measles, influenza, and other pathogens, disproportionate death rates among older Indians severely undercut the resilience of native culture. Religious leaders, masters of oral tradition, artisans, experienced hunters and military leaders, diplomats, and expert agriculturalists were all laid low in high numbers, often before being able to pass on their particular, specialized knowledge or skill. This is an unmitigated disaster in an oral culture.

Unlike beaver pelts, deerskins could be used in numerous practical ways by Europeans. High-quality hides were turned into luxury gloves and the rougher ones for book binding in the burgeoning publishing industry. Deerskins of lesser quality were used for pants, harnesses, and saddles. Historically, cattle hides had been used for these purposes in Europe, but a series of disease epidemics in the early 18th century dramatically reduced the size of the French herds, and England prohibited the importing of diseased French cattle or hides for fear of the spread of contamination. As England was the primary producer of leather goods, its merchants were forced to find a new resource and turned to importing deerskins from North America. Indians were willing and even eager to harvest the hides in order to gain access to the variety of goods available from European traders. The desire for enhanced personal status increased tribal efforts to expand their commercial, military, and political power and drove the exchanges, as did the impacts of disease and the native people's needs to obtain weapons to protect themselves against enemies.

Although the Europeans plied them with attractive goods, Indians struck hard deals when they could, using competition between traders and nationalities to their advantage.

In 1718, 25 deerskins was sufficient for a Creek hunter to obtain a gun; 6 skins earned a duffel blanket or shirt. An axe or hoe went for 4 deerskins. As the deer population decreased later on in the century, 10 to 16 skins was often required for a firearm. Historian Daniel Usner demonstrates the high level of consumption of European goods by Indians in the Lower Mississippi River Valley. Three villages in 1750–51 alone required the following amounts of goods for their deerskins (the items and amounts included here are selected from a longer list but are representative of trade): 90 woolen blankets, 288 pocket knives, 288 hunters knives, 216 combs, 2,000 flints, 60 hatchets, 104 shirts, 800 lbs. of gunpowder, and 1,800 lbs. of gun balls.[5] Not only did Indians provide resources for European manufacturing, but also they constituted an important market for a rapidly growing and developing manufacturing economy on the other side of the Atlantic (see Document 3.3). Indian resources and consumption helped drive the modernization of the European economy in the 18th century. Indians providing goods were an integral part of the Atlantic economy. They harvested and traded highly desired consumer goods such as beaver pelts and useful material such as deerskins and bison hides needed for clothing, manufacturing, and other purposes. They were avid consumers also. An emerging manufacturing economy is nothing without customers, and Indians' avid trading for guns, knives, cloth, hoes, axes, needles, beads, combs, guns, gunpowder, and so on poured money into this economy and helped drive the growth of European and American capitalism.

Axes, combs, and cloth may have appealed to Natives, but they also sought guns for protection. The aggressive pursuit of beaver pelts, deerskins, and slaves by well-armed, expansionist Indian nations necessitated the acquisition of arms, powder, and balls for self-defense. Fighting for control of the beaver trade or access to deer populations and bison herds also increased dependence on weapons and, by extension, European traders and markets. Moreover, Indians' trade with Europeans helped cement relationships that might be needed in times of crisis and war. Historian James H. Merrell discusses the expansion of the Iroquois south, showing how furs and guns and European markets so destabilized Indian cultures:

> [The Iroquois] came south in search of glory, to quiet the cries of those mourning lost relatives, and to bring distant nations under the "*pax Iroquois*." . . . Many were obviously searching for new sources of furs . . . what colonists termed war parties were often bands of hunters come to slay deer and trap beaver. Whether they hunted animals or people, Iroquoians . . . were heavily armed by their French and English allies. When Seneca warriors returned from the southwest with fifty prisoners in 1677, an observer noted that the victims had been outgunned. Senecas had muskets aplenty; these captives "were of two nations, some whereof have few guns; the others none at all."[6]

Clearly, the Iroquois had learned well and adapted since Champlain's pivotal slaying of some of their nation. In the Southeast, Spanish, French, and English colonists paid Indians such as the Creeks, Chickasaws, and others to capture native peoples from other nations for sale and use in the Caribbean sugar economies and in the tobacco and rice fields. Weaker native groups desperate to attain firearms to protect themselves from bondage overharvested deerskins also. If trade goods, alcohol, and weapons for hunting were not enough to compel Indians to overharvest deerskin, the destabilization and warfare arising from European colonialism, pathogens, and economic activities forced native peoples to kill too many deer and other species in order to protect themselves and their families.

When French and English traders were unable to gain more pelts and deerskins with their normal enticements, they turned to the use of another trade good to both ply their customers for favorable transactions and to create the infinite demand so desired. Alcohol increased and sustained the flow of beaver pelts and deerskins in the 17th and 18th centuries and bison hides in later years.

Native peoples adopted alcohol for a number of reasons beyond the generally assumed and simplistic idea of Indian vulnerability to alcohol. Rum, brandy, and whiskey provided shortcuts to visions and spiritual experiences. Also, these beverages could slip into the spot traditionally occupied by the tobacco pipe in councils and discussions, fostering relationships, diplomacy, and discourse. Inebriation also gave license to acts of wildness and violence that violated native sanctions against personal excess. As one Cayuga man stated in the 1660s, "I am going to loose my head; I am going to drink of the water that takes away one's wits."[7] Regardless of the reasons, traders' bartering of alcohol encouraged Indians to harvest an ever-expanding number of beaver pelts and deerskins. This not only severely damaged deer and beaver populations but also took a bitter toll on Indian culture and life as expressed well by an Iroquois chief in the 1750s:

> The Rum ruins. We beg you would prevent its coming in such quantities by regulating the Traders . . . They bring thirty or forty kegs and put them down before us and make us drink, and get all the skins that should go to pay the debts we have contracted for goods bought of the fair traders; by this means we not only ruin ourselves but them too. These wicked whiskey sellers, when they have once got the Indians in liquor, make them sell their very clothes from their backs. In short, if this practice be continued, we must be inevitably ruined.[8]

Prior to Europeans' arrival each Creek family killed approximately 25–100 deer a year. Once the Creek and other Southeastern Indians committed to the deerskin trade for European goods, the rate of killing increased dramatically, to about 200–400 deer per family annually harvested by the Creek. In the late 17th and early 18th centuries, 85,000 deerskins were shipped from ports in Virginia and from Charleston. Charleston exported 120,000 skins in 1707, and by the 1760s the British colonies in North America were handling approximately 400,000 a year. When the harvest of deer for personal use and French and Spanish deer trade are figured in, the annual harvest of deer in the Southeast approached a million per year, causing a precipitous drop in deer numbers and bringing them to the edge of extinction. If colonists were not so aggressively killing predators of deer during this same period, whitetail may have been removed from the region entirely.

The extensive use of credit by traders added to the damage caused by the sale of alcohol. Indians were bound to harvest deerskins or pelts by credit agreements holding their land as collateral. Failure to secure enough of these commodities or careless trading for alcohol would leave Indians at the end of the season with only their land to pay debts. Also, as deer populations plummeted and the beaver neared extinction, Indians had little else to offer for the European trade goods on which they were now dependent other than their land.

When they failed to harvest enough pelts or deerskins, a worsening problem with the collapse of the species, they lost land and livelihood and were displaced and forced west into other Natives' territories. Ironically, this process even facilitated the expansion of European economic interests as Indians forced off their lands into new areas traded their used trade goods with Indians new to the economy, sparking their interest and beginning anew the process of commodification of beaver or deer and the overharvest of these animals for guns, powder, cloth, knives, axes, and so forth.

These highly desired and acquired goods destabilized Indian culture and ecosystems across the continent. One Natchez elder commented in the early 18th century on the impact of these goods: "To seduce our women, to corrupt our nation, to lead our daughters astray; to make them proud and lazy." After commenting that men had to work even harder harvesting deerskins to keep their wives happy, he added,

> "Before the French came into our lands, we were men, we were happy with what we had, we walked boldly upon all our paths, because we were our own master. But today we tread gropingly, fearing thorns. We walk like the slaves which we will soon be, since they already treat us as though we were."[9]

To retain their status, and to perform their traditional roles, they were compelled to deplete the natural resource abundance that had thrived for centuries before the arrival of the Europeans.

Horses, Bison, and Empire

One of the legendary species of the America West and a symbol of both wildness and wanton destruction, the American Bison once blanketed the Great Plains. The number of the great mammal in 1800 likely ranged between 15–27 million. The bison played a critical role as a keystone species in the Great Plains ecosystem. The total number, of course, fluctuated with droughts, hard winters, and summers of high fire activity. Bison are well adapted to the Great Plains environment. Heavy winter coats protect them from the bitter arctic blasts with wind chills down to 70 below Fahrenheit and occasional heavy snows. Their long, spade-shaped heads are an excellent adaptation for the shoveling of snow off of grasses for winter grazing. Moreover, bison (unlike beef cattle) respond to winter storms by walking into the storm instead of away from it. This enables them to more quickly escape dangerous conditions and increases survival rates. In early summer, when the grasses are lush and green, the bison come together in vast seas of animals described in awestruck prose by numerous travelers. One wrote,

> immense herds of bisons, grazing in undisturbed possession and obscuring with the density of their numbers the verdant plain; to the right and left as far as the eye was permitted to rove, the crowd seemed hardly to diminish, and it would be no exaggeration to say that at least ten thousand here burst on our sight in the instant.[10]

While the bison population did extend to the piedmont of North Carolina, to New York, and South Texas, the heart of the bison world was the Great Plains. Bison thrived on the shortgrass prairie of the plains. The majority of these grasses' energy and nutrients go into the root system, allowing these plants to better survive droughts and reducing loss of water via transpiration from leaves to the dry plains air. The dense bundle of roots, much larger than the above-ground plant, spread in a shallow fashion directly under the surface. Because rain is sporadic on the plains and often comes in short, heavy downpours, the grasses evolved this root structure in order to absorb as much of the rarely occurring rain as quickly as possible. Also, during extended dry periods, the short grasses above, only a few inches high, would wither and die, while the roots lay dormant until the next precipitation. These unimpressive, few stubby inches of grass were able to produce bison that could weigh as much as 2,000 pounds (sometimes even more) and hit speeds of 35 miles an hour in short bursts.

Other species thrived because of the bison. The prairie wolf may have numbered as many as 1.5 million as they culled the old, weak, and sick and brought down calves. Ravens, crows, golden eagles, and magpies scavenged bison carrion, as did coyotes and foxes. While grizzly and black bears of the Great Plains consumed mostly berries, roots, and plants, the occasional dead bison added to their diet. A not so obvious beneficiary of the bison were the tens of millions of ducks, geese, terns, and other waterfowl that overflew the Great Plains in what is one of the great migratory bird flight routes in the world. Elliott West explains how bison seeking relief from bugs or seeking to cool themselves helped make this bird-friendly environment. Buffalo wallows began as

> salt or alkali licks or shallow collections of rainwater. Bison and other grazers were drawn to these spots. The subtle depressions deepened as the wind blew away soil loosened by the animals' hooves, and bison further crushed and carried away surface dirt by rolling and wallowing in the shallow water holes to rub off their hair in summer, to cool their bodies, and to coat themselves with mud against biting insects.[11]

As millions of bison regularly used these wallows over thousands of years, they created "a system of water collection in terrain otherwise without drinking places. Water birds found the vast dimpling of wallows ideal for resting on their long flights, so the high plains became one of the earth's great migratory routes."[12]

The bison were the keystone species of a great, thriving Great Plains ecosystem, and to the Indians, Europeans, and Americans they became the cornerstone to empire and economic growth. Bison were pursued by numerous Indian nations prior to the introduction of the horse, but the large mammal remained one part of a complex economic system and was but one of many species pursued. Before the advent of horse-based hunting native peoples showed great creativity in killing bison. They used fire to drive bison off cliffs, hunted them individually on foot, and employed stone piles to make chutes to drive the animal off cliffs or to narrow spots where they could be more easily killed. There were incidents of waste even before Indians' participation in the expanding market economy. Indians sometimes squandered a significant portion of bison kills, at least those from surrounds and bison jumps. They could not control how many of these animals charged over a cliff to their death once a stampede was fully in progress, and they could only harvest and process so much meat and hide at any given time. The testimony of European and American traders from the late 18th century as well as evidence from kill sites and jumps shows that on occasion large amounts of meat and hides were left behind to rot. Explorer Meriwether Lewis wrote,

> the remains of a vast many mangled carcases [sic] of Buffalow [sic] which had been driven over a precipice of 120 feet by the Indians and perished; the water appeared to have washed away a part of this immence [sic] pile of slaughter and still there remained the fragments of at least a hundred carcases.[13]

The description of another surround by the Piegans, one branch of the Blackfeet, confounds our notion of the bison hunt, as it is so different from the popular images of Indian hunting. Peter Fidler of the Hudson's Bay Company wrote the following description in 1792:

> The young Men kill the rest with arrows, Bayonets tyed [sic] up on the end of a Pole, &c. The Hatchet is frequently used & it is shocking to see the poor animals thus pent up without any way of escaping, butchered in this shocking manner, some with the stroke of an axe will open nearly the whole side of a Buffalo & the poor animal runs some times a considerable while all thro' the Pound with all its internals dragging on the ground trod out by the others, before they dye [sic].[14]

Some Indians had a strong taste for drowned bison meat, allowing the meat to age until it turned green, so ripe and tender that it was not necessary to cook the meat for long. Mandans buried animals over the winter and dug up the rotted meat to partake of it in the spring. Some Natives drove bison into rivers to drown and later recovered and consumed the rotting meat further downstream. Inevitably some of this meat was lost.

Indians preferred to harvest cows over bulls because hides were thinner and robes lusher. Also, the flesh of cows tasted better and was more tender. This might help explain the butchering sites and enclosures noted by European and American observers where bulls were barely butchered and left to rot. Also, some have noted the preference for tongues, humps, fetuses, and young calves by Plains Indians. Besides the issue of taste and culinary tradition, over the course of winter and during the lean part of the year in the early plains spring of April and May, the bison used up fat reserves merely trying to survive the difficult winters and were therefore remarkably lean. Meat with high protein content but with little fat increases metabolism, offsetting the caloric benefits of the meat, and was of limited nutritional value in the Plains Indians' diet. Moreover, the consumption of too much high-protein, almost no-fat bison meat could cause the build-up of toxins in the body. Cows have a much thicker layer of back fat than did bulls, and humps, fetuses, and calves were all higher in fat content than an adult bull in winter or early spring. This exculpatory argument only goes so far, however, when one considers other examples of Indian excess and waste. Anthropologist Shepherd Krech, who closely examines the issue of Indian bison harvest and waste, offers some examples of excessive spoilage in the late 18th and early 19th century, before waste could simply be blamed on market forces and dependency. The Scottish explorer Alexander McKenzie observed Gros Ventres Indians killing large numbers of bison and harvesting only the tongues, leaving the rest to "rot in the field." He also noted Cheyenne killing for tongues and leaving 250 cows to rot and Blackfeet harvesting fetuses from cows and abandoning the rest.[15] To simply argue that Indians never harvested or wasted to excess is inaccurate. The fact that they did not commodify these animals in a market economy limited the amount of exploitation, however. Moreover, the fact that the bison Manitou did not turn against them after such incidents of waste proved to them that they did no wrong in leaving the carcasses to feed other animals and rot. This would change with arrival of the horse and European trade goods.

With the introduction of the horse, the hunting of bison developed into the now familiar image, the Indian bestride his best buffalo runner using a short bow to drive one or two arrows directly into the great animal's heart. In this manner the Plains Indians could quickly harvest a large number of the animals and could continue pursuing bison day after day; the bison could no longer find sanctuary from their hunters. Not only were hides turned into tepee walls, flooring, and blankets, but natives put the bison to use in other ingenious ways. The bladder was used for a medicine bag, the stomach as a storage container. Bones could be converted into tools, war clubs, and pipes. Muscles were turned into thread and glue, while tendons made excellent bowstrings. Horns of bison were employed as cups, ladles, and, later, powder horns, while the liver was needed for tanning the hide. In short, they used the bison

Figure 3.2 In this classic Catlin painting of Indians in pursuit of bison, the advantage provided by the Old World animal is evident. The bison could no longer outrun humans; Indians could get much closer for a killing shot and were also able to harvest many more bison at one time. The other advantage came from native peoples' ability to now cover much greater distances in a day. They could pursue the bison on a regular basis. The horse made it possible to harvest bison so effectively that many Indians transformed themselves into a horse and bison culture. Courtesy of the Smithsonian American Art Museum, Gift of Mrs. Joseph Harrison, Jr.

quite comprehensively until European and American manufactured goods appeared, then the focus over time became the hide. Hides, like deerskins in Europe, were converted to leather for a number of uses. As Plains Indians committed to harvesting bison hides, more of the bison was left behind to rot, the animal were hunted in greater quantities by native peoples, and women's status in the tribes deteriorated as they spent a greater proportion of their time in the fleshing and treating of hides for shipment to market.

The Comanche and Cheyenne

The rise of the Comanche is one of the most interesting stories of the American West and of Indian transformation in American history. They built an empire that affected the course of history for Spain, Mexico, the French, and Americans, not to mention native peoples, and it was done from the back of an Old World organism: the horse. Originally from the Great Basin desert, they came from the same language group as the Aztecs; the Indians that became the Aztecs went south. Over centuries the Comanches migrated to the northern Rockies and crossed the mountains to the Front Range on the eastern side. Like most tribes around the Plains they walked onto the grassy steppes to their east to hunt bison, but they did not remain there for long stretches of time. Venturing east in pursuit of bison rendered them vulnerable to raids by Apaches, the dominant tribe of the Southern Plains in the days before the arrival of the Europeans and their biota.

In the headwaters of the Arkansas River they came into contact with the Ute Indians. Their subsequent friendship and alliance transformed the Comanche and the history of the Plains and the Southwest dramatically. From the Utes the Comanches learned to become horsemen, a skill they adroitly mastered. They shared the territory of northern New Mexico and southern Colorado with the Utes but expanded as they transformed. They also positioned themselves close to the Taos Pueblo, the northernmost reach of New Spain in the Southwest, in order to gain access to their horses, the existing slave trade, agricultural products, and European trade goods.

The Pueblo Indians had always grown corn, beans, and squash and traded with more bison-dependent tribes such as the Apache even before the arrival of the Spanish. There was a regular exchange of protein and carbohydrates on the edges of the plains. The Pueblos adopting of melons, peaches, and other European foods made them even more attractive trading partners and raiding targets to the Utes and Comanches. This intensive agricultural production, in addition to the introduction and expansion of the horse, made Comanche specialization in bison possible, and the horse allowed them to pursue and aggressively harvest bison while also making it practical and necessary to conduct war against other tribes to gain access to their territory, grass, and bison. Carbohydrate production by the Pueblos in New Mexico made the specialization possible because bison meat and hides could be traded for those items and rendered agriculture and gathering of wild plants unnecessary. Another benefit accruing from the horse was the ability to haul much heavier loads of goods to and from market than was possible with dogs. This not only increased the harvest of this species but also the accumulation of more European trade goods for further trade and personal edification. Also, the greater consumption enabled by horses is exemplified by the expansion of teepees. When using dogs only, teepees were generally made with 6 or 7 bison hides.

With the arrival of the horse, teepee size essentially doubled to 12 or in some cases tripled to 20 hides. This not only made for more capacious living, particularly welcome during the cold spells of winter when much time was spent around the fire in the teepee, but it also made it possible to accumulate and store more personal items. But, if the development of the Comanche into a specialized horse and bison culture seems foreordained or obvious, consider the decisions of the Utes and the Apaches.

The Utes did not make the complete transformation with the Comanches even as they allied with them on raids against Apaches, Pueblos, the Navajo, and Spanish. They remained in their traditional territory practicing a mixed economy of hunting various species, including bison, and gathering plants. They would eventually rue their alliance with the Comanche and their failure to transform and expand along with them when the Comanche turned against them. The Apaches were heavily bison dependent before the arrival of the Spanish and the horse, but they followed a different path than the Utes and the Comanches. Their choices also reflected the impact of the Spanish and their organisms. There were numerous

Figure 3.3 In this George Catlin painting, Comanches can be seen practicing the arts of war. As the Comanche became increasingly expansionist in the 18th century in pursuit of slaves, horses, and bison, they perfected the art of horse-based warfare. Boys began training at a young age in the use of weapons, shields, and horses, and in order to accrue status and prestige, they were expected to show valor in combat. Courtesy of the Smithsonian American Art Museum, Gift of Mrs. Joseph Harrison, Jr.

smaller uprisings against the Spanish prior to the revolt in 1680, and Pueblos, particularly from Taos, fled Spanish persecution by seeking refuge among the Apache to their east on the plains. In so doing, they introduced agriculture to the Apache. So, even while the Comanche were remaking themselves into a specialized horse and bison dependent culture, the Apache reduced their dependence on bison and began practicing agriculture in the river bottoms of the Southern Plains. They too would rue their decisions because when the Comanche decided to expand into Apache territory in the 1720s, their crops made the Apache vulnerable and easily defeated on multiple occasions. The Comanche drove them off their land into central and then southern Texas.

The Comanches through the 18th century built and extended Comanchería from Taos, New Mexico, to approximately Natchitoches, Texas, with the heart of trade located in the Big Timbers section of the Arkansas River in southern Colorado, in the approximate area of La Junta, Colorado, today. Trade goods, bison hides, meat, and agricultural goods

Figure 3.4 This photograph of Taos Pueblo by Ansel Adams in the 1930s depicts a community that appears little changed in appearance from over hundreds of years. But the quietness of this image belies the bustling energy of the Taos community when the Comanche were at their peak in the late 18th and early 19th century. It anchored the western end of the Comanche area of trade and control that extended east to Louisiana. Pueblo Indians and Spanish settlers of New Mexico benefited from the rise of the Comanche because of the market for their agricultural goods. Indians and Spanish residents of the region also suffered from regular raids by the Comanche in pursuit of carbohydrates, European manufactured goods, and horses. National Archives.

flowed into and out of this zone, including Apache slaves sold into Louisiana and shipped to the Caribbean. Raiding and military campaigns by the Comanches, along with diplomacy, enabled this expansion. By the 1820s and 1830s Comanche expeditions reached deep into Mexico, tearing through the states of Coahuila y Tejas and Chihuahua. These sometimes lasted several months and focused on the seizure of Mexican horses, cattle, and children from the small communities and haciendas scattered across this sparse landscape. The horses and cattle secured in northern Mexico were funneled north into the desperate hands of Anglo-American settlers in Louisiana and East Texas who were experiencing a shortage of those animals of labor. In this way, the horse- and bison-bound Comanche contributed to American expansion while weakening Mexico even as they themselves were dwindling as a culture with the collapse of the bison in the 19th century.

The adoption of the horse, the pursuit of the bison, and the overdependence on both species created a general culture of expansion, conquest, and warfare amongst several Indian tribes. Much of this conflict was based on the pursuit of bison and the need to control vast areas of territory where bison could be found and harvested. Like the Comanche, the Sioux, Cheyenne, and other tribes moved relentlessly onto other Indians' territory in pursuit of the hooved resource. The Sioux and Cheyenne had originally been farming, hunting, and gathering peoples in the woodlands of Minnesota and Wisconsin, growing corn, beans, and squash while harvesting numerous species of game and plants. Disease, warfare, and environmental conditions drove them onto the Plains in the late 17th century, where they became a different people. The Sioux then dislodged the Crow from the Powder River country of northeastern Wyoming to the western side of the Big Horn Mountains in the early 19th century. Similarly, the Cheyenne removed the Kiowa from the Black Hills region of South Dakota before they were also forced south by the Sioux. The Southern Plains became a veritable battlefield because of the competition for bison and because of the introduction of firearms by Europeans and Americans. Historians in recent years have better understood the wealth and power of these plains tribes and have tread new ground, demonstrating native peoples' negative impact on the environment.

The commitment to the horse and bison created both negative environmental and cultural impacts for native plains peoples. Cheyenne Indians needed at least six horses per person due to the need to transport hides, meat, tepees, poles, and various personal goods. Individuals could have much larger herds, in some cases numbering in the hundreds, as accumulations of wealth and symbols of status and rank. While horses provided "a short-cut" to the stored energy of the vast millions of acres of plains grassland that earlier could only be tapped via the meat of prey, by the mid-1800s they depleted this energy source, overgrazing the grass as the tribes and their horses traveled the landscape. The heavy consumption of grass forced the Indians to move on a more regular basis, and in winter camp they were compelled to strip leaves and branches from cottonwoods and other trees to feed the horses even as the ravenous animals themselves consumed young trees. This led to deforestation along plains rivers and streams and a reduction of grass with corresponding erosion in some cases. Also, the millions of horses owned by Indians consumed grass needed by bison, undermining the ability of those herds to reproduce themselves even as they came under increased hunting pressure from both Indians and American market hunters. Historian Daniel Flores estimates that for every 1 million horses, there were 1 million fewer bison. Southern Plains tribes owned approximately 2.5 million horses.

Figure 3.5 Catlin depiction of a Sioux village shows the nature of women's labor in the bison economy. Women previously had many more responsibilities in growing and preparing crops that gave them higher status in Sioux culture. As plains tribes increased their dependency on the bison, and particularly as they required increasing amounts of hides for guns, ammunition, and other goods, women spent most of their time in the strenuous labor of preparing bison hides for trade. This involved removing flesh and fat from the hide; working a mix of brains, bone marrow, and other ingredients into the hide to tan; and drying it. Women did much of this work on their knees as depicted in the picture, and the hours of labor a day fleshing and working material into the hide was painful and debilitating, particularly for women's hands. Courtesy of the Smithsonian American Art Museum, Gift of Mrs. Joseph Harrison, Jr.

The dependence on horses undermined the cohesive identity of the Cheyenne. The Cheyenne relate the story of a warning to tribal leaders by Maheo (the primary god of the Cheyenne) when the Comanche introduced the horse to them when they still lived in earthen homes somewhere along the Missouri River:

> You will have to move around a lot to find pasture for your horses. You will have to give up gardening and live by hunting and gathering, like the Comanches. And you will have to come out of your earth houses and live in tents . . . You will have to have fights with other tribes, who will want your pasture land or the places where you hunt. You will have to have real soldiers, who can protect the people. Think, before you decide.[16]

At one time, during their exodus onto the Great Plains under the leadership of the prophet Sweet Medicine, the "called-out people" shared a distinct, unified identity as a people. The needs of horses broke them apart, forcing the camps of several hundred into smaller groups or bands of 40 to 60 people to accommodate large horse herds. These approximately 40 bands sought to maintain their shared unity through one group called the Council of Forty-Four, with representatives from each band and through the warrior societies whose membership spanned across the various bands. But conflict intensified between the fragmented Cheyenne groups as they were also weakened in the face of American westward expansion and onslaughts of epidemic disease.

The Comanche suffered a similar fate as their even larger horse herds competed with an ever-dwindling bison population. An expansionist and warring empire, the Comanche depended on the guns, powder, and bullets making their raids possible. Bison hides paid for these implements of war. The ideal hides were those of bulls in the approximate three-year range, and the overharvest of this particular group added to the problem of overharvest by lowering reproduction. Their large horse herds devastated riparian habitats and overgrazed the grasses of the Southern Plains, taking away food from the needed bison. By the 19th century captives were increasingly put to work stripping back bark and leaves from trees to feed the horse herds. Moreover, the Comanche increased their rate of polygamy, with some wealthy leaders having as many as a dozen wives. These wives and some young men were also pressed into labor supporting the horses and tending to hides.

Not all cultural and historical change should be chalked up to environmental processes, but in the past the opposite has been assumed—that the environment itself was a minor player. The Comanches transformed into a specialized horse and bison culture and in the process also become aggressive expansionists that celebrated warrior culture. In so doing, they effected dramatic political, cultural, economic, and political change while weakening other Indian nations, Spanish colonial hopes, and Mexican control of its northern regions. The horse made this possible, but their overreliance on it and the bison undercut their own economy, reducing their strength as Americans began settling and expanding across Texas.

Furs from the Pacific

The sea otter is a charismatic animal, one that is beautiful and endearing in its cuteness. Females carry their pups on their stomachs while floating on their back, and otters also use tools for hunting and accessing food. The species also almost went extinct with an estimated population of 1,000 to 2,000 individuals in the early 20th century. An animal that spends its entire day and life in the ocean, the sea otter ranges between 30 to 100 pounds, and an older male can reach 5 feet in length. They eat prodigiously because of their high metabolism, consuming between a quarter to a third of their bodyweight every day. Sea urchins, clams, mussels, abalone, and some fish are their primary food sources. Sea otters' consumption of sea urchins protects kelp beds from destruction by that species. Following the near extermination of the sea otter, many kelp beds were damaged by large sea urchin population increases. Adding to their appeal, they are one of the few animals to use tools. Sea otters dive to the bottom of the ocean to harvest clams, mussels, abalone, and sea urchins, and they store them in a chest pouch to carry back to the surface and eat. Also in that chest pouch is a stone the sea otter uses to hammer open shellfish while floating on its back. Sea otters also employ the stones to pound on abalone, knocking them loose from their tight hold on underwater boulders. The mammals' lack of fat requires not only high food consumption but also a remarkably thick, warm pelt. With up to 150,000 hairs per square centimeter, the sea otter boasts the densest fur of any animal. Its beauty, warmth, and ability to shed water drove the desire of Europeans, Americans, and Asians, leading to the death of approximately 1 million of these animals and their near demise as a species.

The fur trade in the Pacific Northwest began in earnest in the 1780s after British Explorer Captain Cook's crew (the captain had just been killed in Hawaii) profited highly from the sale of sea otter pelts in China, making a profit of almost 2,000 percent. The crewmembers were almost mutinous in their demands to return to Nootka Sound for more sea otter pelts. While Spain and England both contended for control of this region, traders of both nationalities, as well as Russians and Americans, arrived in the hundreds in the late 18th and early 19th century. Nootka Sound sat at the heart of this trade economy and is located on the eastern side of Vancouver Island, a body of saltwater separating the island from what is now mainland British Columbia. The trade extended through Puget Sound and the Olympic Peninsula of Washington and south to northern California. Also, the Russians had by this point already established a fur economy in Alaska, enslaving Aleut Indians to hunt sea otters. This item was highly desired, particularly in China, for its insulating qualities, luxurious coat, and its ability to shed water easily. The coat remained prime year round also, so trade was continuous. Russians mostly used the sea otter pelt ornamentally.

The leaders of Northwestern peoples sought to control trade and the distribution of goods to reinforce their power and to preserve stability. They did this through their political authority in the tribe, through women intermediaries who married or entered into relationships with traders, and by lying to the Europeans and American ship captains, telling them the residents of other villages were vicious and that the sea otter was not available beyond their area. As with other natives, this control broke down over time as traders sought out individuals and some Indians ignored hierarchy and rushed to the new economic opportunities. Europeans proffered the typical trade items: axes, guns, copper and iron pots, blankets, knives, chisels, cloth, and so forth. Indians of this region were fond of woodwork, making totem poles, masks, and building houses and longhouses with performance stages. Hence, carpentry tools such as saws and carpenter planes were quite popular. Looking glasses and thimbles were coveted as well. Knowing too that the traders considered the darkest pelts of the highest quality, some shrewd Indians conducted business at dusk in the dying light and rubbed charcoal into lighter toned fur to increase their trade value. Northwest Indians were choosy with a finite demand for goods even after they chopped pots into jewelry. Desperate to acquire more pelts, traders sometimes set up blacksmith shops near villages to produce goods directly conforming to Indian needs and desires. However, the Indians did harvest and trade the desired pelts with such alacrity that they needed the cloth and coats of the traders to replace the otter pelts as clothing. As with other native peoples across the continent, once an animal could be commodified and used to gain access to desired trade goods, they overharvested the species. At the same time, they grew increasingly dependent on European and American goods and agricultural commodities. The pursuit of pelts and skins, while temporarily offering benefits and advantages, over the long-term weakened native peoples by damaging their land and resource base, increasing conflict, and making them increasingly dependent on Europeans and Americans.

In the Northwestern fur trade women played a central role as they did across the continent. French traders had been encouraged to marry native women and have children because it aided trade but also because religious and political leaders believed it would lead to more rapid acculturation of Indians. They were disappointed when the opposite was often true; the men many times adopted Indian behavior, language, and dress. The British did not encourage relationships, but the traders engaged in them anyways, and American fur traders and trappers did as well. First and foremost, women served as intermediaries or cultural brokers between villages and tribes and Europeans and Americans. Women translated and provided cultural and political knowledge, and through them traders were immediately embedded in extended kinship networks that increased their odds of economic success. These relationships not only served the interests of the traders, it also benefited the community of the wife or partner because they then had greater access to desired trade goods and could control distribution to some degree, temporarily enhancing the tribe and tribal leader's status and political power. These women provided a number of services, from translating to collecting, storing, and preparing food to also participating in trade negotiations.

Native Americans pursued their opportunities where they could. Hammered and devastated by Old-World pathogens that scoured their populations, they lost much of their oral culture and traditions while grasping for advantages from the Europeans and, later, Americans. Trade goods improved life in many ways, and guns offered some protection or even the possibility of greater power and expansion versus other nations, but at a cost that was not understood in those days of chaos and change. Similarly, trade goods and certain organisms such as the horse allowed many Indians to gain unprecedented power and wealth. However, the consequences of these changes were dramatic. Native groups were annihilated in war with other Indians. Traditional culture was undermined and weakened, while ecosystems were devastated with Indians losing both food supplies and land even as the Europeans and Americans continued their steady progress west. The losses of the Indians benefited the colonizers. The continental extraction of a once abundant storehouse of pelts, skins, and hides increased the growth and prosperity of European nations and the United States while leaving native peoples vulnerable to settlement, conquest, and removal from their ancestral lands.

Document 3.1 Letter from *Father Claude-Jean Allouez,* 1670

Claude-Jean Allouez was an early French Jesuit missionary and explorer in the western regions of New France in North America in the mid to late 17th century. He established Catholicism among numerous native peoples and traveled, preaching, through what is now Canada, Wisconsin, Michigan, Indiana, and Illinois. Allouez was on a mission among the Potawatomi Indians of Wisconsin when he wrote the following excerpt.

1. How active are Gods in the creation of the landscape, and what affect did they have?
2. How do you read this story in terms of Indians seeking an understanding of French interest in their country?

They say that it is the native Country of one of their Gods, named Michabous—that is to say, "the great Hare," Ouisaketchak, who is the one that created the Earth; and that it was in these Islands that he invented nets for catching fish, after he had attentively considered the spider while she was working at her web in order to catch flies in it. They believe that Lake Superior is a Pond made by Beavers, and that its Dam was double,—the first being at the place called by us the Sault, and the second five leagues below. In ascending the River; they say, this same God found that second Dam first and broke it down completely; and that is why there is no waterfall or whirlpools in that rapid. As to the first Dam, being in haste, he only walked on it to tread it down; and, for that reason, there still remain great falls and whirlpools there.

This God, they add, while chasing a Beaver in Lake Superior, crossed with a single stride a bay of eight leagues in width. In view of so mighty an enemy, the Beavers changed their location, and withdrew to another Lake, Alimibegoung [Nipigon],—whence they afterward, by means of the Rivers flowing from it, arrived at the North Sea, with the intention of crossing over to France; but, finding the water bitter, they lost heart, and spread throughout the Rivers and Lakes of this entire Country. And that is the reason why there are no Beavers in France, and the French come to get them here. The people believe that it is this God who is the master of our lives, and that he grants life only to those whom he has appeared in sleep. This is a part of the legends with which the Savages very often entertain us.

Document 3.2 From "The Extermination of the Bison," *William Temple Hornaday,* 1889

Willliam Temple Hornaday was an important leader in the early conservation movement, a zoologist, and for several years served as the first director of the New York Zoological Park, later renamed the Bronx Zoo. He worked with Teddy Roosevelt to save the bison from extinction. Hornaday acquired and protected 40 of the animals at the Zoological Park, and they were the basis for later reintroductions in the West in the early 20th century. He is probably most famous for his book *Our Vanishing Wildlife: Its Extermination and Preservation*. Through his book and other efforts, he was a strong voice for wildlife protection and hunting restrictions. The piece from which the following passage is excerpted helped build popular support for protecting and restoring the bison. The first, quoted section, describes a hunt by mixed-blood Indians of Sioux, Ojibwa, and Blackfeet descent.

1. Analyze the structure and organization of the bison hunt as described below. What does this tell you about the importance of the bison and the relationship of the hunt with their faith?
2. How is the serious nature of the bison hunt reflected in these formal arrangements and in the punishments for violations of the rules?

"After the start from the settlement has been well made, and all stragglers or tardy hunters have arrived, a great council is held and a president elected. A number of captains are nominated by the president and people jointly. The captains then proceed to appoint their own policemen, the number assigned to each not exceeding ten. Their duties are to see that the laws of the hunt are strictly carried out. In 1840, if a man ran a buffalo without permission before the general hunt began, his saddle and bridle were cut to pieces for the first offense; for the second offense his clothes were cut off his back. At the present day these punishments are changed to a fine of 20 shillings for the first offense. No gun is permitted to be fired when in the buffalo country before the "race" begins. A priest sometimes goes with the hunt, and mass is then celebrated in the open prairies.

At night the carts are placed in the form of a circle, with the horses and cattle inside the ring, and it is the duty of the captains and their policemen to see that this is rightly done. All laws are proclaimed in camp, and relate to the hunt alone. All camping orders are given by signal, a flag being carried by the guides, who are appointed by election. Each guide has his turn of one day, and no man can pass a guide on duty without subjecting himself to a fine of 5 shillings. No hunter can leave the camp to return home without permission, and no one is permitted to stir until any animal or property of value supposed to be lost is recovered. The policemen, at the order of their captains, can seize any cart at night-fall and place it where they choose for the public safety, but on the following morning they are compelled to bring it back to the spot from which they moved it the previous evening. This power is very necessary, in order that the horses may not be stampeded by night attacks of the Sioux or other Indian tribes at war with the half-breeds. A heavy fine is imposed in case of neglect in extinguishing fires when the camp is broken up in the morning.

In sight of buffalo all the hunters are drawn up in line, the president, captains, and police being a few yards in advance, restraining the impatient hunters. "Not yet! Not yet!" is the subdued whisper of the president. The approach to the herd is cautiously made. "Now!" the president exclaims; and as the word leaves his lips the charge is made, and in a few minutes the excited half-breeds are amongst the bewildered buffalo."

. . . The "great fall hunt" was a regular event with about all the Indian tribes living within striking distance of the buffalo, in the course of which great numbers of buffalo were killed, great quantities of meat dried and made into pemmican, and all the skins taken were tanned in various ways to suit the many purposes they were called upon to serve.

Mr. Francis La Flesche informs me that during the presence of the buffalo in western Nebraska and until they were driven south by the Sioux, the fall hunt of the Omahas was sometimes participated in by three hundred lodges, or about 3,000 people all told, six hundred of whom were warriors, and each of whom generally killed about ten buffaloes. The laws of the hunt were very strict and inexorable. In order that all participants should have an equal chance, it was decreed that any hunter caught "still-hunting" should be soundly flogged. On one occasion an Indian was discovered in the act, but not caught. During the chase which was made to capture him many arrows were fired at him by the police, but being better mounted than his pursuers he escaped, and kept clear of the camp during the remainder of the hunt. On another occasion an Omaha, guilty of the same offense, was chased, and in his effort to escape his horse fell with him in a coulée and broke one of his legs. In spite of the sad plight of the Omaha, his pursuers came up and flogged him, just as if nothing had happened.

Document 3.3 "Memories of Sea Otter Hunting," *Recollections of Mowachaht Nuu-chah-nulth elder, August Murphy*

August Murphy was an elder of the Mowachaht Nuu-chah-nulth peoples (otherwise known as the Nootka, an incorrect name given Indians of the area by Captain Cook). This Indian nation lived along the western coast of Vancouver Island in what is now British Columbia, Canada, and traveled widely by canoe. An active fishing and hunting tribe, they harvested large amounts of sea otters for trade and later for cash. Murphy wrote an account of his experience hunting sea otters as a young man.

1. How organized and regulated is this hunt? What are some of the keys to preparation?
2. What does the organization of this hunt tell you about Pacific Northwest Coastal Indian culture and the importance of the sea otter pelt trade?

The best time for chase sea otter had to be very calm and smooth sea and no wind at all. The purpose choosen the calmest weather was for tracing the little bubbles coming up when sea otter is living underwater and they claim that the otter is more lively with brezze [sic] and little ripples on the surface of the sea.

There were some very good weather-men to watch the signs for good and bad weather. These men were praying and watching closely for the best calm weather for going out. When they were sure of the calm weather, before dawn they went to all the houses of hunters knocking on sides of their houses to wake them up, and bring the good news of fine weather for the hunt.

About the year of 1896 or 1897, at that time there were 24 small canoes, special made for sea otter hunt. Each canoe had 2 or 3 men. When they were all up and every men went down on the beach and helped each other to lift haul down their canoes to the water's edge.

The boss We-yak gave orders "We shall go to Wa-ka-ta [Bajo Reef] and work slowly toward west. You all know the rules when you sight a sea otter. Both men shall lift his paddle above his head and wave. All canoes shall in haste pull towards the canoe [gaven] signal. When we come to Wa-ka-ta all canoes shall get on line up from shore to off sea, spread approsemately [sic] 200 yards apart from each other. Travel slow. Watch for otter and signal."

My aunt's husband took me along. One of the canoes near the Bajo point reefs gave us the waving of their paddle signal. All canoes turned and paddled towards the gaven signal.

No time lost, the signal canoe was in the middle of a big circle of 23 canoes and the leader We-yak was appointed for watching the otter's bubble, and when pointed his paddle toward the course of the bubbles, leads, and canoe that direction was then ready with their bows and arrows.

When poor sea otter came up surface to take in his breath, rain of arrows was pouring down at the otter. All the shooters were watching their arrows to see if he hit the otter. When one of the hunters hit it, the steermen stands up and shouts on top his voice, he mentions his hunter's name. His arrows has hit the otter.

In that season, 12 otters was caught. Large size was worth $400.00 medium size, $300.00 Small $200.00 very small $150.00. . . . a person catch one otter he made money.

Notes

1 Eric Jay Dolin, *Fur, Fortune, and Empire: The Epic History of the Fur Trade in America* (New York: W.W. Norton, 2010), 21, 22.
2 Quoted in ibid., 48.
3 Quoted in Daniel K. Richter, *The Ordeal of the Long-house: The Peoples of the Iroquois League in the Era of European Civilization* (Chapel Hill: The University of North Carolina Press, 1992), 61.
4 Ibid., 62.
5 Daniel H. Usner, Jr., *Indians, Settlers, & Slaves in a Frontier Exchange Economy: The Lower Mississippi Valley Before 1873* (Chapel Hill: The University of North Carolina Press, 1992), 265.
6 James H. Merrell, *The Indians' New World: Catawbas and Their Neighbors from European Contact Through the Era of Removal* (New York: W.W. Norton & Company, 1989), 42.
7 Quoted in Richter, *The Ordeal of the Long-house,* 86.
8 Quoted in Dolin, *Fur, Fortune, and Empire,* 107.
9 Quoted in Shepard Krech III, *The Ecological Indian: Myth and History* (New York: W.W. Norton & Company, 1999), 158.
10 Quoted in Andrew Isenberg, *The Destruction of the Bison* (New York: Cambridge University Press, 2000), 23.
11 Elliott West, *The Contested Plains: Indians, Goldseekers, and the Rush to Colorado* (Lawrence: University Press of Kansas, 1998), 41.
12 Ibid., 41.
13 Quoted in Krech, *The Ecological Indian,* 133.
14 Ibid., 131.
15 Ibid., 134, 135.
16 Quoted in Isenberg, *The Destruction of the Bison,* 41.

Further Reading

Bartram, William. *Travels: Through North & South Carolina, Georgia, East & West Florida* (1791). New York: Penguin Books, 1988.
Cronon, William. *Changes in the Land: Indians, Colonists, and the Ecology of New England.* New York: Hill and Wang, 1983.
Dolin, Eric Jay. *Fur, Fortune, and Empire: The Epic History of the Fur Trade in America.* New York: W.W. Norton & Company, 2010.

Hamalainen, Pekka. *The Comanche Empire.* New Haven, CT: Yale University Press, 2009.
Isenberg, Andrew C. *The Destruction of the Bison: An Environmental History 1750–1920.* New York: Cambridge University Press, 2000.
Kirk, Ruth. *Tradition and Change on the Northwest Coast: The Makah, Nuu-chah-nulth, Southern Kwakiutl and Nuxalk.* Seattle: University of Washington Press, 1986.
Krech III, Shepard. *The Ecological Indian: Myth and History.* New York: W.W. Norton & Company, 1999.
Merrell, James H. *The Indians' New World: Catawbas and Their Neighbors from European Contact Through the Era of Removal.* New York: W.W. Norton & Company, 1989.
Richter, Daniel K. *The Ordeal of the Long-house: The Peoples of the Iroquois League in the Era of European Colonization.* Chapel Hill: The University of North Carolina Press, 1992.
Silver, Timothy. *A New Face on the Countryside: Indians, Colonists, and Slaves in South Atlantic Forests, 1500–1800.* New York: Cambridge University Press, 1990.
Usner, Daniel H., Jr., *Indians, Settlers, & Slaves in a Frontier Exchange Economy: The Lower Mississippi Valley Before 1783.* Chapel Hill: The University of North Carolina Press, 1992.
Van Kirk, Sylvia. *Many Tender Ties: Women in Fur Trade Society, 1670–1870.* Norman: University of Oklahoma Press, 1983.
West, Elliott. *The Contested Plains: Indians, Goldseekers, and the Rush to Colorado.* Lawrence: University Press of Kansas, 1998.
White, Richard. *The Middle Ground: Indians, Empires, and Republics in the Great Lake Region, 1650–1815.* New York: Cambridge University Press, 1991.
———. *The Roots of Dependency: Subsistence, Environment, and Social Change among the Choctaws, Pawnees, and Navajos.* Lincoln: University of Nebraska Press, 1983.

Timeline

British Royal Proclamation Stopping Expansion West of the Appalachian Mountain Crest	1763
Virginia Instituted a Four-Year Deer Hunting Moratorium	1772
The Declaration of Independence	July 1776
The End of the American Revolutionary War	1783
The War of 1812	1812–1814
John Taylor Publishes *Arator*	1813
The Battle of Horseshoe Bend	1814
The Treaty of Fort Jackson	1814
Creation of the Albemarle Agricultural Society	1818
The Missouri Compromise	1820
The Indian Removal Act	1830
Edmund Ruffin Began Publishing the *Farmer's Register*	1833
Collapse of Cotton Prices	1837
The Treaty of Guadalupe Hidalgo	1848
The Compromise of 1850	1850

A Great Farming Nation 4

The popularity of tobacco in 17th-century Europe is hard to comprehend and practically impossible to exaggerate. Europeans avidly consumed the leaves, spending exorbitant amounts of money on Spanish and Chesapeake tobacco because of their enjoyment of the buzz from smoking it, chewing it, or inhaling it as snuff. They also believed that it provided a number of health benefits such as increased vigor, reduction of fevers, stronger lungs, and longevity. King James I of Great Britain was so disturbed by the new addiction that he issued a 1604 denunciation titled, "A CounterBlaste to Tobacco." The monarch inveighed against the leaf and British use on numerous grounds. One point he argued is

> this tobacco is not simply of a dry and hot quality but rather hath a certain venomous faculty joined with the heat thereof which makes it have an antipathy against nature as by the hateful nature thereof doth well appear. For the nose being the proper organ and convoy of the sense of smelling to the brains, which are the only fountain of the sense, doth ever serve us for an infallible witness, whether that odor which we smell be healthful or hurtful to the brain!![1]

It took almost four centuries to prove King James right about the dangerous health impacts of tobacco. He also took issue with the belief that the plant improved health in numerous ways:

> [A]re you therefore no wiser in taking Tobacco for purging you of distillations than, if for preventing cholic, you would take all kind of windy meats and drinks; and for preventing of the stone, you would take all kind of meats and drinks that would breed gravel in the kidneys. And then when you were forced to void much wind out of your stomach, and much gravel in your urine, that you should attribute the thank, therefore, to such nourishments as breed those within you that behooved either to be expelled by the force of nature, or you have to burst at the broadside, as the Proverb is.[2]

The angry and sarcastic monarch's dissection of beliefs about tobacco's health benefits is followed by an attack on the attitudes and behaviors that helped make smoking, chewing, and sniffing the plant all the rage. In noting this he was observing one of the great driving forces of a capitalist economy, a process that helped lift his nation to preeminence among the world's nations for three centuries.

> For such is the force of that natural self-love in every one of us, and such is the corruption of envy bred in the breast of every one as we cannot be content unless we imitate every thing that our fellows, and so prove ourselves capable of every thing whereof they are capable, like apes counterfeiting the manners of others to our own destruction.[3]

Tobacco was the first widely popular exotic import in Europe and was followed soon by tea and coffee. The European hunger for these crops and others such as indigo, rice, and cotton helped drive the rise of the colonial economies, expansion in the New World, and dramatic ecological transformation through the 17th, 18th, and 19th centuries.

From the initial stages of land acquisition, beginning agriculture for subsistence as well as the market, and the introduction of livestock, regional agricultural economies developed around central staples that thrived in their climates and soils and produced profits in the markets of Europe, the Caribbean, and other colonies in the New World. The development of these agricultural economies necessitated the creation and expansion of labor systems ranging from wage-labor to indentured servitude and slavery. Moreover, in some cases the landowners initiated even more dramatic, directed transformations of the land in order to systematize production of rice, tobacco, and other commodities. Extracting and creating abundance was the central goal, and the production without limits enabled by constant movement onto new lands temporarily came to an end in the late colonial period. While some planters sought to slow down expansion in order to preserve their communities, the temporary barrier to expansion caused by the Proclamation of 1763 and treaties with Indians on the western side of the Appalachian Mountains contributed to tensions leading to the Revolutionary War.

By the early and mid-18th century, distinct agricultural communities were developing along the eastern seaboard. Grain and livestock production for subsistence and markets were hallmarks of New England agriculture, while tobacco was the dominant cash-crop of the Middle Atlantic colonies, an export commodity that tied the planters into a broad and complex Atlantic trade economy. In the Georgia and Carolina lowlands, planters moved earth and water to produce a remarkable rice economy and culture. Even as these agricultural economies developed and farmers and planters sought to make these economies sustainable, the westward trek into the piedmont of Virginia and the Carolinas and the Berkshires of Massachusetts continued. The processes of land use and transformation described in chapter two essentially continued on the westward edge of European society as the expansion continued, setting in motion the frontier experience that for a long period of American history was celebrated as a key component in the development of the American character. Historians today challenge the simple notion of a frontier line, emphasizing numerous frontiers, borderlands, and points of contestation.

Tobacco, the Plant of "Infinite Wretchedness"

Tobacco saved the Virginia Colony in the 1620s, emerging as the staple crop *par excellence* for the Chesapeake region. Economic and land conflicts arose from the production of this popular plant as was evidenced in Bacon's Rebellion. This colonial civil war rose directly from the need for more land for tobacco agriculture and resulted in instituting African

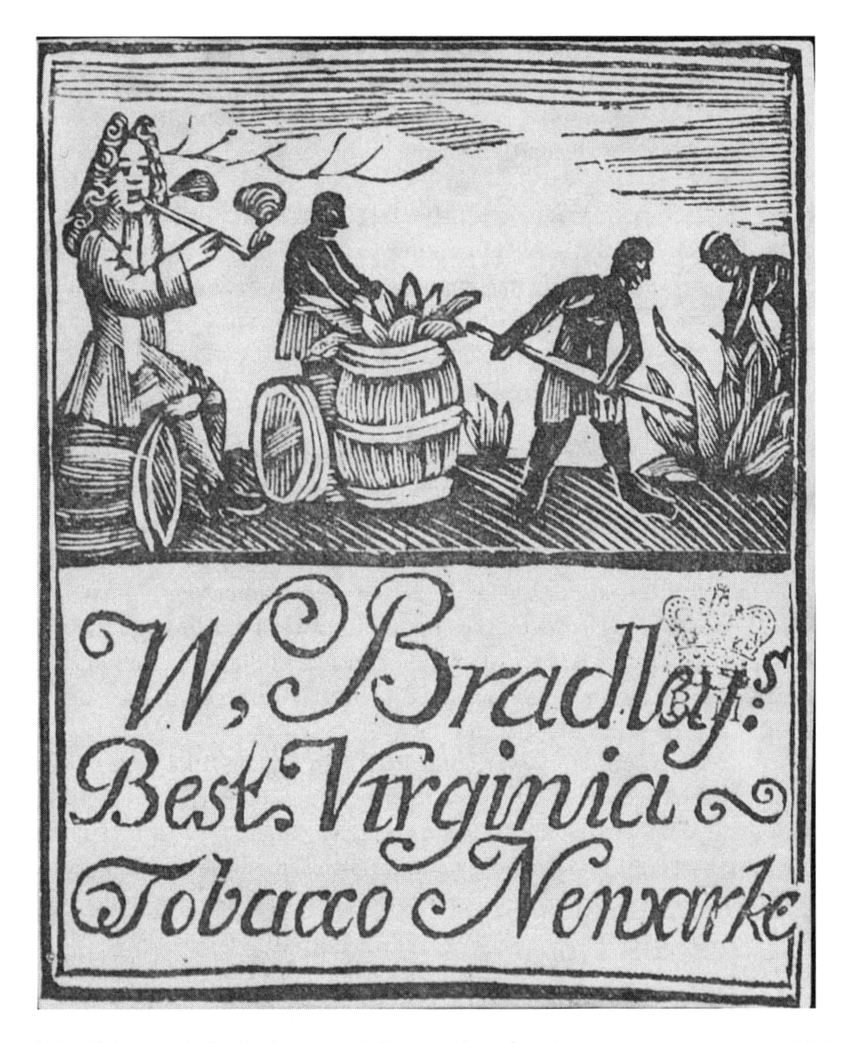

Figure 4.1 Tobacco Labels promoted the product for European consumers with images of colonial life in America. They typically showed a planter at his leisure and slaves working the tobacco crop. Courtesy of The Jamestown Yorktown Collection, Williamsburg, Virginia.

slavery as the primary labor force for southern agriculture in America. Tobacco continued to consume land even as it exhausted the sinew and muscle of hundreds of thousands of men and women.

The author of *American Husbandry* observed tobacco agriculture in the early 1760s and 1770s and described the painstaking and ongoing labor necessary for successful production of the plant. He explained that the seed is first planted in fertile soil, then transplanted when attaining the height of 4 or 5 inches. Ideally, these plants would be set in rich, black soil from where forests had been cleared by fires.

> Sometimes it is so badly cleared from the stumps of trees, that they cannot give it any ploughings; but in old cultivated lands they plough it several times, and spread on it what manure they can raise. The negroes then hill it; that is, with hoes and shovels

> they form hillocks, which lie in the manner of Indian corn, only they are larger, and more carefully raked up: the hills are made in squares, from six to nine feet distance, according to the land; the richer it is the further they are put asunder, as the plants grow higher and spread proportionally.

After a month the tobacco plants reached about a foot in height. It was necessary then to top and prune the plants in the unrelenting summer heat. The bottom leaves had to be cut off to leave only seven or eight per stalk in order to create a decently sized leaf. While the tobacco was growing to a height of 5 to 7 feet over the following six weeks, the slaves were working hard, pruning off suckers, weeding hillocks and rows, and picking worms off the plants.

> When the tobacco changes its colour, turning brown, it is ripe and they then cut it down, and lay it close in heaps in the field to sweat one night: the next day they are carried in bunches by the negroes to a building called the tobacco house, where every plant is hung up separate to dry, which takes a month or five weeks; this house excludes the rain, but is designed for the admission of as much air as possible. They are then laid close in heaps in the tobacco houses for a week or a fortnight to sweat again, after which it is sorted and packed up in hogsheads; all the operations after the plants are dried must be done in moist or wet weather, which prevents its crumbling to dust.[4]

Slaves in the tobacco economy were organized into a gang-labor system of 10 to 12 men per gang, closely supervised by an overseer or owner. Hoeing, pruning, deworming, and harvesting, the men worked the crops for a long day under a hot sun while constantly prodded to work harder. With leaves drying in the sheds, then there were more forests to clear, more ground to break, and stumps to remove. Competition with the rice economy in the 18th century kept the price of slaves imported from Africa or the Caribbean high, and tobacco planters were compelled to treat their slaves better than rice plantation owners. The lower prices for tobacco meant they could not simply work a slave to death and replace him or her with another as was the case in the sugar economy and had been true of the rice economy for a period in the early 1700s. Pregnant slaves were given more provisions and time to rest from labor. While kindness may have played some small part in this treatment, the economic motivation to produce more slaves because of high prices was the more important factor in this decision.

As hard on land as it was on people, the leafy plant stripped soil of its nutrients within three years, necessitating constant movement into new areas and using rotation as a secondary option. In the 17th and early 18th century, that meant exhausted soils were abandoned to return to pines and oak. When trees had reconquered an "old field," it was seen as a sign that fertility had returned and one or two years of production of notably reduced yields were typical before the land was then abandoned for good. The increasing population growth of whites and slaves and the exhaustion of soil pushed the tobacco economy west into the piedmont and foothills. Many exhausted farms were sold or abandoned as the population worked its way westward. The whole notion of westward expansion and the contestation of a frontier landscape arose from European market demands and the careless agriculture of tobacco agriculture.

Figure 4.2 This image from 1800 depicts the many stages involved in tobacco production and the labor intense nature of this cultivation that necessitated indentured servitude and slavery. Library of Congress.

Trying to stop this exodus, planters of the Chesapeake country in the mid-18th century emulated a British agricultural model known as the Norfolk system. The goal of this land-use strategy was to create productive, permanent farms, eliminating the practice of constant abandonment of fields for newly opened Indian land and forest that characterized tobacco agriculture in this region for over a century. Wide-ranging, wandering livestock were to be corralled behind permanent fencing and crops rotated through a five-field system. They varied production with crops such as wheat and corn, while committing more land and labor to grass production and manure collection. Anxious agriculturalists sought to conserve and use manure, fencing livestock with a vengeance, in some cases even building movable corrals to target infertile soil with nutrient-rich cattle waste. Some farmers also began growing clover as a cover crop. Historian Brian Donahue writes of clover that

> there may be nothing magic about manure, but there is deep magic in clover. Rotating arable crops with grass and legume crops is in most cases the most productive, ecologically stable form of mixed husbandry known. Legumes such as clover fix atmospheric nitrogen into a farm that other plants can use, and many also have deep foraging roots that bring mineral nutrients from the subsoil.[5]

Speculation, Tobacco, and Revolution

This unstable agricultural economy with its fluctuating prices and the need for ever more land was inherently unstable. As a result many planters turned to land speculation as a way to either enhance profits or manage debt burdens. The tensions that resulted from British government efforts to control westward expansion and the corresponding impact on speculating elite planters played a small but important role in the increasing tensions that eventually turned into calls for revolution.

Speculation in land was an essential part of the colonial economy, and Virginia planters, as well as economic elites in other colonies, came to rely on profits from these land transactions to offset low tobacco prices and help pay their escalating debts to British merchants for imported manufactured goods. Such renowned Virginia figures as George Mason, Thomas Jefferson, Patrick Henry, and George Washington engaged in extensive land speculation (see Document 4.1) and, to varying degrees, relied on these ventures to reduce their debt burdens. In 1768 and 1769 Jefferson was involved in an effort to obtain title to 7,000 acres on the western side of the Appalachian Mountains intended for sale only. Arthur Lee, a Virginia planter and one of those later demanding revolution, tried to find a British official in London that would allow the transfer of a 2,500,000 acre parcel of land to a company created by Lee, his brothers, and several Virginia elites. Great orator of liberty Patrick Henry also suffered economic pressure from British action as he had purchased more than 3,000 acres of land on the other side of the Appalachians.

The belief that veterans of the French and Indian War would receive land bounties triggered another speculative frenzy. A former officer in that war and future Continental Army commander and president of the United States and his brother began buying up bounties from veterans. According to historian Woody Holton, George Washington

> advised his brother Charles to approach veterans "in a joking way, rather than in earnest at first" in order to "see what value they seem to set upon their Lands." If Charles Washington could attain the veterans' grants for seven pounds or less per thousand acres, he was directed to do so.[6]

Jefferson involved himself in two other speculative efforts in 1769, and the reality is this sort of mad pursuit of land on the other side of the mountains was typical of this class and time. These efforts were made in spite of treaties with powerful Indian nations claiming sovereignty to the land on the west side of the mountains and the Proclamation of 1763 banning migration into Indian land on the western slope of the Appalachians.

Virginia planters found Parliamentary interference frustrating and economically damaging. Their criticism of the proclamation was relatively muted but certainly hurt them economically and tested their loyalties. Holton points out that,

> the abolition of land grants was surely a major complaint for Virginia's leading revolutionaries because it hurt almost all of them. George Mason, who would write the constitution for the new Commonwealth of Virginia, had watched the Proclamation of 1763 destroy first his beloved Ohio Company and then his hopes of obtaining fifty thousand acres of Kentucky land using headrights. Richard Henry Lee, who would introduce the motion for Independence at the Continental Congress, had seen his Mississippi Land Company's hope of obtaining 2,500,000 acres of Indian land disappear behind a double barrier: the Proclamation of 1763 and the Quebec Act of 1774.[7]

Washington was also a member of the Mississippi Company and had purchased thousands of acres of land that he subsequently lost. The other man with equal standing as a pivotal leader in the independence movement, Thomas Jefferson, stood to gain 17,000 acres of land in multiple deals but lost it to the British government policy. He, of course, wrote the Declaration of Independence. And the firebrand who issued the Virginia Resolves that helped inflame the argument against the British Parliament and declared "Give me Liberty or Give me Death," Patrick Henry, had participated in several land speculation ventures from which he would lose all of his money if British policy remained uncontested.[8]

It wasn't simply access to land but the crop they were so dependent on that helped drive dissent toward war and new government. Tobacco planters were furious at merchants' control of the economy and the low prices planters received versus what the merchants received after processing it through England and shipping the tobacco back out to other markets. Stymied by mercantile policies, planters' anger and demands for greater economic independence grew. The desire to expand profits from tobacco production, as in other areas of American colonial economic life, fanned the flames of independence rhetoric. For the Virginia planters, the low prices for tobacco were one factor in the planters' increasing debt. Their economic burden was higher than that of northerners, and when efforts were made to impose and enforce the Sugar Act, the Stamp Act, and Townsend Duties, southern planters felt that they were being doubly taxed. One might argue that they were trussed in a trap of their own crafting, that they had created their own dependence on tobacco production, but by the 1760s there was an increasing emphasis on mixed crops and local food production. In fact, the House of Burgesses (the Virginia Colonial assembly) had petitioned the Privy Council for a dramatic increase in the duties paid on imported slaves. This had been done for several reasons, one of which was to make slaves so expensive that planters and farmers would be compelled to move away from overreliance on tobacco production. The Privy Council rejected the request, increasing resentment on the part of Virginia elites.

As a nation we take great pride in the pursuit of liberty and the creation of a new form of representative government following the American Revolution. It is going too far to assert that anger over policies regarding access to land, tobacco pricing, and the importing of slaves caused the American Revolution. But better understanding the role of speculation, anxiety over economic opportunity, and frustrations associated with British policy, land availability, and tobacco prices, provides a more complete understanding of how environmental factors and land-use practices contributed to the swirling pool of motivations and complaints that eventually led to a revolution and an independent America.

Having won their independence, the planters and now political leaders of Virginia still faced many of the same ecological, economic, and cultural problems plaguing them prior to the conflict. The exhaustion of soil and failing crops engendered concern in Virginia in the post-Revolutionary War years through the antebellum period. The anxiety arose from both fear of economic failure and because of the massive out-migration to lands further west. Like the leaders of many rural small towns in America today, farm reformers sought to keep their youth on the land. One observer wrote about the exodus,

> Year after year we behold the anxious struggling crowd, pressing forward through sunshine and through storms, over mountains and valleys, in long continuous crowds of carriages and waggons, [sic] rich and poor, young and old, white and black, master and slave, hastening with impetuous ardor and zeal to this fancied El Dorado and Elysium of the West, till we seemed as we behold the stream, to be left desolate and alone, amid the depopulated and abandoned scenes of our youth.[9]

According to historian Lynn Nelson,

> stung by the sight of abandoned fields, covered with gullies and weeds, anxious planters called for farmers to take a longer view when trying to intensify their state's agricultural environment—to stop "wasting" its productivity and thereby ensure sustainable gains in farm production. Gathering under the banners of "agricultural reform," "improvement," and "high farming," these men preached that the decline of the state could be stopped if planters would conserve the ecological wealth they already possessed.[10]

John Taylor, a Virginia landowner and veteran officer of the Revolutionary War, noted the limitations of even these strategies and argued vociferously through numerous publications for the increased use of green manure. This refers to the growing of cover crops such as clover, corn, and other plants and then tilling them directly into the soil to enhance fertility rather than running them through the digestive tracts of cattle first. Clover was also beneficial as a cover crop because it would slow down erosion, conserving soil. Taylor also promoted deep plowing, believing this would break up clays below the surface, bringing them to the surface and improving the soil overall. Deep plowing provided the added benefit of ridges that would control erosion.

Another southern reformer took up the mantle of soil improvement and cultural preservation. Edmund Ruffin is best known as the long-haired, wild-eyed states' rights advocate who lit the fuse on the first cannon shot at Fort Sumter by the Confederacy, thereby launching the Civil War. Not as well known are his contributions to soil science, pioneering that field in the South and the United States. Ruffin despaired over the diminishing quality of the soil on his James River plantation. Studying Taylor's work, he applied his innovative ideas, but to limited affect. Erosion continued to eat away at his land, and the quality of the soil did not improve. He recognized that the ideas promoted by Taylor, built on English agricultural reforms, were not a good fit for the complex ecosystems of the Chesapeake and Piedmont regions. Moreover, Ruffin's studies of organic chemistry led him to the central weakness in Taylor's reforms: the failure to counter acidity in tidewater soils with lime.

With the aid of an elderly slave, Ruffin discovered beds of marl on plantation. Marl is a mix of mud and decomposing shells. This calcium carbonate rich soil, a ready and available supplement for planters and farmers of that region, was rich in lime—the material required to offset acidification. He put his slaves to work excavating marl and working it into the fields. Ruffin advertised "marling" as the solution to planters' woes and saw his crop yields increase substantially. However, other planters did not experience the same consistent success. The pH levels of fields ranged quite a bit, as did marl supplements themselves. Marling could push the soil too far the other direction and make it overly alkaline. Moreover, this method of amending the soil was labor intensive and did not preclude the other work necessary to building soil fertility—collection and deposition of livestock manure, planting and plowing of green manure, and the use of erosion control strategies. When farmers in the Chesapeake moved away from an overreliance on tobacco agriculture to growing multiple crops, there was a corresponding transition from complete dependence on slave labor to a wage-labor economy. This made marling too expensive for many planters. As a result, Ruffin's solution did not gain popular adherence, and Southern farmers eventually turned to imported guano to fertilize and keep their land productive.

The creation of the Albemarle Agricultural Society in 1818 by Charlottesville, Virginia, planters represented an organized effort to staunch the bleeding of population through conserving and restoring the soil. Retired president and planter James Madison was asked to serve as the first president of the organization and, speaking before the group, gave his "Address to the Albemarle Agricultural Society," a speech anticipating later conservationist thought, Madison emphasized the abundance of diversity of nature, commenting on the need for greater balance between man, nature, and agriculture and arguing that farming every inch of the land would bring negative results (see Document 4.2).

A primary point of the address was Madison's emphasis on efficient soil conservation and the need to effectively recycle nutrients back into the soil. He argued for intelligent use of good land and leaving marginal land out of production. Madison also emphasized the need for the collection and distribution of manure. In this address that is so focused on finding a more sustainable way of using land, Madison's passion originates from his belief that successful agriculture is clearly the foundation of America's greatness. When abundance is threatened, it undermines the possibilities of American life. The very aspirations of the young Republic were built on an edifice of natural wealth and abundance, and he and the other planters did not want to see their society fail because of wasteful land use.

Making a Hydraulic Landscape for Rice

As tobacco dominated the landscape of Virginia, the lowlands of the Georgia and South and North Carolina colonies were transformed to grow rice. The creation of what historian Mart Stewart calls a "hydraulic landscape" also dramatically altered the development of slave culture in the rice economy, increasing deaths from disease while also providing them, for a time, with greater autonomy and the ability to preserve their culture and identity. Slaves were also able to build families and communities to better protect themselves against the horrors of slavery.

Great labor and knowledge were required to grow rice. West African slaves were highly valued for their knowledge and experience from African rice fields. Before skill came into play, brute force was applied to the land. Trees were cut down and stumps rooted from the ground: habitats stripped away to begin the movement of soil to control water. Temporary and permanent banks were shoveled and piled to separate and hold water. Cross or check banks were then built separating rice paddies into squares of 20–25 acres. Trunk lines were built to move water into and out of the fields. According to historian Mart Stewart,

> Originally, the trunk may have literally been a tree trunk, hallowed out to serve as a conduit from river to field. Trunks were commonly used for this purpose in hydraulic agriculture in West Africa; much of the tidal system for growing rice came from the slaves. By the late eighteenth century the trunk was essentially a large square culvert, usually made of heavy planks and timbers, with gates on either end. It was the main conduit, on a well-regulated plantation, between the water within and the water outside, and it was also the nodule of control in this hydraulic system.[11]

Slaves then dug ditches and drains, seeking to level the squares. African slaves moved remarkable quantities of soil in this endeavor. According to Stewart,

> a plantation of one-eight square mile, when it was completely "improved," had two and a quarter miles of exterior, interior, and check banks, and twelve or thirteen miles of canals, ditches, and quarter-drains. In other words, slaves working with shovels in ankle-deep mud and water had to move well over thirty-nine thousand cubic yards of fine-grained river swamp muck to construct an eighty-acre plantation, in addition to clearing the land and leveling the ground in the fields.[12]

Historian Donald Worster argues that the control of nature leads to the control of men and often a brutalizing of their condition. Slaves' identities were stripped away from them, and they were given classical ancient European names like Cato or Nero to destroy their African culture and mock them; their independence was also limited. The ability of slaves to travel between plantations and to even serve in the militia was ended in the early 1700s. A dramatic increase in the use of violence accompanied the increasing physical labor performed by slaves. Whipping, burning, and hangings of troublesome slaves became commonplace. Historian Ira Berlin writes that

Figure 4.3 This image of rice agriculture from *Harper's Weekly* in 1867 shows the various tasks required for successful cultivation. Men are shown digging ditches and shooting at birds to drive them away. The image also depicts African-Americans weeding the rice as well as the complex machinery of water control required to manage the hydraulic landscape. Library of Congress.

> even punishment for "small faults" took on a monstrous quality, as with the planter who placed slave miscreants in a "coffin where they are almost crushed to death," keeping them "in that hellish machine for Twenty Four hours." Such atrocities disturbed a few clerics, but planters believed terror to be a critical element in sustaining their dominion over a people, who, as one slaveholder professed, were "created only for slavery."[13]

As they wrought their will on the land, they did the same on human bodies, even destroying and crippling bodies like they did the lowland habitats of Georgia and the Carolinas.

The work involved in raising and harvesting rice in this hydraulic landscape was difficult. In its early years slaves worked in closely supervised groups under the gang-labor system developed in the tobacco economy. Ten to twelve slaves, mostly men, working under the close supervision of an overseer, felt constant pressure and experienced periodic punishment compelling them to labor strenuously at all times. Sowing of rice seed occurred from April through June, and the slaves used their bare feet to press the seeds into saturated soil. For the rice to germinate the fields needed to be flooded; the most skilled slaves, many bringing their knowledge from Africa, governed the flow of water in and out via the trunks. With germination, the slaves worked the muck, hoeing away weeds. Summer was filled with a busy schedule of alternating flooding and draining of fields while actively driving birds

Figure 4.4 African-Americans working on a Cape Fear River Plantation in North Carolina. This engraving was produced in 1866 for *Frank Leslie's Illustrated Newspaper*. Library of Congress.

away. Weeding remained a regular feature of daily life, and slaves performed these tasks in mud up to their ankles or higher, in stagnant water. Their labors were made miserable by the possibility of water moccasin bites and the swarms of biting insects like flies and mosquitoes whose numbers exploded with the increased standing water on the land. As a result, slaves in the early 18th century were decimated by malaria and yellow fever; white owners fled to the cities in the summer to escape the diseases. After harvest in September the crops were processed. Slaves then returned to other tasks, repairing canal and drainage banks, extending waterways, clearing more land, and numerous other onerous jobs. The high profits of the rice economy made the overworking of slaves possible in the middle of the 18th century. Theirs was truly a life of constant toil, transforming the land, maintaining the flow of water, and producing the staple crop so popular in the Atlantic trade economy.

Disease devastated the slaves but also changed the nature of their bondage. Malaria coupled with difficult labor and low rations inflicted heavy losses on slaves for many years. Frederick Law Olmsted is better known for his work as a landscape architect and is probably most famous for his contributions to the design of New York City's Central Park and others. He worked as a journalist early in his career, and traveling through the South, providing dispatches for the *New York Daily Times* between 1852 and 1857, he provided valuable glimpses of southern life. This description of a lowland plantation illustrates the impact of malaria:

> In the upper part of this pine land is a house, occupied by his overseer during the malarious season, when it is dangerous for any but Negroes to remain during the night in the vicinity of the swamps or rice fields. Even those few who have been born in the region, and have grown up subject to the malaria, are said to be generally weakly and short-lived. The Negroes do not enjoy as good health on rice plantations as elsewhere; and the greater difficulty with which their lives are preserved, through infancy especially, shows that the subtle poison of the miasma is not innocuous to them.[14]

Malaria was likely introduced early in the colonial period, although the first type of malaria in the colonies was a mild form originating in Europe. A parasite carried and introduced by mosquitoes, *Plasmodium vivax,* travels to the liver once introduced into the body and mostly attacks young blood cells. This early malaria inflicted a death rate of approximately 5 percent. Those not killed by malaria remain more vulnerable to other diseases, as is the case with all forms of malaria. Fevers and chills beset the malaria victim, and after recovery the malaria will return again in less than a year; this process will repeat regularly for several years until the victim finally builds immunity. A more dangerous form of malaria, most likely from West Africa and carried by slaves brought to the New World, arrived in the southern colonies after 1650. *Plasmodium falciparum* is much more lethal, killing up to 25 percent of its victims by attacking both young and mature red blood cells and causing blood clotting and death. This new form of malaria became a scourge at least partially because the hydraulic landscape provided a perfect environment for the spread and growth of mosquitoes carrying the disease. In comparison, in New England, malaria disappeared as wetlands were drained and converted into pasture or farmland.

This new lethal malaria resulted in an unexpected benefit for slaves of the hydraulic landscape. Plantation owners and their families fled the rice plantations during the malaria season, residing in coastal cities such as Savannah and Charleston to escape the parasite. Because they returned to the plantations when the threat of malaria abated and usually were only a day's journey from their crops and slaves, the rice plantations did not become as brutal (by way of plantation conditions and treatment by the plantation owner) as the absentee landlord sugar plantations of the West Indies. This is also because prices dropped after the early 18th century, and slaves became too expensive to replace. Because of the need to keep slaves alive and due to the existence of the new, more lethal breed of malaria, slaves gained more latitude in their daily lives. With greater autonomy and power to petition and lobby owners, they leveraged greater control of their time and lives. The shift to a task labor system, gave them freedom to pursue other activities. Since owners were now absent during the most crucial part of the rice-growing season, it was in their interest to negotiate with slaves because of the possibility of sabotage to crops and equipment or simply lackadaisical or intentionally slow labor in the fields resulting in lower yields. According to Berlin,

> under the task system, a slave's daily routine was sharply defined: so many rows of rice to be sowed, so much grain to be threshed, or so many lines of canal to be cleared. Such a precise definition of work suggests that city-bound planters found it difficult to keep their slaves in the field from sunup to sundown and conceded control over work time in return for a generally accepted union of output, especially when it could be measured from afar.[15]

While the hydraulic landscape required and expanded the control of human labor, it unexpectedly created a pathogenic region that provided slaves some relief from their subjugation.

Rice economy slaves were able to create a limited world of their own. Because of the size of plantations and the ability to travel the countryside between operations, it became easier to marry and produce families as well as build communities on plantations and between them. Families and communities improved slaves' quality of life and provided them important psychological strength and comfort within an evil system. The physical independence of slaves of the Georgia and Carolina low-country translated into some cultural and religious autonomy. The slaves of this region resisted cultural assimilation and conversion to Christianity longer than African slaves of any other region in the United States. The persistence of African languages, cultural practices, and religions that slaves used to bring order to their world of enslavement is at least partially the result of the malaria parasite. If pathogens hadn't devastated white slave owners and their families, they would have remained in closer contact with their slaves and compelled more rapid assimilation. Slaves found some comfort in this brave new world.

Because they had more time, they were able to produce more food for themselves and to sell in local markets. Doing so allowed them to enrich their diets, improve their health, and gain income to purchase desired or needed objects and, in a few cases, their own freedom. They worked their "little piece of land . . . much better than their master . . . they rear hogs and poultry, sew [sic] calabashes, etc. and are better provided for in every thing than the poorer white people with us."[16] Low-country slaves not only grew more food, but they also introduced new species into the American diet that are integral to southern food culture. For example, the black eyed pea, purple hull pea, and crowder pea, all of the cow pea family, originate in Africa and were a part of slaves' diet there. They arrived in American via the digestive tracts of cows or were intentionally brought by slaves. Seen by whites as food fit only for consumption by cows, slaves made these peas a major part of their diet, and they were adopted over time by poor whites. Similarly, the sweet potato originated in South America and the Caribbean and was grown by some Indians in the American South. It entered American diets in a number of ways, but one key route was via slaves brought from the Caribbean with a fondness for the nutritional crop. Along with cow peas, sweet potatoes became an important crop in slaves' diets. This tuber was particularly useful because of its ability to grow anywhere; the ease of storing, cooking, and carrying it; and its nutritional benefits and high carbohydrate count. By 1900 the per capita annual consumption of sweet potatoes by Americans was 60 pounds a year. The spread of malaria not only allowed slaves to improve the amount and diversity of their diets but also enriched American foodways.

With their "free time" slaves also hunted, fished, and ran trap lines. The deer, squirrel, rabbit, bass, catfish, and other species they harvested were incorporated into their meals and also bartered and sold among slaves and whites in small town and urban markets.

Figure 4.5 Gullah developed in the South Carolina Sea Islands and the lowland from the Cape Fear River of North Carolina to Georgia. It represents the strongest persistence of native culture and the rise of a distinct culture incorporating African, Portuguese, Caribbean, and American elements. Gullah has its roots in the low-country rice economy where the presence of endemic diseases such as yellow fever and malaria enabled African slaves to hold onto traditional African folk beliefs, foodways, language, and other cultural forms while blending them with European, American, and Caribbean language and cultural forms to create a distinct culture that persists to the present day. This painting, called "The Old Plantation" and produced in 1790, shows African slaves dancing and playing musical instruments like the banjo, which originated in Africa. The women are playing a traditional Sierra Leonean instrument called the shegureh. Courtesy of the John D. Rockefeller, Jr. Library, Colonial Williamsburg Foundation.

In due time, slave owners imposed stricter controls on the men and women working the rice fields. But for a while, at least, the introduction of malaria from Africa and the creation of a hydraulic landscape made it possible for slaves to mitigate their conditions through controls on labor and punishment by enriching their diet; preserving their culture, language, and religions; and through building families and communities. In historical events and cultural developments such as these, we can clearly see the impact of environmental change.

When Cotton Was King

Cotton was grown through the colonial period in the American South, but it wasn't until a technological innovation, the creation of the cotton gin in 1793, that the fluffy, white boll came to dominate the land and culture of Dixie. Until that invention slaves picked seeds from the cotton by hand; a typical pace was about a pound per day. The cotton gin created an exponential increase of 50 pounds per day. This drove the price of cotton down and made cotton the primary cloth of the textile industry, resulting in Great Britain and then New England's dependence on the crop and increasing the wealth of southern planters.

With the extreme profits to be made from these plants, land formerly used for grain, corn, and tobacco in the Southeast was converted to cotton farming. More importantly, the desire to extract wealth from cotton and the diminished quality of soil in the southeastern states after long, intensive use caused bellicose expansionist rhetoric and demands for access to Indian lands in the early 19th century. Much of the lobbying for the War of 1812 emanated from western and southern states seeking to seize Spanish land in Florida and Indian land in southern Georgia, Alabama, and Mississippi, where a number of tribes, including the Choctaws, Chickasaws, and Creeks, still controlled millions of acres of potential cotton farms. An excellent example of how the war opened land for cotton farmers is the Treaty of Fort Jackson. Forced upon the Creek Indians after the defeat of the Red Stick band of the Creeks at the Battle of Horseshoe Bend in 1814 by General Andrew Jackson, this treaty seized 23 million acres of Indian land, much of it river and creek bottom land well suited for cotton agriculture. Removal of British and Spanish interests from the region and the diminishment of Indian power and status enabled southerners to flood into the black-soiled cotton belt running through Alabama and Mississippi. The war made possible the continued expansion that was so integral to sustained American economic growth and the production of abundant resources and agricultural goods.

Mississippi became the heart of the cotton kingdom for a number of reasons: deep fertile soil along rivers, loess blown from the Midwest on the hills paralleling the Mississippi River, and ready access to streams and rivers for the easy transport of cotton to the market in New Orleans. The deep, loamy soil of river and creek bottoms replenished for numerous centuries of flooding and deposition were much more fertile than soils of the Southeast. Indian practices had sustained fertility, so their seized and abandoned fields were farmed first. A land rush in the 1830s occupied lands made available through the removal of remaining Choctaws and Chickasaws and because of easy credit from banks. Also driving this mass migration were high prices for cotton in the 1830s. By the end of the decade cotton bolls dotted the landscape, easy credit was gone, and prices were sinking.

As in the colonial period, land speculation figured prominently in the process of gobbling up as much acreage as possible. One investment group with 62 members gained control of 1.5 million acres in northern Mississippi, and another, with a large portion of capital from outside the region, the New York and Mississippi Land Company, gained title to 206,000 acres. The national was made local with the creation of the Pontotoc and Holly Springs Land Company, which purchased approximately 50,000 acres from the New York and Mississippi group. Speculation drove land prices up, and farmers and planters often started their new ventures burdened by debt. In the economic crisis following 1837, many would lose their land, unable to service the debt with depressed cotton prices. The need to service debt through participation in the market also meant that farmers and planters had to maximize their cotton production immediately if they hoped to hold onto their land. This resulted in faster expansion and more intensive use of the landscape.

The difficulty and variety of cotton agricultural labor overwhelmed white cotton planters, from the poor man scratching his plot from deep forest to the planters' sons bringing slaves with him from South Carolina or Georgia. Early Anglo-American settlers moving into this region brought with them a complex set of attitudes and land-use practices created over the previous 200 years of westward expansion. Natural history essayist Edward Hoagland describes the first Anglo-Americans wandering into this region:

> The first settler straggled into the wilderness with a single-shot rifle, leading a couple of mules, with a crate of brood hens on the back of one and two piglets in a sack to balance the load, some seed corn, potatoes and soldier beans, and dragging a long-suffering cow with a half-grown bull at her heels which the fellow hoped might manage to freshen her again before he butchered it.[17]

Driven by a belief in the rightness of expansion as part of God's mission for America, and a commitment to the idea that wilderness was merely an unfinished landscape, they put axe and fire to good use in converting the forests and thickets to productive farmland as quickly as possible.

Bringing sunshine to plants was the first, most important step. Using a system practiced since colonial New England, these intrepid pioneers cut down enough trees to bring light on a garden plot as they created their first patches in the landscape. Using axes and saws they brought down these trees, burning them to introduce nutrients into the soil, a practice of Indians across North America. With enough land opened for a kitchen garden and with aspirations of market farming, they then began girdling as many trees as possible. This was a labor efficient system for clearing land over time. The cutting of a strip of bark completely around the circumference of the tree stopped the flow of sap, the lifeblood of trees, leaving them to die. They would then decay, collapse, or be easier to cut down than a green tree. In the very beginning, however, it was essential to burn as quickly as possible, and they did so in stages, burning trees, cutting them down, piling them up, and burning them again to get both sunlight on the soil and nutrient rich ash into the ground.

The imperative from the beginning was to produce a crop that was both usable for personal consumption but also had value in the market, in order to barter for other needed goods (such as sugar and coffee) or even sell for cash. In many regions of the United States, generally to the north, wheat was favored for its cash and consumable value. Corn was generally a close second. Corn's cash value was lower, but it is tougher than wheat and less likely to be lost to a hail storm or problems like sprouting. In fact it was consistently the pioneer's first crop of choice. According to cultural geographers Terry Jordan and Matti Kaups,

> The preeminence of corn as the backwoods grain crop had an ecological basis. It often produced a thousandfold increase . . . corn produced four times as much food per unit of land as wheat from a tenth the seed and came to harvest in a shorter time span.[18]

Planted in mounds around stumps, the ground did not have to be tilled or plowed in preparation and corn required few tools and stored easily. As soon as the soil was cleared, and even while still building lean-tos or primitive cabins, corn was quickly planted. Some of these settlers were so cash poor they did not have steel tools and, like the Indians, used sharp sticks or crafted tools out of wood and other materials. Corn could be bartered or sold in the market. It could also be ground daily to be cooked in corn bread or Johnny cakes, a standard staple of 19th-century poor southern farmers. If there was a surplus of corn produced—good harvests in the region could result in 60–70 bushels an acre, at least in the first few years when there was still plenty of nitrogen in the soil—then it could be bartered and sold. If the market for corn was bad, then corn was fed to pigs with the pigs serving as a form of bank account or a way to transform the corn into something more valuable, such as sausage, bacon, and ham. Other foods grown for the family diet, what is commonly known as kitchen garden produce, were sweet potatoes, potatoes, pumpkins, turnips, melons, radishes, and peas of various sorts, including purple hull and black eyed peas. Potatoes, turnips, and radishes were easily stored in root cellars and could be used throughout the year.

It was hard to make a living from this land as was evidenced by dietary habits. Solomon Wright described a typical southern meal for poor whites:

> After a hard day's ride I stopped at a house near the road for supper and shelter for the night. About fifteen minutes after my arrival my host announced supper was ready. I cast my eyes over the anticipated meal. My digestive organs, after the inspection of the supper spread before me, rebelled and contracted. The following is the bill of fare complete: Cornbread, very fat bacon, and clabber [a product of unpasteurized milk allowed to go solid like yogurt and acquiring a sour taste]. As I am not fond of clabber, I did not eat it. My host called his daughter and said: 'Emma Jane, bring this man some water.' My heart felt sick within me to think I could not get a cup of coffee.[19]

Poorer whites were often limited to more marginal land: clayey soils, hill country, and the like. They were compelled to pursue their living in multiple ways. While images of American history and the frontier are often dominated by bucolic pictures of cabins nestled on the edge of forest and pasture, smoke curling upwards, and cattle and swine cozy in their fenced lots and fields, the reality was, in fact, much more messy and ecologically damaging. They brazenly killed a wide range of wildlife throughout the region. Deer, rabbit, opossum, alligator, quail, and bear fell before their gun or, as in the case of the pigeons, under the club as they roosted on trees at night. The protein of these animals was a necessary addition to the limited diets of the early settlers. Their wanton destruction and harvest of game reflected a cultural view of resource use developed in the Carolinas and carried throughout the South by people that traced their lineages and attitudes about farming and nature back to piney woods regions of South and North Carolina. Game was not only killed for protein and fat, which could also be used for cooking oil or lamp oils (bears were particularly valued in this regard), but also for barter or for sale in local towns. The hides of animals were put to use in a number of ways—beds, window and door coverings, and clothing. Furs were trapped for trade and sale, and, of course, deerskins were still eagerly harvested. Birds such as herons, ivory-billed woodpeckers, and egrets were killed for feathers to sell to the millinery trade. While historians have noted deforestation and agriculture as key contributors to the decline of game species across the North American landscape, market hunting as a factor has not been given its due.

As in the colonial period, some animals were hunted because of the competition they created with domesticated animals. This repeats patterns established earlier in American history, but it is important to understand a culture of land-use conquest and transformation that is integral to the American pioneering and frontier process. Wolves, panthers, and bears were all extirpated from the region because of their killing of livestock, goats, and pigs. The black bear is omnivorous, and, contrary to fierce depictions of teeth and claws capable of rending flesh, a typical black bear's diet is 80 percent bugs, plants, roots, and carrion. Mast (hardwood nuts such as acorns decomposing on the ground) constituted a major part of the black bear diet, as it did for deer, squirrel, and pigs. As pointed out earlier, pigs were allowed to run loose, and they fattened on mast. When settlers needed pork or to barter pig meat for medical care, tools, sugar, or coffee, they could harvest a mast-fattened pig. With feral pigs such an important part of the settlers' diet and economy, bears could not be allowed to compete and were hunted aggressively. By the early 20th century the black bear was essentially extinct throughout the South.

The red wolf was another species that both frightened and competed with settlers. Hoagland writes,

> the settlers had good reason to be afraid of wolves . . . wolves digging under a dead man's cairn to wolf down his spoiling remains, wolves disemboweling the family cow, feeding on her thighs and abdominal fat, burying their heads inside her, although her entrails lay unbroken and she was still alive and watching them. When wild game was no longer available, wolves killed the new livestock prodigiously—such stupid, lavish, feasty beasts presented to them on a tray.[20]

Wolves were also hunted aggressively and driven extinct across southern states.

By patching together a variety of activities and ways to earn money from farming, trapping, hunting, and other means, some of these poor whites were able to generate enough revenue to begin acquiring slaves. This population of slaveowners, farmers with a handful of slaves, constituted the majority in the South. They acquired a slave or two and built from there, either from increased profits from cotton sales or from natural reproduction. Slaves owned by small landowners performed field labor almost entirely, and because their owners operated on a thin economic line, there was much unpredictability in the life of these slaves as they were more likely to be sold themselves or see children or spouses "sold down the river" never to be seen again.

While great variety existed in slaveholding patterns based on location and soil fertility, the plantation model dominated, and most slaves were owned by large operations of this type. Slaves were almost immediately organized under a gang-labor system, although some of the planters originating from the rice economy tried to preserve the task labor system for a short period. Gang labor returned slaves to a more restricted work environment where they labored under constant supervision, criticism, and punishment. The ideal plantation was from 1,000 to 1,500 acres, and since each field slave could work approximately 10 acres, it was common for these planters to own 150 slaves or more. Out of a white population of 8 million in 1860, less than 400,000 were slave owners; approximately 340,000 of that group owned fewer than 20 slaves. But "the peculiar institution" defined the culture and economy of the South that many of those whites who did own slaves aspired to. This was a brutal system that provided no legal rights to human property, and the only variation in treatment arose from the benevolence or malevolence of the owner. The whip was applied liberally, particularly on plantations with overseers, the most abusive people

Figure 4.6 While this photograph of African-Americans picking cotton was taken in 1917, it bears strong similarities to cotton slavery. The white overseer or supervisor sits his horse while workers harvest the cotton. Following the Civil War, plantation owners wanted to hire former slaves to work their fields, but Freedmen sought greater autonomy and instead helped create the sharecropper and tenant farmer system. In this labor system the land owner provided land, tools, housing, and other supplies, and the sharecropper or tenant farmer only received a share of the harvest as payment. Tenant farmers typically owned their own homes and tools and thus received a greater share of the harvest. While this perpetuated poverty, it also gave African-Americans greater freedom from constant oversight and interference by whites. Library of Congress.

in the slave economy. Women suffered rape on a regular basis from slave-owners and their sons, and families were regularly fragmented by the sales of spouses and children to other plantations.

Of the approximately 4 million southern slaves, almost 2 million of them worked on cotton plantations. They worked from "can't see to can't see" (from before sunrise to after sunset) and often into the evening as well. While slaves in the cotton economy did not experience the autonomy that those in the rice economy did for a period, many of them tended small gardens and raised chickens. This was an important supplement to the steady diet of salt pork and corn bread provided by their owners. Some hunted or fished, but they typically had to slip away to do that and ran the risk of being viewed as trying to escape, and suffering the severe punishments doled out for that offense.

Deep in the land of cotton it was common to see one plantation after another, producing nothing but the fluffy white staple.

> Planters crowded upon every spot of their fields with cotton plants, and sent their ox teams to the Mississippi River for every grain of corn consumed upon their plantations. You might have travelled all day without seeing a corn field of any importance.[21]

The prices for cotton in the early 1830s were so high planters decided to abandon mixed agriculture for monoculture. One group of planters felt compelled to record the reasoning behind this narrow focus:

> At a time when cotton commanded prices from 15 to 20 cents per lb., and corn could be purchased at 37½ or 50cts. per bushel, pork at 3 or 4 cents per lb. and other necessaries at a rate proportionate, the planter, living convenient to market where the trouble of wagoning is not too great, was pursueing [sic] a true economy to devote exclusive attention to cotton and purchase those articles with the proceeds of its sale, which thus cost him less labor than would have been necessary to their domestic production.[22]

The logic of the market trumped mixed agriculture and even resiliency against market fluctuations.

The early stages of cotton production in the old Southwest in the early 19th century resembled land-use practices from the opening stages of the various agricultural economies. Poor whites working their own land and slaves laboring for planters immediately cleared forest to gain access to the soil. They converted forest to farmland much more quickly, in a more intensive manner, than had been the case earlier in American history, cutting and removing trees, digging up stumps and roots, and clearing as much forest as possible for the first planting. Cotton could not be planted around stumps as had been the case with the first stage of corn and even tobacco planting; this agriculture was much more land intensive from the start. Soil was broken and pushed into rows for successful cotton agriculture. Therefore great labor was expended clearing the land. In this first stage, the practice of clearing land, farming until soil lost its fertility, and then rotating onto newly cleared land held firm. But once the land was all possessed and prices on cotton dropped in the late 1830s, farmers and planters looked to conserve soil and enhance cotton production through a number of measures.

The precipitous drop in cotton prices was ecologically beneficial because the crisis triggered efforts to conserve soil and create a more sustainable cotton economy. One of the first responses was a return to mixed agriculture. Planters put more land back into corn and produce such as cowpeas, fruit, and sweet potatoes. While cotton prices remained low, interest in pig and cattle husbandry increased. Cotton farming caused damaging erosion, with cropland quickly turning into a landscape of "frightful precipices and yawning chasms" after heavy rains. Cotton planters focused on conserving topsoil. Some planters began planting rows horizontally to the slope. This reduced the stress for beasts of burdens pulling plows, and the rows helped contain the flow of water downhill, slowing erosion and holding soil. These first efforts were of limited benefit because of the crudeness of leveling tools. Rows were uneven in height and length, so pooling would develop beyond some rows, leading water to flow around the end and then channel down the slope. By the 1840s, leveling instruments had improved to such a degree that rows could be planted in consistent height and length to prevent pooling and erosion. Drainage ditches were also used to siphon away water. The first ones ran vertical to the field and could also lead to erosion, so planters began digging carefully sloped ditches in a diagonal line across the fields. Further improvements included lining ditches with boards and wood to prevent soil from washing away and, for well-capitalized planters, the use of buried pipes to drain water. All of these measures helped to conserve soil on the plantation. An advocate of ditching and soil conservation wrote, in a tone reminiscent of the concern of Virginia tobacco planters in the late 18th and early 19th century, that "hillside ditches well run will in a measure stop the spirit of emigration. The sunny South can be saved by them I know."[23] In short, sustainable cotton agriculture would preserve southern culture.

While holding soil was important, sterile dirt was useless. Like farmers in other regions, cotton planters also made crop rotation integral to their land-use strategy in order to restore fertility. They discovered that cotton crops thrived in fields previously used for corn and cowpeas. Cowpeas like black eyed peas and purple hull peas fix nitrogen in the soil. The cotton planters adopted a rotational strategy where they followed the corn and cowpea fields with cotton for two or three seasons, then succeeded cotton with another round of corn and cowpeas. They also planted these imports from Africa with the corn, between rows, and even among the grain. In mixing crops like this they unknowingly imitated the Choctaws and Chickasaws whose lands they now occupied, and were moving toward a more sustainable form of agriculture. Edmund Ruffin referred to cowpeas as "the clover of the South." After harvest, pigs and livestock were allowed to eat corn stalks and cowpea vines, and the remaining debris was left on the fields over winter to provide some protection against rain and erosion. In spring, they tilled any remaining vegetative matter directly into the dirt. Excess cottonseed was used as fertilizer and laid in rows with fresh seed. Unlike northern farmers and tobacco planters, southern planters did not adopt manuring, as in the Northeast and Southeast, complaining that there was too much land versus too little cattle for that measure to be economically worthwhile. Much of their meat was imported anyways, so manure simply wasn't available in large amounts.

Once cotton farms and plantations were established, the work of producing cotton remained complex and strenuous. In the spring, the ground was broken by plows, digging 2 or 3 inches into the soil, and followed by harrows breaking up clods of dirt. Seeds were originally broadcast by hand, but over time that was replaced by the planting of seeds in a shallow trough. Slaves plowed ditches between rows when the plants were an inch or two in height, then plowed in reverse one week later, throwing loose dirt on the sides of the rows, giving roots more space to grow. These plowings also killed weeds and grass by covering and choking the competing plants. As the cotton grew taller, slaves worked their way down the rows, thinning the plants to 2 to 3 feet apart and removing weeds. At the same time, plowing between the rows continued through the summer to suppress competing plants. Improvements in farming technology led to a scraper pulled by oxen or horses that effectively removed competing plants between rows. The rows themselves could only be kept clear by dint of constant vigilance from slave crews wielding heavy hoes.

Cotton agriculture became heavily mechanized with a corresponding dependence on horses and oxen and ongoing study and adaption of new equipment. Still, there were ongoing problems—the biggest of which was the tendency to persist in shallow plowing even in the 1850s after decades of the deep plowing gospel. Not all planters could afford the high-quality plows needed to break the soil deeply—much of their capital was tied up in slaves. Certainly poor and yeoman cotton farmers would have struggled to acquire that equipment as well. Also, old habits died hard for planters and slaves alike, and there was no real incentive, other than the lash of the whip, for slaves to bury the plow deeper in the soil than they were accustomed. By the 1850s cotton planters were again realizing large profits with the increased demand from the booming textile industry in Europe and New England. By 1860 the South produced more than 5 million bales of cotton a year with the largest slave-holding states producing a disproportionate amount of that. With money flowing in, they again abandoned mixed agriculture and food production, increasing their dependence on northern food production. Their ability to plant and harvest cotton sustainably, their control of black slave labor, and resulting high profits led to a jaunty confidence on the part of Southerners.

Having built their economy primarily around a staple crop and constructed a system of labor and control dedicated to the production of cotton, southerners had concerns about possible limits on slavery's expansion and increasing tensions between the North and the South, particularly after 1850. Their ace in the hole was that both northeastern and British textile manufacturers were almost completely reliant on cotton. Possibly not reflecting enough on their own dependence and lack of economic diversity, Southerners argued that "King Cotton" would preserve Southern culture. From their perspective the North would never act against them or go to war because of its need for cotton. In the case that proved wrong, British dependence on the crop would inevitably lead to their support of southern states. Confident of their future, they continued to fill the land with cotton plantations as the nation steadily moved toward schism through the 1850s.

Rebuilding the Soil in New England

Farmers in New England, like their southern country cousins, realized the necessity to institute better land-use practices to create an agricultural system less dependent on constant expansion onto new lands. Many of their practices reflect innovations in agriculture in England but are also direct adaptations to the challenges and benefits of the variety of soils and elevations of New England. One of the most important was the increased incorporation of manure into farm practices. Building barns close to the fields made transport of manure less burdensome—some farmers let the cattle out onto farmland in the offseason for the manure deposits and in some cases even used fencing to move cattle and fertilizer into desired locations. One New Hampshire farmer rotated his cattle and dung through his land by fencing and enclosing a new section each year. In fact, this may have been more common than the historical record shows. Manure hauled into fields was used to dress corn and in some cases sown directly with the seeds.

Instituting a pattern of mixed use of the varied landscape, farms were laid out in a fairly consistent pattern. Kitchen gardens worked by women were situated close to the home, 6 to 12 acres of land were put into tillage just beyond the kitchen garden (depending on where the best land for agriculture was located), and a barn was located close to the farm fields to ease transport of manure. Uplands were used for pasture, woodlots, and apple orchards. The production of apples and reliance on cider as an integral part of the agricultural cycle developed as an agricultural adaptation in the Concord, Massachusetts, area. The apple trees growing small, sour, and tough crab apples were productive on rocky high ground that could not be used for corn or wheat, and cider almost completely replaced beer as the beverage of choice in that community. Being nearly sustainable in one place required greater versatility and varied use of different parts of the landscape. Reaching the limits of natural resources, farmers became better stewards of the land in this region.

The key to more sustainable, successful farms was a more careful, planned transformation of the land than was practiced before. Meadows constituted an essential component of this mixed land-use system because of their importance in providing graze and cut hay for cattle. Farmers drained marshes and wetlands through a complex ditch system that dried the soil so the grasses cattle preferred could thrive over the undesired sedges and cattails common to wetlands. Ditches had to be cleaned, maintained, and even re-dug on a regular basis. This necessitated cooperation and shared labor between farmers. This system worked effectively for at least a century. Farmers also built small dams to store water for release on the meadows in winter to stimulate grass growth. Pastures and woodlots were situated on to the least ideal land. Sandy soil, rocky terrain, and moist bottomlands were ill suited for tillage but could be worked for pasture. Farmers extended their reach across the landscape, acquiring pastures and woodlots separated from the original farm site. Woodlots were purchased and maintained for regular production of pine and oak timber for firewood and building construction. Farmers embracing sustained land use showed a remarkable adaptability in the New England region and transformed the land into one of multiple agricultural uses.

This organization of the land into patterns of production that together constituted a viable agricultural community resulted in benefits for wildlife. There was still no place for predators such as bear, cougar, wolf, and bobcats, and the elk and moose could not compete with guns and livestock in this agricultural mosaic. But others species were able to take advantage of the mix of grasslands in meadow and pasture, the orchards, the woodlots, and the tilled fields themselves. Deer began their recovery in this region in the 18th century, aided also by the eradication of predators. Turkey, rabbits, squirrels, songbirds, crows, ravens, and a variety of other insects, small mammals, and other species found room for themselves on these farmers' lands. Striped bass, shad, alewives, and salmon continued to spawn up the rivers that had not been devastated by erosion and, on many watersheds, constituted an important part of farmers' diets as well as an important economic commodity.

Even as farmers in New England struck a balance between economic growth and land use that was more sustainable than earlier practices, the population of the region grew too quickly for the land to support. Both immigrants seeking their own opportunities and the children of native farmers, whose property could not be further divided, were compelled to move west, north, and to higher elevations in order to obtain needed land. Farming in the foothills and mountains of New England, they brought even more land into production and began growing new crops better suited for marginal soils, such as potatoes. As these farmers gained elevation, they faced more challenges. Thinner and rockier soil, earlier and later crop-killing frosts, and more problems with erosion made their lives difficult, and these farms were largely abandoned over time. Of course, the migration of colonists to the always moving frontier was an established response to land shortages and soil exhaustion and a dynamic of American settlement and expansion from the early 17th century into the late 19th century. Even as farmers in New England learned to diversify, adapt to ecological limitations, and use the land in a more sustainable pattern, ongoing rapid population growth pushed against the environmental limits of the region so that expansion continued into the Ohio River Valley and farther.

The 18th and early 19th centuries were a period of ongoing transformation of ecosystems and expansion of agriculture throughout the colonies. Farmers, planters, and leaders of agricultural communities sought to reduce community fragmentation and economic distress by conserving soil, introducing new, innovative ways to amend soils, and even speculating in land far beyond the reach of settlement to prevent bankruptcy. Even as regions such as New England shifted to a sustainable model of agriculture, tobacco planters were compelled to embrace a more complex agricultural economy, and the construction of a hydraulic landscape in the Georgia and Carolina lowlands dramatically transformed ecosystems of that region. Population growth, worn out soil, and economic stress continued to exert pressure to expand geographical space and land use. Implicit in the continued expansion west and the implementation of agricultural reforms were two key assumptions.

The first was that more land was always needed for the growth and health of the United States. An abundance of land and the inherent opportunities seemed to reduce or prevent conflict, although, ironically, the acquisition of vast territory in the West after the defeat of Mexico in the Mexican-American War would launch and intensify the debate over the expansion of slavery that ultimately culminated in the Civil War. An ideology of abundance was developing in America in this era—a belief that Americans enjoyed a special relationship with God who had provided for them by creating the vast reservoirs of natural capital and millions of acres of fertile soil that were theirs for the taking. This natural abundance was there to ensure the success and expansion of the nation not only from "Sea to Shining Sea" but, in the minds of some, even into Cuba and Latin America.

The second implicit assumption was that of dominion over nature, African slaves, and native peoples. Dominion means never having to say you're sorry. On a more serious note, the assumption of dominion over nature and other peoples absolved Americans of careful consideration of the impacts of their actions on nature, of the meaning of their destruction of ecosystems and cultures as they transformed the land. Certainly they understood their mission to subdue the earth, but Americans also believed in the divinity of creation. Subduing clearly took priority over protection of God's creation in this era. Similarly, Americans owned slaves throughout the colonies, but after the Revolutionary War, slavery became a Southern institution. The desire and will to build an entire agricultural economy on an edifice of human flesh and enslaved labor emanated from the belief in dominion but was also justified by the ideology of abundance and a determination to produce as much agricultural commodities and wealth as naturally and humanly possible. Southerners did not carefully consider the right or wrong of owning humans; dominion and a belief in abundance as a wellspring of American success prevented careful reflection on their actions and led instead to thousands of pages of books, articles, and sermons justifying slavery and its expansion.

But this nation of crops, fields, and grazing animals was already changing as mills, dams, and factories began rising and canals and railroads were dug and laid across the land. The rise of a manufacturing economy, the expansion of the market, and ever-increasing abundance and prosperity would transform the nation in ways no one could anticipate.

Document 4.1 Letter from George Washington in Opposition to the Royal Proclamation Line, September 21, 1767

George Washington made much of the money of his early career as a surveyor and land speculator. This was common in the late colonial period. Planters and businessmen acquired land west of settlement for sale and rent. The passage of the Royal Proclamation by the British Parliament in 1763 drew a line down the crest of the Appalachian Mountains beyond which British citizens of the colonies were not allowed to travel. This not only proved frustrating for traders and migrating farmers but also stymied the economic plans of Virginia planters and other speculators. Washington, Thomas Jefferson, Patrick Henry, and numerous other Virginia elites were deeply involved in land speculation schemes when this proclamation was announced.

1. What are the characteristics Washington stipulates for land purchase? What is his intent in acquiring this land?
2. Can Washington's attitude about speculation and the British Government's efforts to stop westward expansion via the Royal Proclamation be determined from this letter? How does Washington view these issues, and how does he characterize potential competition for land speculation?

I then desired the favor of you (as I understood rights might now be had for the lands which have fallen within the Pennsylvania line,) to look me out a tract of about fifteen hundred, two thousand, or more acres somewhere in your neighborhood, meaning only by this, that it may be as contiguous to your own settlement as such a body of good land can be found. It will be easy for you to conceive that ordinary or even middling lands would never answer my purpose or expectation, so far from navigation, and under such a load of expenses as these lands are incumbered [sic] with. No; a tract to please me must be rich (of which no person can be a better judge than yourself), and, if possible, level. Could such a piece of land be found, you would do me a singular favor in falling upon some method of securing it immediately from the attempts of others, as nothing is more certain than that the lands can not remain long ungranted, when once it is known that rights are to be had.

The mode of proceeding I am at a loss to point out to you; but, as your own lands are under the same circumstances, self-interest will naturally lead you to an inquiry.

I offered in my last to join you in attempting to secure some of the most valuable lands in the King's part, which I think may be accomplished after awhile, notwithstanding the proclamation that restrains it at present, and prohibits the settling of them at all; for I can never look upon that proclamation in any other light (but this I say between ourselves) than as a temporary expedient to quiet the minds of the Indians. It must fall, of course, in a few years, especially when those Indians consent to our occupying the lands. Any person, therefore, who neglects the present opportunity of hunting out good lands, and in some measure marking and distinguishing them for his own, in order to keep others from settling them, will never regain it. If you will be at the trouble of seeking out the lands, I will take upon me the part of securing them, as soon as there is a possibility of doing it, and will, moreover, be at all the cost and charges of surveying and patenting the same. You shall then have such a reasonable proportion of the whole as we may fix upon at our first meeting; as I shall find it necessary, for the better furthering of the design, to let some of my friends be concerned in the scheme, who must also partake of the advantages.

By this time it may be easy for you to discover that my plan is to secure a good deal of land.

I recommend, that you keep this whole matter a secret, or trust it only to those in whom you can confide; and who can assist you in bringing it to bear by their discoveries of land. This advice proceeds from several very good reasons, and, in the first place, because I might be censured for the opinion I have given in respect to the King's proclamation, and then, if the scheme I am now proposing to you were known, it might give the alarm to others, and, by putting them upon a plan of the same nature, before we could lay, a proper foundation for success ourselves, set the different interests clashing, and, probably, in the end, overturn the whole.

Document 4.2 "Address to the Agricultural Society of Albemarle," *James Madison*, 1818

After an illustrious career spanning the debates over responses to Parliament prior to the American Revolution, the creation of the Constitution, and, of course, his service as president of the United States for two terms, James Madison returned home to his own Montepelier plantation in the piedmont of Virginia. The Agricultural Society of Albemarle was organized by planters to create and implement strategies of soil conservation in order to support their own lifestyles and slow the flood of young people migrating west for better opportunities. President Madison was asked to serve as the organization's president and to address the group.

1. What does Madison have to say about the diversity and abundance of nature? Is there a foreshadowing of the ecological thought that would develop later in the 19th century? Is there an emerging ethos of restraint and conservation in this speech?
2. Is there a moral good or evil associated with proper and improper land use? How does Madison evaluate this? How does proper land use relate to the functioning of democracy?

. . . Agriculture, once effectually commenced, may proceed of itself, under impulses of its own creation. The mouths fed by it increasing, and the supplies of nature decreasing, necessity becomes a spur to industry; which finds another spur in the advantages incident to the acquisition of property, in the civilized state. And thus a progressive agriculture, and a progressive population ensue.

. . . The earth contains not less than thirty or forty thousand kinds of plants; not less than six or seven hundred of birds; nor less than three or four hundred of quadrupeds;

to say nothing of the thousand species of fishes. Of reptiles and insects, there are more than can be numbered. To all these must be added, the swarms and varieties of animalcules and minute vegetables not visible to the natural eye, but whose existence is probably connected with that of visible animals and plants.

On comparing this vast profusion and multiplicity of beings with the few grains and grasses, the few herbs and roots, and the few fowls and quadrupeds, which make up the short list adapted to the wants of man, it is difficult to believe that it lies with him so to remodel the work of nature as it would be remodelled, by a destruction not only of individuals, but of entire species; and not only of a few species, but of every species, with the very few exceptions which he might spare for his own accommodation.

Such a multiplication of the human race, at the expense of the rest of the organized creation, implies that the food of all plants is composed of elements equally and indiscriminately nourishing all, and which, consequently, may be wholly appropriated to the one or few plants best fitted for human use. Whether the food or constituent matter of vegetables be furnished from the earth, the air, or water; and whether directly, or by either, through the medium of the others, no sufficient ground appears for the inference that the food for all is the same.

. . . Could it, however, be supposed that the established system and symmetry of nature required the number of human beings on the globe to be always the same; that the only change permitted in relation to them was in their distribution over it: still, as the blessing of existence to that number would materially depend on the parts of the globe on which they may be thrown; on the degree in which their situation may be convenient or crowded; and on the nature of their political and social institutions, motives would not be wanting to obtain for our portion of the earth its fullest share, by improving the resources of human subsistence, according to the fair measure of its capacity. For in what other portion of equal extent will be found climates more friendly to the health, or congenial to the feeling of its inhabitants? In what other, a soil yielding more food with not more labor? And above all, where will be found institutions equally securing the blessings of personal independence and social enjoyments? The enviable condition of the people of the United States is often too much ascribed to the physical advantages of their soil and climate, and to their uncrowded situation. Much is certainly due to these causes; but a just estimate of the happiness of our country will never overlook what belongs to the fertile activity of a free people, and the benign influence of a responsible Government.

The error first to be noticed is that of cultivating land, either naturally poor or impoverished by cultivation. This error, like many others, is the effect of habit, continued after the reason for it has failed. Whilst there was an abundance of fresh and fertile soil, it was the interest of the cultivator to spread his labor over as great a surface as he could. Land being cheap and labor dear, and the land co-operating powerfully with the labor, it was profitable to draw as much as possible from the land. Labor is now comparatively cheaper and land dearer. Where labor has risen in price fourfold land has risen tenfold. It might be profitable, therefore, now to contract the surface over which labor is spread, even if the soil retained its freshness and fertility. But this is not the case. Much of the fertile soil is exhausted, and unfertile soils are brought into cultivation; and both co-operating less with labor in producing the crop, it is necessary to consider how far labor can be profitably exerted on them; whether it ought not to be applied towards making them fertile, rather than in further impoverishing them? or [sic] whether it might not be more profitably applied to mechanical occupations or to domestic manufactures?

. . . we ought to yield to change of circumstances, by forbearing to waste our labor on land which, besides not paying for it, is still more impoverished, and rendered more difficult to be made rich. The crop which is of least amount gives the blow most mortal to the soil. It has not been a very rare thing to see land under the plough not producing enough to feed the ploughman and his horse; and it is in such cases that the death-blow is given. The goose is killed, without even obtaining the coveted egg.

There cannot be a more rational principle in the code of agriculture, than that every farm which is in good heart should be kept so; that every one not in good heart should be made so; and that what is right as to the farm, generally, is so as to every part of every farm. Any system, therefore, or want of system, which tends to make a rich farm poor, or does not tend to make a poor farm rich, cannot be good for the owner; whatever it may be for the tenant or superintendant, who has transient interest only in it. The profit, where there is any, will not balance the loss of intrinsic value sustained by the land.

The evil of pressing too hard upon the land has also been much increased by the bad mode of ploughing it. Shallow ploughing, and ploughing up and down hilly land, have, by exposing the loosened soil to be carried off by rains, hastened more than any thing else the waste of its fertility. When the mere surface is pulverized, moderate rains on land but little uneven, if ploughed up and down, gradually wear it away. And heavy rains on hilly land, ploughed in that manner, soon produce a like effect, notwithstanding the improved practice of deeper ploughing. How have the beauty and value of this red ridge of country suffered from this cause? And how much is due to the happy improvement introduced by a member of this society, whom I need not name, by a cultivation in horizontal drills, with a plough adapted to it?

Document 4.3 From *Notes on the State of Virginia, Thomas Jefferson*, 1787 (First written in 1780–1781, revised and published later)

As both a statesman and politician, Thomas Jefferson struggled with the relationship between land and labor, particularly with how the nature of labor liberates man or reduces his humanity. Vociferously opposed to industrialism because he believed that factory labor made humans inferior and rendered them incapable of participation in a democracy, Jefferson celebrated the yeoman farmer, arguing that this group constituted the backbone of American democracy. As is evidenced in the document below, he is uncomfortable with the impact of slave labor on African slaves.

In this section of *Notes on the State of Virginia* Jefferson seeks to explain the differences between blacks and whites in America, with a strong emphasis on physical distinctions. In fact, this marks the beginning of efforts to define racial categories per se and ascribe to specific races certain physical characteristics, spiritual tendencies, intellectual skills, tendencies, and so on.

1. How does Jefferson deal with the issue of labor and bondage in regards to the condition of slaves? How would you explain his different conclusions on the qualities of white farmers versus black slaves and farm laborers?
2. Evaluate Jefferson's efforts to equate the physical with the mental and spiritual. How does he devalue the rights and abilities of African-Americans? What role might labor, subjugation, and his own status contribute to these conclusions?

[on Whites]

Those who labour in the earth are the chosen people of God, if ever he had a chosen people, whose breasts he has made his peculiar deposit for substantial and genuine virtue. It is the focus in which he keeps alive that sacred fire, which otherwise might escape the face of the earth. Corruption of morals in the mass of cultivators is a phaenomenon of which no age nor nation has furnished an example. It is the mark set on those, who not looking up to heaven, to their own soil and industry, as does the husbandman, for their subsistance, depend for it on the casualties and caprice of customers. Dependance begets subservience and venality, suffocates the germ of virtue, and prepares fit tools for the designs of ambition.

[on Blacks]

Why not retain and incorporate the blacks into the state, and thus save the expence of supplying, by importation of white settlers, the vacancies they will leave? Deep rooted prejudices entertained by the whites; ten thousand recollections, by the blacks, of the injuries they have sustained; new provocations; the real distinctions which nature has made; and many other circumstances, will divide us into parties, and produce convulsions which will probably never end but in the extermination of the one or the other race.—To these objections, which are political, may be added others, which are physical and moral. The first difference which strikes us is that of colour. Whether the black of the negro resides in the reticular membrane between the skin and scarf-skin, or in the scarf-skin itself; whether it proceeds from the colour of the blood, the colour of the bile; of from that of other secretion, the difference is fixed in nature, and is as real as if its seat and cause were better known to us. And is this difference of no importance? Is it not the foundation of a greater or less share of beauty in the two races? Are not the fine mixtures of red and white, the expressions of every passion by greater or less suffusions of colour in the one, preferable to that eternal monotony, which reigns in the countenances, that immovable veil of black which covers all the emotions of the other race? Add to these, flowing hair, a more elegant symmetry of form, their own judgement in favour of the whites, declared by their preference of them, as uniformly as is the preference of the Oran-ootan for the black women over those of his own species. The circumstance of superior beauty, is though worthy attention in the propogation of our horses, dogs, and other domestic animals; why not in that of man? Besides those of colour, figure, and hair, there are other physical distinctions proving a difference of race. They have less hair on the face and body. They secrete less by the kidnies, and more by the glands of the skin, which gives them a very strong and disagreeable odour. This greater degree of transpiration renders them more tolerant of heat, and less so of cold, than the whites. Perhaps too a difference of structure in the pulmonary apparatus, which a late ingenious experimentalist has discovered to be the principal regulator of animal heat, may have disabled them from extricating, in the act of inspiration, so much of that fluid from the outer air, or obliged them in expiration, to part with more of it. They seem to require less sleep. A black, after hard labour through the day, will be induced by the slightest amusement to sit up till midnight, or later, though knowing he must be out with the first dawn of the morning. They are at least as brave, and more adventuresome. But this may perhaps proceed from a want of forethought, which prevents their seeing a danger till it be present. When present, they do not go through it with more coolness or steadiness than the whites. They are more ardent after their female: but love seems with them to be more an eager desire, than a tender delicate mixture of sentiment and sensation. Their griefs are transient. Those numberless afflictions, which render it doubtful whether heaven has given life to us in mercy or in wrath, are less felt, and sooner forgotten with them. In general, their existence appears to participate more of sensation than reflection. To this must be ascribed their disposition to sleep when abstracted from their diversions, and unemployed in labour. An animal whose body is at rest, and who does not reflect, must be disposed to sleep of course. Comparing them by their faculties off memory, reason, and imagination, it appears to me, that in memory they are equal to the whites; in reason much inferior, as I think one could scarcely be found capable of tracing and comprehending the investigations of Euclid; and that in imagination they are dull, tasteless, and anomalous. It would be unfair to follow them to Africa for this investigation. We will consider them here, on the same stage with the whites, and where the facts are not apocryphal on which a judgment is to be formed. It will be right to make great allowances for the difference of condition, of education, of conversation, of the sphere in which they move. Many millions of them have been brought to, and born in America. Most of them indeed have been confined to tillage, to their own homes, and their own society; yet many have been so situated, that they might have availed themselves of the conversation of their masters; many have been brought up to the handicraft arts, and from that circumstance have always been associated with the whites. Some have been liberally educated, and all have lived in countries where the arts and sciences are cultivated to a liberal degree, and have had before their eyes samples of the best works from abroad. The Indians, with no advantages of this kind, will often carve figures on their pipes not destitute of design and merit. They will crayon out an animal, a plant, or a country, so as to prove the existence of a germ in their mind which only wants cultivation. They astonish you with strokes of the most sublime oratory; such as prove their reason and sentiment strong, their imagination glowing and elevated. But never yet could I find that a black had uttered a thought above the level of plain narration; never see an elementary trait of painting or sculpture. In music they are more generally gifted than the whites with accurate ears for tune and time, and they have been found capable of imagining a small catch. Whether they will be equal to the composition of a more extensive run of melody, or of complicated harmony, is yet to be proved. Misery is often the parent of the most affecting touches in poetry.—Among the blacks misery is enough, God knows, but no poetry. Love is the peculiar oestrum of the poet. Their love is ardent, but it kindles the senses only, not the imagination. Religion indeed has produced a Phyllis Whately, but it could not produce a poet. The compositions published under her name are below the dignity of criticism. The heroes of the Dunciad are to her, as Hercules to the author of that poem. Ignatius Sancho has approached nearer to merit in composition; yet his letters do more honour to the heart than the head . . . The improvement of the blacks in body and mind, in the first instance of their mixture with the whites, has been observed by every one, and proves that their inferiority is not the effect merely of their condition of life . . . in this country the slaves multiply as fast as the free inhabitants.

. . . Whether further observation will or will not verify the conjecture, that nature has been less bountiful to them in the endowments of the head, I believe that in those of the heart she will be found to have done them justice. The disposition to theft with which they have been branded, must be ascribed to their situation, and not to any depravity of the moral sense.

Notes

1 King James I of England, "A Counterblaste to Tobacco," 1604.
2 Ibid.
3 Ibid.
4 Anonymous, *American Husbandry. Containing an Account of the Soil, Climate, Production and Agriculture, of the British Colonies in North-America and the West-Indies* (London, England: J. Bew, 1775), 233, 234.
5 Brian Donahue, *The Great Meadow: Farmers and the Land in Colonial Concord* (New Haven, CT: Yale University Press, 2004), 70.
6 Woody Holton, *Forced Founders: Indians, Debtors, Slaves, & the Making of the American Revolution in Virginia* (Chapel Hill: The University of North Carolina Press, 1999), 11.
7 Ibid., 36, 37.
8 Ibid.
9 Quoted in Lynn A. Nelson, *Pharsalia: An Environmental Biography of a Southern Plantation, 1780–1880* (Athens: The University of Georgia Press, 2007), 67.
10 Ibid., 69.
11 Mart Stewart, *"What Nature Suffers to Groe": Life, Labor, and Landscape on the Georgia Coast, 1680–1920* (Athens: University of Georgia Press, 1996), 102.
12 Ibid., 104.
13 Ira Berlin, *Many Thousands Gone: The First Two Centuries of Slavery in North America* (Cambridge, MA: Harvard University Press, 1998), 150.
14 Frederick Law Olmsted, *The Cotton Kingdom: A Traveller's Observations on Cotton and Slavery in the American Slave States*, Vol. I (New York: Mason Brothers, 1862), 235.
15 Berlin, *Many Thousands Gone*, 153.
16 Quoted in Ibid., 164.
17 Edward Hoagland, "Lament the Red Wolf," in *Out Among the Wolves*, edited by John A. Murray (Seattle, WA: Alaska Northwest Books, 1993), 58, 59.
18 Terry G. Jordan and Matti Kaups, *The American Backwoods Frontier: An Ethnic and Ecological Interpretation* (Baltimore, MD: The Johns Hopkins University Press, 1989), 115.
19 Robert S. Maxwell and Robert D. Baker, *Sawdust Empire: The Texas Lumber Industry, 1830–1940* (College Station: Texas A&M University Press, 1983), 13.
20 Hoagland, "Lament the Red Wolf," 58.
21 Quoted in Olmsted, *The Cotton Kingdom*, 15.

22 Quoted in John Hebron Moore, *The Emergence of the Cotton Kingdom in the Old Southwest: Mississippi, 1770–1860* (Baton Rouge: Louisiana State University Press, 1988), 15.
23 Olmsted, *The Cotton Kingdom*, 33.

Further Reading

Berlin, Ira. *Many Thousands Gone: The First Two Centuries of Slavery in North America*. Cambridge, MA: Harvard University Press, 1998.

Dattel, Gene. *Cotton and Race in the Making of America: The Human Costs of Economic Power*. Chicago: Ivan R. Dee, 2009.

Donahue, Brian. *The Great Meadow: Farmers and the Land in Colonial Concord*. New Haven, CT: Yale University Press, 2004.

Fenn, Elizabeth A. *Pox Americana: The Great Smallpox Epidemic of 1775–82*. New York: Hill and Wang, 2001.

Holton, Woody. *Forced Founders: Indians, Debtors, Slaves & the Making of the American Revolution in Virginia*. Chapel Hill: The University of North Carolina Press, 1999.

Kirby, Jack Temple. *Mockingbird Song: Ecological Landscapes of the South*. Chapel Hill: The University of North Carolina Press, 2006.

Kulikoff, Allan. *Tobacco and Slaves: The Development of Southern Cultures in the Chesapeake, 1680–1800*. Chapel Hill: The University of North Carolina Press, 1986.

Moore, John Hebron. *The Emergence of the Cotton Kingdom in the Old Southwest: Mississippi, 1770–1860*. Baton Rouge: Louisiana State University Press, 1988.

Nelson, Lynn A. *Pharsalia: An Environmental Biography of a Southern Plantation, 1780–1880*. Athens: The University of Georgia Press, 2007.

Stewart, Mart. *"What Nature Suffers to Groe": Life, Labor, and Landscape on the Georgia Coast, 1680–1920*. Athens: University of Georgia Press, 1996.

Timeline

Northwest Land Ordinance	1787
Edisto River Dam Conflict	1787
Louisiana Purchase	1803
Congress Finances the National Road	1806
Beginning of Erie Canal	1819
Land Act	1820
Completion of Erie Canal	1825
Indian Removal Act	1830
Kennebec Dam Protest	1834
The National Road Completed	1839
Publication of *Walden*	1854
Publication of *The Maine Woods*	1864

"A Newer Garden of Creation" 5

New York Governor Dewitt Clinton had suffered his share of slings and arrows when he announced plans for the construction of the Erie Canal and what detractors referred to as "Clinton's Ditch." With the 364-mile waterway connecting Lake Erie to the Hudson River and New York City now complete, it was already producing commerce and revenue far beyond expectations. So no one could blame Clinton's and other politicians' and businessmen's grandiose celebration of the canal's completion in November 1825.

The festivities opened with a parade in Buffalo, and Clinton and his Lieutenant Governor boarded a small boat called the *Seneca Chief.* On board the boat were two barrels of Lake Erie water destined to be poured into the Atlantic, and the *Seneca Chief* was pulled by four decorated horses. Two paintings decorated the interior: one a straightforward composition of Buffalo joined with the canal and the other a much more ornate depiction of Clinton in a heroic, god-like mode. Clad in Roman clothing, Clinton directs Hercules's and Neptune's attention to the lock; they are both shocked by such a feat of human engineering. As Clinton's boat left, others followed, including one named *Noah's Ark,* which contained western animals, including a bear and two eagles, as well as some Indian boys in traditional clothing. Upon leaving Buffalo the flotilla's launching was announced with a volley of fire from small arms and a cannon. People along the shoreline cheered as the boats moved slowly downstream. Other communities were organizing celebrations and events, so cannons were pre-positioned for the whole 364-mile route within hearing range of each other to communicate through firing the movement of the flotilla downstream. When they arrived at their final destination of Sandy Hook, New Jersey, the volley of celebration there then launched a reverse round of cannon fire all the way up the canal back to Buffalo, letting them know that Clinton and company had completed their journey safely.

As they traveled downstream giving speeches, joining celebrations and feasts on shore, and cheering back at the crowds lining the canal banks, it was clear to the promoters of this great work that the residents of New York celebrated this unprecedented American accomplishment (the Aztec, Inca, and Maya might not have been impressed) and held great hopes for the changes it would bring in their lives.

Both the nation and the economy dramatically expanded from the late 18th to mid-19th century. Federal and state policies distributing land, funding roads and canals, and forcing the relocation of native peoples supported both geographical expansion and economic growth. At the same time the nation reached west, the dramatic economic expansion known as the Market Revolution was transforming the land, the economy, and resource use practices. In the North the revolution was more dramatically evident with the rise of widespread manufacturing, but it influenced the southern economy as well with the demand for cotton and the construction of dams for lumber and grain milling. The building of roads and canals and the increased use of steamboats, combined with increased capital investment and the expanding reach of the market economy, all contributed to the rise of market agriculture and industrial manufacturing. As this economy muscled up against geographic limits, railroads then expanded the reach of the market even further with corresponding expansion and growth.

Economic growth and changes to the land created internal dissension among Americans while forcing outright conflict and resistance by Indians desperate to hold onto their lands. At the same time, a few Americans began to question some of the governing assumptions of American life. Most prominent among them was Henry David Thoreau. In a time of unbridled prosperity and a seeming cultural consensus that prosperity and social mobility were birthrights of Americans, he suggested otherwise; he questioned the strengthening ideology of abundance and foreshadowed the critiques of a century later.

The course of American history is framed around a number of familiar events such as the American War of Independence and the Civil War. The less well-known intervening

Figure 5.1 Steamboats preceded the railroad, but their heyday was quickly eclipsed by both the speed and geographic versatility of railroads. Romanticized in Mark Twain's work, they remained crucial to economies along the Mississippi and Missouri rivers. National Archives.

period, the Market Revolution, is as important to American history and environmental change as any of the pivotal and transformative wars. This era spanned the first half of the 19th century and occurred primarily in the Northeast; it was characterized by the explosive growth of transportation infrastructure, such as improved roads, canals, steamboats, and, most importantly, railroads. This grid of transportation networks enabled an unprecedented economic expansion. Moreover, this era and place marked the edge of an expanding capitalist economy that was sweeping the world, transforming communities, economies, and ecosystems along the way. Americans strived to find their place in this burgeoning economy. Businessmen, farmers, and laborers alike had to be nimble, flexible, and aware of and responsive to change as the world transformed around them. Furthermore, they were compelled to quickly grasp and respond to both opportunities and threats. Market principles reorganized society and continued to transform the American landscape. Although the Market Revolution provided more jobs, created great wealth, and laid the industrial foundation of America's emerging modern capitalist economy, this came at great ecological cost. The wealth of nature, the accrued capital of large trees in uncut forests, fertile soil, great fish runs, and other stored natural wealth subsidized this great economic growth. The abundance of the American landscape, so quickly accessed and exploited with the Market Revolution, created an abundance of manufactured goods and prosperity and helped propel the United States forward as a powerful economy and nation.

New York and the Hudson River Valley

Merchants in New York and other major urban centers invested in roads specifically to gain access to agricultural goods in their city's hinterland and to sell merchandise to farmers. There were numerous ways to improve this reach into rural areas. Plank roads greatly increased the amount of weight that could be hauled and reduced transit times. The creation of turnpikes also made transport of goods more cost effective, as did canals, the increasing use of rivers, and, finally, the construction of a web of railroads across the Northeast and then the nation. These merchants and boosters also encouraged farmers practicing mixed agriculture to reorient themselves to market thinking. In an 1833 issue of the *Albany Argus,* in an article titled "Hints to Farmers," agriculturalists were told that

> undertak[ing] to raise all kinds of crops upon one kind of soil misapplies [the farmer's] labor. He had better confine himself to those which make the best return—sell the surplus, and buy with a part of the proceeds that for which his neighbor's soil is better adapted than his own.[1]

Standing on a rural highway in Iowa and gazing on mile upon mile of corn, or viewing the rolling hills of wheat in southeastern Washington, we can see the end point of this logic—specialization in agriculture is the key to success in the marketplace.

In the Hudson River Valley this meant an increased emphasis on grass and livestock. By 1800, river valley farmers had worn out the soil from farming. In the 1820s, as the Market Revolution picked up momentum, farmers in this region began implementing some of the strategies employed in other areas, planting more clover and timothy grass to restore the pasture, as well as rotating fields and distributing manure. They increased hay and oats production to serve the needs of the New York market; in one area of the Hudson Valley, hay yields grew by 24 percent between 1845 and 1855. Much of this was shipped downstream to Albany and New York City to feed cattle, dairy cows, and horses. The increased emphasis on hay and grass production also led to increases in cattle herds in the valley with a corresponding proliferation in dairy products such as cheese and butter produced for city consumers. The introduction of sheep for wool into the local ecosystem and economy was not so much an adaption to the limits of ecology, such as in the West, but an effort to maximize participation in the growing market economy. The high demand for wool along with the existence of unused forested hills made this a natural choice for Hudson Valley farmers. Historian Martin Bruegel writes,

> this development constituted a critical moment in the extension of capitalist relations to hill areas because farmers were not merely making up, or simply substituting a new product, for a small market crop lost to lower prices. Sheep raising brought them into the extralocal market to a significantly greater extent than they had been before.[2]

But in order to create pasture for sheep in the unused hills, the forests had to be removed. A British observer noted,

> An *ax* is their tool and with that tool, at *cutting down* trees *or cutting them up,* they will do *ten times* as much in a day as any other man that I ever saw. Set one of these men upon a wood of timber trees, and his slaughter will astonish you.[3]

Of course, Americans had spent the last two centuries perfecting their use of the ax.

More forests fell in service to another economic activity tied to the Market Revolution, the hide tanning industry. With the creation of a tanning process requiring hemlock bark, the vast stands of hemlock trees in the river valley, in combination with the ease of transport via the Hudson River, attracted investment in tanning factories in the valley. They processed hides from South America into leather then exported the product. In this way, not only were the citizens of this region tied to the urban markets of the east coast but even to the broader Atlantic economy. Although this industry only lasted approximately three decades in this region, it was more than long enough for forests to be destroyed.

The rise of hay and grass production for urban markets as well as dairy products boosted the Hudson River economy and provided a way for local farmers to specialize in agricultural production while improving the quality of their fields. Sheep herding and the tanning industry diversified the economy, increased the local participation in larger market economy, and, of course, led to changes to the landscape, particularly the removal of forests for the creation of pastures and meadows or simply their destruction for the harvest of bark. By 1850 the landscape of this valley had been transformed from one of heavy forests to one of mixed agriculture, grazing, and hillside pastures interspersed with patches of remaining forest.

Expansion and Improvement

The Federal Land Ordinance of 1785 and the Northwest Ordinance of 1787 ended the troublesome land speculation in unsettled lands of the colonial and early Republic and also stopped efforts by states to engage in great western land grabs. They provided a system for surveying land and organized it into a grid pattern of 640-acre sections broken down into smaller subsections down to 80 acres. A clear process of denoting land title and easing distribution of land to settlers was created. The Northwest Ordinance specified the manner in which territories were to be organized and how to pursue a statehood. In addition to creating much needed revenue for a government saddled with onerous foreign debts from the Revolutionary War, these laws created an orderly system of political and geographical expansion that was stable and predictable and within which Americans of varied economic status could be ensured of gaining access to new land.

When President Jefferson negotiated the Louisiana Purchase in 1803, he added 828,000 square miles to the nation, approximately doubling it size. For Jefferson to do this was to violate his original arguments for a strict interpretation of the Constitution, but one reason to change his interpretation was the opportunity to provide more abundance of land and resources for the growing country. Opposed to industrialization because he believed that such work debased humans and made democracy impossible, Jefferson desired a United States of yeoman farmers that would grow the food for an industrializing Europe. As noted in his *Notes on the State of Virginia,* he also believed that farmers represented the best of humanity and America's best chance. More land would support all of those goals and his vision.

The Land Act of 1820 further democratized land ownership as the federal government disposed of allowing credit to be used to purchase large sections of public land. Now requiring cash on the barrelhead, the price was lowered from $2.00 an acre to $1.25, and the minimum acreage sale was reduced from a quarter section of 160 acres to 80 acres. This played a key role in convincing 5 million Americans to cross the Appalachian Mountains into the Ohio River Valley and other areas by the 1840s.

Networks of Travel and Commerce

Seeking to facilitate faster, more efficient trade and travel, many states in the late 18th and early 19th century passed laws and created charters supporting the building of transportation infrastructure. Private companies were given charters for building better turnpikes and charging a toll for their use. Built with gravel and better maintained, these roads were less likely to become muddy and unusable during rains like the rough dirt tracks that often connected country and town. Also, states financed the building of short canals for bypassing waterfalls and other transportation barriers and provided funding for dredging of rivers for improved navigation. In 1806 Congress financed the National Road with the intent of fostering settlement expansion into the Ohio River Valley and the Midwest, as well as the flow of commodities east into urban markets. Completed in 1839, the gravel road ran from western Maryland to Vandalia, Illinois. As these roads, canals, and dredged rivers reached further west, settlers moved with them, buying up land within easy traveling distance of these transportation routes with the specific intent of raising commodities for urban markets. These arteries of transport and trade became then tools of expansion, economic growth, and ecological change.

The most famous transportation project of this period is the Erie Canal. The water route designed to connect New York City via the Hudson River to Lake Erie, would extend the eastern city's reach far into the heartland of America. It was completely unprecedented, and New York Governor Dewitt Clinton was roundly mocked and ridiculed when he announced the project and predicted its economic importance. One reason for the criticism was the exorbitant cost of $7 million, representing a third of New York's capital in the banking and insurance industries. Investments were originally small and included money from Europe and even China. As the canal grew and began generating revenue, the amount and size of investments grew. Constructing it required thousands of laborers, mostly Irish immigrants. The canal was 364 miles long, 40 feet wide, and 4 feet deep. Workers had to clear forest for the canal; most of it was dug by hand, and dirt was hauled away with oxen and horses. Rock had to be quarried, and workers constructed dams to build the lock system and dams to create reservoirs to control water flow. Even with the high costs

Figure 5.2 This painting of the Erie Canal is one suggesting a harmonious balance between human use and nature. Painted by George Harvey in 1837, "Pittsford on the Erie Canal—A Sultry Calm," it suggests the nation's divine favor by depicting a peaceful process of progress in an idealized pastoral setting. Little of the ecological devastation that accompanied the construction of the Erie Canal and the corresponding economic development is shown here. Courtesy of the Memorial Art Gallery of the University of Rochester.

and difficult labor, when the first section opened for business in 1819, it paid for itself within a year. Within a year of the canal's completion in 1825, 7,000 boats were in use. The revenue for tolls far outstripped the annual payment for interest on bonds, and the debt for the entire canal was paid off within 11 years. Boosters for the canal projected upon completion it would handle 250,000 tons of shipping a year; it managed more than 300,000 tons in the first year. Governor Clinton and his investors' vision was vindicated as commerce exploded along the canal and at both ends.

All along the canal, even as it was being dug, farmers quickly converted from subsistence economies to producing for the market. Those settling uninhabited areas cleared forest and put land to work intending from the get-go to fully participate in the expanding market economy. The grain mills of Rochester, New York, produced 26,000 barrels of flour for export in 1818; that number exploded to 200,000 barrels by 1828 and 500,000 by 1840. A wheat belt 150 miles long and 40 miles wide soon surrounded the canal.

The success of the Erie Canal triggered numerous other canals and imitators, none of them as successful as the Erie, but all contributing to the expansion of the market economy. By 1850 approximately 3,700 miles of canal had been dug and blasted up and down the Atlantic Seaboard.

The cumulative impact of these internal improvements was dramatic in a number of areas. Agricultural production was the very paragon of abundance. In 1839 the nation's corn crop was just less than 378,000 bushels. That number reached 838,793,000 in 1859. The wheat harvest in 1839 was a measly 84,823,000 bushels compared to the approximately 173,000,000 bushels harvested 20 years later. Because so much land in the South was devoted to cotton, the vast majority of these food crops were grown in the North. This would be an important factor in the conduct of the approaching Civil War. While agriculture boomed, so did industry. The entire route along the Erie Canal was marked with cities producing cotton, cloth, iron, and other manufactured goods. This contributed to Northern economic growth, and the rise of New York City as the center of economic investment in the nation is in large part due to the prescient planning of Governor Clinton and others and the flood of timber, agricultural, and manufacturing goods that flowed into and through the city.

Indian Removal

These improvements and federal policies encouraging migration brought Americans into contact and conflict with native peoples time and again in the Ohio River Valley, the Midwest, and the South. By the early 19th century this process was already an old story for Americans but one that was shocking for native peoples seeing their land being taken and game driven away for the first time. When they rose up to protect their land, militia and federal troops invariably defeated them and compelled them to sign treaties "guaranteeing" them much reduced territory but continued existence in their traditional homelands. But even these straitened circumstances and small inholdings were too much in the face of insatiable American land hunger.

The last land not settled by Americans in the East, Indian occupied territory protected by federal treaties, was opened up for settlement by the Indian Removal Act of 1830. An ardent expansionist and a man who built his military career and financial wealth on the basis of displacing Indians from their land, President Andrew Jackson strongly backed this legislation. When the John Marshall Supreme Court declared the law unconstitutional, Jackson ignored the decision and forced the relocation of the five "civilized" (see Document 5.1) tribes anyway. The Creek, Choctaw, Seminoles, Chickasaw, and Cherokee were forced to migrate to a newly created "Indian Territory" in the area that is now Oklahoma. The Seminole rose up in fierce rebellion but were finally defeated and removed. The Cherokee particularly suffered as the civilian contractors responsible for their relocation skimmed from the funds paid them and fed the Indians low quality provisions and did not provide enough food. Approximately a quarter of the 16,000 Cherokee died of starvation or illness on The Trail of Tears. Northern Indians were relocated as well by a federal government determined to keep its citizenry satisfied with regular supplies of new, abundant land. The Pottawatomie, Fox, Sauk, Winnebago, and others were forced to relocate from the Ohio River Valley and upper Midwest to locations in what is now Kansas and Nebraska in order to provide more land for Americans pushing west. Indian resistance by some of these tribes also failed. These forced removals and violations of federal treaties and the constitutional system of checks and balances are a testament to the reality that American society could only thrive as long as more land, more resources, and more abundance could be provided.

Harnessing Rivers for Prosperity

Dams proved integral to the Market Revolution because they converted the flow of water into mechanical power for textile mills, sawmills, grain mills, and foundries. Important instruments of economic production, dams also created economic and environmental problems. The conflicts generally reflected the tensions between a market-oriented, manufacture-driven economy versus traditional subsistence, barter, and pre-industrial capitalist practices. An ongoing debate about the best uses of the environment erupted during this period of rapid economic growth, anticipating later conservation thinking and discourse.

Dam Tensions in the North

The right to a river and its fish was codified in law in the colonial period, and this interpretation continued into the early part of the Market Revolution. The law so favored the traditional uses of the river that disgruntled fishermen and farmers traditionally had the right to remove dams prior to a determination by a court in cases of threatened fish runs. Although colonial law supported the continued use of rivers for numerous subsistence activities, it also codified the rights of mills to build dams to generate mechanical power and cause some upstream flooding in service of that goal—requiring only that they provide suitable fish passageways, a requirement that acknowledged multiple uses of and rights to the river. As long as mills served as an integral part of the local economy and were generally small in size, conflict was rare. The production of blast furnaces and cotton textile mills, with their larger scale and the greater need for more water for greater power production, led to increased conflicts over dam construction. One key moment indicating the shifts to come in the economy is found in the fight over the Furnace Unity dam in 1748.

Built on the Blackstone River in Rhode Island, upstream residents complained that the dam blocked the passage of spawning fish and convinced the local justice of the peace to order the dam opened for fish passage. The dam owners, with local support, were able to make their case in court and preserve the dam without providing fish passage. This constitutes an early example of manufacturing gaining precedence over subsistence or commercial activity based on harvest and sale of fish.

Conflict over dams arose in other locations throughout New England and in the South. In the first half of the 19th century farmers upstream of the Bilerica Dam on the Concord River in eastern Massachusetts used the courts to protest the building of that dam and later efforts to raise the height of the dam to enlarge the upstream reservoir. Important hay-producing meadows were flooded; this caused a variety of other environmental impacts, including softening of the soil, making it harder to harvest from, and a switch from the beneficial feed plants in these upstream pastures to riparian vegetation less palatable to cattle. Their protests, except for temporary victories, failed to block the expansion of the dam as the Northeastern courts increasingly favored industrial development over the traditional economy.

The proposal to build a dam on the Kennebec River of Maine triggered strong opposition in 1834 and 1835. Dam proponents sought mechanical power to support the creation of textile mills, emulating the uses of rivers for such purposes in economic powerhouses such as Lowell, Massachusetts. Investment from outside Augusta, Maine, and from business leaders of that community drove this effort to consolidate control of a river resource that most Kennebec Valley residents viewed as a commons, a shared set of river-based resources to be used by all members of the community. The most important feature of this particular commons was the millions of fish, including striped bass, shad, herring, sturgeon, and Atlantic salmon, that spawned in the river, providing both subsistence and commodities for barter and sale in the local economy. Opponents wrote and submitted petitions to the Maine legislature objecting to the planned dam for its likely environmental and economic impacts.

One petition addressed the negative impact of the dam on the movement of logs downstream, hence, potentially damaging the lumber economy. Heavy logging along the Kennebec River caused severe ecological damage, including forest habitat loss, erosion, and destruction of river bottoms from log drives, rendering almost useless an integral part of the river habitat. These dam opponents were fighting not to preserve the ecology of the Kennebec River for its own sake but rather for the river as a tool for their own economic gain. However, a petition from Greenfield, Maine, also mentioned the impact on salmon, warning of the dam's potentially devastating impact on the salmon fishery and stating that an important economic resource of "$3000 to 5000" would be lost. These opponents also wrote that the salmon "afford to a class of people at one season a cheap living, and are a source of considerable profit to a portion of our citizens. But this business must vanish and give way to the 'march of improvement.'"[4] This last lament against development reveals petitioners' concern with the damage that drastic economic change and the damming of the river would do to their traditional lifestyle. Accordingly, the dam opponents argued for the preservation of more traditional economic uses of the river as well as subsistence, in short, a preservation of the commons. They also provided a limited critique of industrial development, their rhetoric foreshadowing the later conservation movement with their argument for the preservation of fish for their economic benefits and the necessity for equity in the preservation and distribution of natural resources.

Those dependent on shad, salmon, and alewives for subsistence and commercial sale realized the danger of the proposed dam. Their petitions did not so much question progress as beseech the state to intercede on their behalf to protect their economic interests and subsistence needs. This was a fight over how nature could be best used, and by whom, for economic gain. Furthermore, this was a conflict over control of the commons, in this case, the Kennebec River and its fisheries. Many settlers in New England and Maine sought to balance subsistence activities with economic participation in the expanding market. This allowed them to maintain independence from wage labor while trying to pursue a yeoman ideal and strive to advance themselves economically. For this strategy to work they had to maintain access to common resources such as lumber, fish, and game species. Efforts to retain control over or access to the commons were a regular occurrence across the Northeast in this period.

Capitalists favored intensive development and extraction of resources for capital accumulation. Private logging companies sought to clear land completely, removing resources for locals while also destroying habitat that both game and fish needed for survival. Mill owners needed to control rivers in order to harness power for the conversion of logs to lumber and capital and convert cotton into textiles and capital. This intensive development necessitated exclusive control of the commons and brought dam proponents and opponents face to face in a conflict over the best use of the river. The supporters of intensive private development won control of the Kennebec River in Augusta, as they did throughout the region, and the dam ensured that many of those trying to pursue a mixed economic lifestyle would lose that choice and become wage laborers in the expanding market economy. Not only were the options of lower-class citizens restricted, but the anadromous fish runs went into serious decline with corresponding negative impacts on predators and bugs, birds, and others dependent on bass, shad, alewives, herring, and other fish.

River Conflict in the South

Outrage was sparked by a dam on the Edisto River of South Carolina in 1787, and protestors announced "they are totally cut off from availing themselves of the common Rights of Mankind."[5] This marked an early conflict in an ongoing point of struggle over dams in the southeastern states of Virginia, North Carolina, South Carolina, and Georgia over the next 70 years. Colonists had learned to catch and prepare shad from local Indians, and that fish, along with alewives and herring, was an integral part of the subsistence economy. As planters expanded their farming operations with ever increasing amounts of slaves, their need for more ground cornmeal and the growing commercial demand for lumber and flour engendered the construction of increasing amounts of mills and dams, directly imperiling yeoman famers. Whereas northern dams powered grist mills, foundries, and textile mills propelling the growth of the early manufacturing economy in the late 18th and first half of the 19th century, dams in the South powered lumber mills and ground wheat for export and corn for slaves. Although not often noted, dams and mills were integral to the southern economy because they helped feed the slaves required for cotton, sugar, and other crops.

Upcountry yeoman farmers in Georgia, Virginia, North Carolina, and South Carolina worried that full-scale commercial agriculture would trap them in debt, costing them their land and independence. In a society built on slave labor, landless, poor whites had few options, and slaves even considered them lower in status. Working the land not only preserved independence, but it was proper labor and denoted status in this agricultural society. They therefore tried to preserve a pioneer-subsistence economy in the face of this powerfully expanding market economy. Even with the radical differences between North and South, rural middle-class farmers of both regions shared many of the same anxieties about changes in the economy and their future status.

As they fought to protect their access to shad, a "gift from God" and key to preserving their liberty, arguments successfully made in petitions time and again during the colonial period and in the 19th century often resulted in local as well as colonial and state statutes requiring that decrepit dams be removed and passage be built through functioning dams. But, even with these compromises, fish runs declined because of widespread erosion from the fields of planters and yeoman farmers. The increased turbidity of rivers and streams destroyed fish eggs, killed fish, and kept them from returning to spawn. The development of a commercial shad fishing economy, primarily in coastal estuaries for the market and to feed slaves, pushed the fish runs to complete collapse. In the end, it wasn't the mill dams as much as ecological change and overharvesting that deprived the yeoman farmers of their fish, forcing them to even more farming and a greater reliance on pork.

Railroads and the Second City

While the growth of the canal system was impressive, it was a sluggard compared with the pace of railroad track construction and the speed of movement. Appropriately enough, the first major rail line was built to parallel the Erie Canal, but rails were not as limited by topography as canals and could more effectively connect rural areas and towns with cities. By 1840, 2,818 miles of rail were lain; that number reached just over 9,000 in 1850, and by the eve of the Civil War in 1860, 30,626 miles of railroad tracks were in operation. The vast majority of those rail lines were located in the North, and the great interior hub connecting vast reaches of the American hinterland to the East Coast and Europe was Chicago.

No city more aptly epitomizes the control and exploitation of nature for economic growth and expansion than does Chicago. Situated at the lower end of Lake Michigan at the mouth of the Chicago River and a small bay, the site offered access to East Coast and European markets by water navigation through the Great Lakes and the recently completed Erie Canal. The millions of acres of deep, mature forests to the north and the rich soils to the south and west suggested this location as a natural center of economic development. What differed between Chicago and other towns that aspired to great metropolis status and failed were the successful efforts by local farmers and merchants to raise funds for an initial rail line connecting the town of Chicago to Illinois farmers, government-funded improvement of the bay, the access to lake-borne commercial trade, and the flow of capital into the town, then city, from the East Coast and Europe.

Railroads were integral to the city's rise and instrumental to the booming industrial capitalism of the second half of the 19th century and the extraction of western resources

Figure 5.3 This painting of Chicago by James Palmatary in 1857 shows a bustling city of economic vigor. Serving as a center of the flow of resources and capital between the Midwest and the West and the East Coast, the city profited and grew from the wheat, lumber, livestock, and numerous industries. Chicago became known as "hogbutcher to the world" as it processed more meat than any other city in the world. These industries gave the city the financial foundation to also become a bustling manufacturing center of great retail importance with the rise of Sears Roebuck and Montgomery Ward in that city. Courtesy of the Chicago Historical Society.

as well. The railroads became the arteries and veins of American capitalism for a number of reasons and also drove economic development. First among many was the issue of speed. Railroad engines and cars moved exponentially faster than did wagons, barges, and carts and substantially faster than steamboats. For example, the amount of time to move goods from New York to Chicago was reduced from approximately 3 weeks to 2–3 days. This pace of shipment meant that crops more prone to spoilage could be moved greater distances, changing what farmers grew in their fields and making participation in the market more appealing to those who were once too far away. Crops could be shipped at much lower prices, encouraging the westward expansion of agriculture at the same time as it reduced the cost of transporting manufactured goods, encouraging farmers to grow for the market, buy manufactured goods, and thereby move away from a local economy to a national and international one.

Speed was not the only way railroads transformed nature and helped the capitalist economy to expand rapidly both geographically and in economic output. The railroads opening of access to land, forests, grazing, and mines provided the natural wealth to keep growth humming along. Rain at the wrong time could turn a dirt road into a morass of deep mud in which a wagon of fragile wheat or peaches might be stuck fast. Wheat caught in rain will quickly sprout, rendering it unsaleable and useless. A farmer hoping to make a nice profit from peaches one year, taking a chance to haul the load a little further to a town where he might realize a higher profit, might now face severely straitened

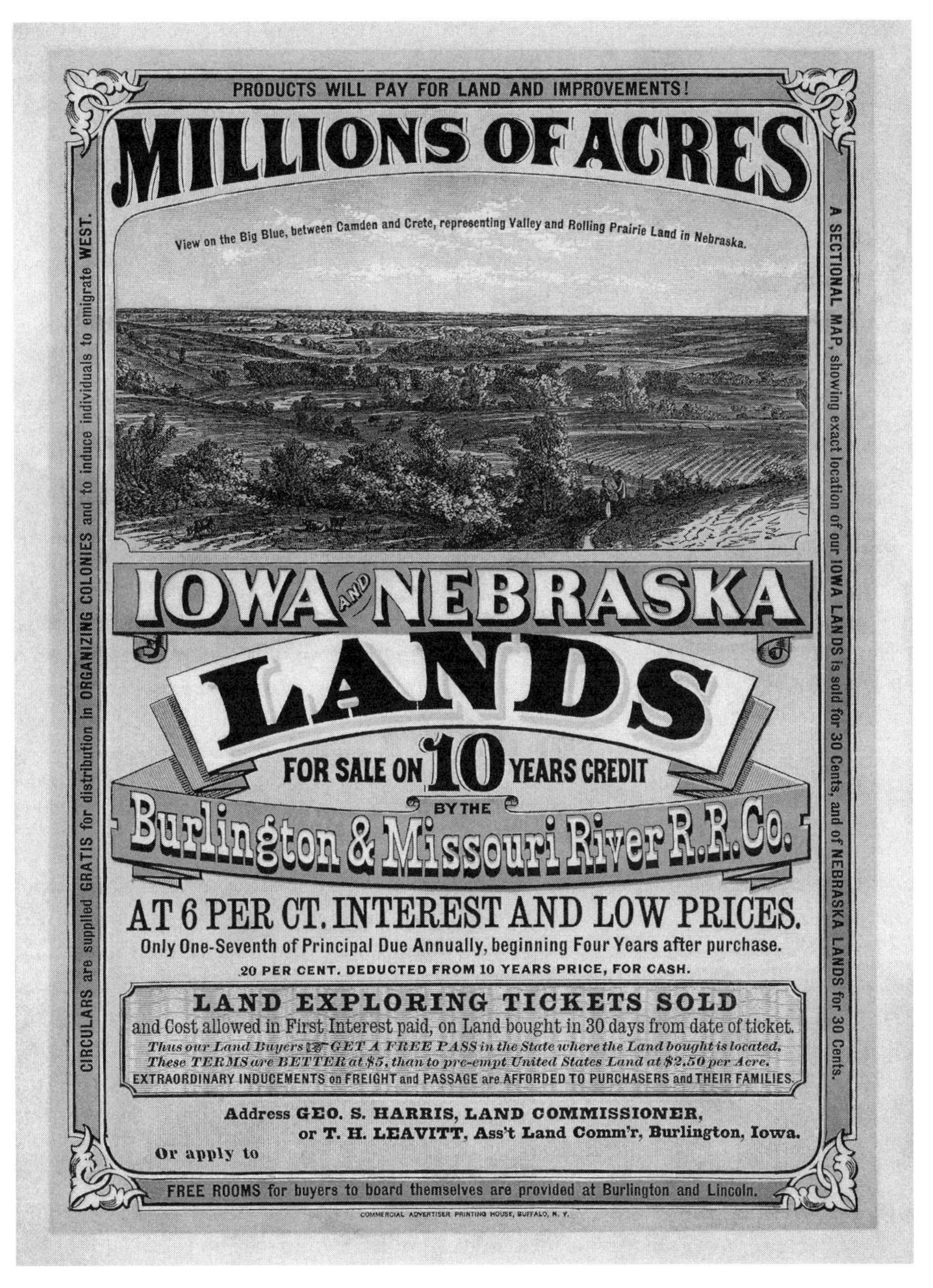

Figure 5.4 The federal government provided land subsidies to railroad companies to push west. In a competitive and cutthroat industry with rapid spikes and collapses, these corporations sold the land for needed cash. They did so by advertising across the country and in Europe and by offering payment plans and even covering travel costs for some aspiring western settlers. These companies often exaggerated the quality of the land and the types of crops that could be grown. The result was a rapid settling of the West, then a process of depopulation of the region as failed farmers and ranchers moved to urban communities. Library of Congress.

circumstances or bankruptcy as his harvest now rotted in a trapped wagon. Similarly snow and ice locked up many roads in the North in winter months and also stopped Great Lakes navigation. Railroads operated through rain, and ice and snow was cleared from one set of tracks so that the cars kept moving. This meant, for the first time, year-round, predictable, dependable transportation and travel for farmers, merchants, and capitalists. This made it easier to commit to and less risky to participate in markets farther and farther away from crops and factories. The fact that telegraphs accompanied the rails made participation in the market even less risky and more predictable as it grew easier to determine current prices for commodities such as wheat and beef in distant markets rather than rely on 2–3 week old information.

Railroads provided a logic to market transportation infrastructure that did not exist before. Railroad lines did not have to follow rivers or even follow topography to remain as flat as possible. Tracks could be laid in straight lines, directly from an urban center such as Chicago to the heart of wheat or beef country. With innovations in explosives by the 1870s, railroad lines traveled through mountains, extending their reach to the West coast and creating a continental transportation system.

The increasing use of railroads necessitated access to vast reservoirs of stored energy. This constitutes a threshold moment in environmental history and the history of capitalism as American and European societies, and following them Latin American, Asian, Indian, and African economies, shifted from transport and labor primarily based on animal and human labor, or biological sources of power, to machines powered by ancient, stored non-living energy; fossil fuels such as coal and, in the 20th century, petroleum. Railroad engines in America first used timber, and a great deal of deforestation was a result of the voracious appetite of engines as well as the wood needed to build trestle bridges and for railroad ties. But the shift to coal opened up access to seemingly infinite deposits of dense, high-energy fossil fuels, energy stored deep underground for millions of years but now brought to light to drive the industrial capitalist economy. This source of energy was particularly beneficial because it was no longer necessary to acquire food, water, and housing for living animals that could get sick, injured, or die, thereby increasing costs and interrupting predictable, regular work and transport. As a farmer told me once, "I would rather drive a dirt bike than use a horse that never stops eating." This line captures the logic and fundamental flaw in coal and, later, petroleum use. It is seemingly cheaper because the cost of use never factors in the costs of air and water pollution, costs pushed back to later generations to contend with.

The excavation of coal, also needed for the rapidly growing steel industry and for manufacturing and home heating, was shockingly destructive. Deforestation, erosion, and the filling of watersheds with rock, gravel, and slurry degrade and still damage the natural landscapes of coal communities. That environmental and social cost has not been included in the calculus of "fuel costs." The ecological damage, the loss of clean air, water, fish, and game, eroded farm soil, and so forth, constitute economic externalities to be absorbed or paid for by others: the government, community members, future generations, and nature itself. The atmosphere functions as a sink for the pollution and greenhouse gases created by coal use and other fossil fuels. Moreover, the economic boom and consumption, the culture of abundance made possible by these fuels, further create externalities only really accounted for in the late 20th century and later. Certainly the industrialists of the late 19th century cannot be indicted for failing to see the warming of the earth, but fossil fuels inflicted costs they did not pay.

"What May Avail a Crowbar against the Bilerica Dam?"

By the 1840s, a republic for only 60 years, the United States was bustling with energy and economic growth and constantly pushing up against its geographical and ecological limits. Commercial ventures, religious and political freedom, and the desire for social mobility were powerful forces in the rise and expansion of this nation. Nature played a significant role in the nation's development, but little thought, beyond conserving soil, had been given to the role on nature in one's life or how nature may be a place of respite, sanctuary, or even sacred space. The creation of Transcendentalist philosophy and thought offered a new way to view nature and man's relationship with it. For two centuries of North American history, unruly wilderness was a place of danger and threat, dense forest and swamp from which bears, panthers, or Indians might emerge. The early colonial settlers and their offspring subdued and transformed this landscape, turning forests and swamps into pasture and farms. As wilderness became a memory, a few sought to reimagine the landscape. One of these, Henry David Thoreau, is much loved by naturalists and environmentalists, one of his most oft-used quotes being, "in wildness is the preservation of the world."[6] Of course his views of nature were more complex than this and did not simply spring full-grown from the exhausted soil and deforested land of eastern Massachusetts.

The first American school of philosophy is Transcendentalism. It rose among clergy, writers, and intellectuals of New England in response to the increasing materialism of a market society and the perceived constraints of social institutions such as churches, traditional thought, and higher education. They embraced the idea of radical freedom, arguing that man and nature were inherently good but that traditions and institutions stymied people's intellectual and spiritual growth. Many of the ideas of the Transcendentalists originated with the German philosopher Immanuel Kant, but they understood his ideas largely through the writings of Romantic poets and philosophers such as Samuel Coleridge, Thomas Carlyle, and others. For these men landscape was infused with the divine and allowed the reflective individual to escape mere rationalism and scientific thought to return to his natural self.

For the father of Transcendentalism, Ralph Waldo Emerson, nature surpassed the church as a place to find a better relationship with God, to learn more about oneself, to cast off the binds of tradition and dogma. While he celebrated the beauty of nature and its ability to inspire awe, bringing one closer to God, he also enthused about putting nature to work by turning forests into parks, creating abundant crops, and harnessing rivers to alleviate man's labors. Nature could provide solace and respite, but Emerson was not interested in protecting and respecting nature for its own sake—a position that would have been extreme for even one of America's most radical thinkers of the mid-19th century. Daniel G. Payne sums up Emerson's view of nature, writing, "although nature and nature study figure prominently in Emerson's essays, his studies of nature were undertaken less for what they would tell him about natural history than for what they might tell him about himself."[7]

WALDEN;

OR,

LIFE IN THE WOODS.

BY HENRY D. THOREAU,

AUTHOR OF "A WEEK ON THE CONCORD AND MERRIMACK RIVERS."

I do not propose to write an ode to dejection, but to brag as lustily as chanticleer in
morning, standing on his roost, if only to wake my neighbors up.—Page 92.

BOSTON:

TICKNOR AND FIELDS.

M DCCC LIV.

Figure 5.5 While most Americans know *Walden* over Henry David Thoreau's other work, his budding conservationism is more evident in other works such as *The Maine Woods* and *A Week on the Concord and Merrimack Rivers*. But *Walden* makes an important contribution to environmental thought in its questioning of the assumptions of capitalism and materialism, suggesting ways of living that reject consumption and the pursuit of wealth. Library of Congress.

While this great philosopher took a stand in the fight over slavery, he never moved to a specific critique of the uses of nature or to a defense of the land.

Thoreau looked to experience, simplicity of living, and the natural world to help him reconcile the life of the mind and the physical self. Donald Worster writes,

> by themselves, Emerson's moral doctrines could not sustain Thoreau for long, for they were aspiring branches that had no roots to support them. They were ideas that were not *soiled* enough. For himself, he would have the roots of a perennial plant: thick, heavy tubers that could support and nourish the most soaring vision of human possibilities.[8]

One of these possibilities was a conscious effort to reconstruct our relationship with nature. His ideas about nature and society are grounded in ongoing interactions with and observations of wildlife, trees, and plants ranging from his daily four hours of walking and rambling through forest and meadow to his charting the depths of Walden Pond, notes on nature in his journals, and his own journeys down rivers, through forests, and to the tops of mountains. As a result he goes much further than Emerson in his critique of American culture, particularly consumption and materialism, and in his embrace of the natural world and formation of an incipient conservationist frame of mind with intimations of later deep ecology thought.

While *Walden* remains his most popular and widely read book, it is within his journals and two other books, *The Maine Woods* and *A Week on the Concord and Merrimack Rivers,* that his naturalist perspective is clearly developing. Moreover, work he was conducting at the end of his life on seed dispersal and forest regeneration indicates an increasing knowledge of the natural world, an early version of ecological thinking, and a seeking for solutions to a damaged environment. One of the Market Revolution impacts that Thoreau observes, comments upon, and actually acts to correct is the impact of the Bilerica Dam on fisheries and upstream farmers whose fields had been inundated by the dam reservoir. He criticized the dam and its negative impacts, noting the lower numbers of fish in the river. He also mused on what he thought could only be a dim possibility.

> Perchance, after a few thousand years, if the fishes will be patient, and pass their summers elsewhere meanwhile, nature will have levelled the Billerica dam, and the Lowell Factories, and the Grass-ground River run clear again, to be explored by new migratory shoals [schools of fish].[9]

He would certainly be surprised to discover it only took about 150 years for some of the dams in New England to start coming down. Thoreau was not only critical of crass materialism but also questioned the obsession with progress and the implicit assumption of progress's inherent good. He, like environmentalists a century later, questioned the benefit of progress if it came at the cost of ecological destruction. In this particular case, he, like the petitioners against the Kennebec dam in Maine, pointed out but one of the many negative consequences of economic development.

His travels through the forest interior of Maine moved him to more extensive commentary on wasteful practices in the lumber industry. Thoreau also questioned the assumed value of nature by society, querying as to the highest value of a tree, for example. Inquiring as to whether a lumberman is the true "friend and lover of the pine" or the tanner who has used its bark and tapped its turpentine, Thoreau asserted,

> No! no! it is the poet; he it is who makes the truest use of the pine, who does not fondle it with an axe, nor tickle it with a saw, nor stroke it with a plane, who knows whether its heart is false without cutting into it, who has not bought the stumpage of the township on which it stands.[10]

Writing that "all the pines shudder and heave a sigh when *that* man steps on the forest floor," Thoreau then opined that,

> No, it is the poet, who loves them as his own shadow in the air, and lets them stand. I have been into the lumber-yard, and the carpenter's shop, and the tannery, and the lampblack factory, and the turpentine clearing; but when at length I saw the tops of the pines waving and reflecting the light at a distance high over the rest of the forest, I realized that the former were not the highest use of the pine. It is not their bones or hide or tallow that I love most. It is the living spirit of the tree, not its spirit of turpentine, with which I sympathize, and which heals my cuts. It is as immortal as I am, and perchance will go to as high a heaven, there to tower above me still.[11]

Thoreau strongly criticized the utilitarian uses of nature as well as a defense of the rights of poets and philosophers to value and use nature in their way, arguably a superior use. He rejected the nearly unanimous assumption in that culture that instrumental uses of nature automatically constituted their best use. Beyond that, the language employed here is suggestive of deep ecology in its assertion that the pine's right to exist is its highest purpose. He vents his spleen about the tendency of loggers to announce the size of a tree by stating that a yoke of oxen had stood on its stump and in so doing offers strong evidence for his pantheistic frame of mind.

> The character of the logger's admiration is betrayed by his very mode of expressing it. If he told all that was in his mind, he would say, it was so big that I cut it down and then a yoke of oxen could stand on its stump. He admires the log, the carcass or corpse, more than the tree. Why, my dear sir, the tree might have stood on its own stump, and a great deal more comfortably and firmly than a yoke of oxen can, if you had not cut it down. What right have you to celebrate the virtues of the man you murdered?[12]

One of Emerson's greatest disappointments in Thoreau was the younger man's unwillingness to participate in the political world or be more of a man of action. But in regards to nature, he actually made specific proposals and took some limited steps. Living simply while rejecting consumption and waste was one way to reduce the damage done to the natural world and is an implicit rejection of the ideology of abundance and pursuit of wealth integral to the American identity. He collected evidence for the opponents of the Bilerica Dam to support their lawsuit to have the dam removed in 1859. Thoreau also studied the best way to regenerate forests, a clear and pressing problem in the increasingly deforested region, and proposed that parks should be set aside as places for retreat and introspection as well as for protecting parts of nature from destruction. In these ways, he pointed a path for future conservationists and environmentalists, and while these ideas were not appreciated in his own lifetime, in later years, environmentalists would turn to his publications and ideas for their own reforms.

Undoubtedly, the citizens of the United States were a long ways from tackling environmental problems. Action requires imagination, and that is what Thoreau provided, but it would also require the passage of time and a sense that the nation was truly running out of resources. The leap from a utilitarian mode of thinking to one that seeks to reframe the discussion of the best uses of nature and the relationship of Americans and nature is pivotal to later efforts to preserve and restore wilderness and ecosystems. But at the time that Thoreau considered who truly values a pine the most, the economy was growing like gangbusters in both the North and the South. Manufacturing in the northern states along with high profits from staple crop production like cotton in the South, built confidence in Americans that the nation was on the right track and prosperity their birthright. The ability of the government and capitalists to create a robust economy and lay the veins and capillaries of settlement and economic expansion furthered the belief that abundance could always be produced or found somewhere else. If that meant displacing Indians to get to soil, forests, or gold, Americans shrugged it off. Even as the nation attained economic heights unimaginable a few decades earlier, the schism between North and South, the debate over the expansion of slavery, was bringing the irreconcilable sections to the cusp of the bloodiest war in U.S. history.

Document 5.1 Letter to Albert Gallatin, *John Ridge,* 1826

The excerpt below is from a letter written by John Ridge to Albert Gallatin in 1826. John Ridge was a Christianized Cherokee, educated in mission schools and fluent in English. He was a political leader among the Cherokee and a wealthy slaveowner. Gallatin, former Secretary of the Treasury under President Thomas Jefferson, was collecting information on Indian tribes to show their progress in assimilating to white culture. Ridge responded to a query for information from Gallatin with this letter. It was written to show the ways the Cherokee Nation (like the Creek, Chickasaw, Seminole, and Creek) had transformed their culture and land use in order to assimilate and be successful amongst the Americans that threatened to swallow them up or, more realistically, steal their land from them. These efforts failed as the Cherokee and many other southern and mid-western tribes were forced west under the terms of the Indian Removal Act 1830.

1. In what ways and examples does the writer of this letter assure Gallatin that Cherokees have abandoned traditional land and resource use for methods learned from the Americans? To what degree do gender issues and race reflect

this effort? Is there a persistence of old methods, and if so, how is that described or explained?

2. How does the embrace of capitalism show in this document? Do issues of class, status, and consumption appear, and if so, what does this indicate about changes in tribal culture and assimilation? Can this be brought back to land use?

The Cherokee Nation is bounded on the North by east Tennessee & North Carolina, east by Georgia, south by the Creek Nation & state of Alabama, & west by west Tennessee. The extreme length of the Nation must be upwards of 200 miles & extreme breadth about 150. At a rough conjecture, it has been supposed to contain about 10,000,000 of acres of land. It is divided into eight districts or Counties by a special act of the National Council, & their boundaries are distinctly designated and defined. A census of the Nation was taken last year (1825) by order of the Council to ascertain the amount of property and Taxable persons within the Nation. The correctness of this may be relied on, and the population proved to be 13,583 native citizens, 147 white men married in the Nation, 73 white women, and 1,277 African slaves, to which we add 400 cherokees who took reservations in North Carolina & who are not included in the census & who have since merged again among us, the Cherokee Nation will contain 15,480 inhabitants. There is a scanty instance of African mixture with the Cherokee blood, but that of the white may be as 1 to 4, occasioned by intermarriages which has been increasing in proportion to the march of civilization. The above population is dispersed over the face of the Country on separate farms; villages, or a community, having a common enclosure to protect their hutches, have disappeared long since, & to my knowledge, there is but one of this character at Coosawattee, the inhabitants of which are gradually diminishing by emigration to the woods, where they prefer to clear the forest & govern their own individual plantations. In this view of their location, it really appears that they are farmers and herdsmen, which is their real character. It is true that there are distinctions now existing & increasingly so in the value of property possessed by individuals, but this only answers a good purpose, as a stimulus to those in the rear to equal their neighbors who have taken the lead. Their principal dependence for subsistence is on the production of their own farms. Indian corn is a staple production and is the most essential article of food in use. Wheat, rye & oats grow very well & some families have commenced to introduce them on their farms. Cotton is generally raised for domestic consumption and a few have grown it for market & have realized very good profits. I take pleasure to state, tho' cautiously, that there is not to my knowledge a solitary Cherokee to be found that depends upon the chase for subsistence and every head of a family had his house & farm. The hardest portion of manual labor is performed by the men, & the women occasionally lend a hand to the field, more by choice and necessity than any thing else. This is applicable to the poorer class, and I can do them the justice to say, they very contentedly perform the duties of the kitchen and that they are the most valuable portion of our Citizens. They sew, they weave, they spin, they cook our meals and act well the duties assigned them by Nature as mothers as far as they are able & improved. The African slaves are generally mostly held by Half breeds and full Indians of distinguished talents. In this class the principal value of property is retained and their farms are conducted in the same style with the southern white farmers of equal ability in point of property. Their houses are usually of hewed logs, with brick chimneys & shingled roofs, there are also a few excellent Brick houses & frames. Their furniture is better than the exterior appearance of their houses would incline a stranger to suppose. They have their regular meals as the whites, Servants to attend them in their repasts, and the tables are usually covered with a clean cloth & furnished with the usual plates, knives & forks. Every family more or less possess hogs, Cattle & horses and a number have commenced to pay attention to the introduction of sheep, which are increasing very fast. The horse is in general use for purposes of riding, drawing the plough or wagon.

Domestic manufactures is still confined to women who were first prevailed to undertake it. These consist of white or striped homespun, coarse woolen Blankets & in many instances very valuable & comfortable, twilled & figured coverlets. Woolen & cotton Stockings are mostly manufactured for domestic use within the Nation. I can only say that these domestic cloths are preferred by us to those brought from New England. Domestic plaids our people are most generally clothed with them, but calicoes, silks, cambricks, &c. Handkerchiefs & shawls &c. are introduced by Native merchants, who generally trade to Augusta in Georgia. The only trade carried on by the Cherokees with the adjoining States, is in hogs & horned Cattle. Skins formerly were sold in respectable quantities but that kind of trade is fast declining & getting less reputable. Cherokees on the Tennessee river have already commenced to trade in cotton & grow the article in large plantations and they have realized very handsome profit. All those who have it in their power, are making preparations to grow it for market & it will soon be the staple commodity of traffic for the Nation.

Document 5.2 Petitions against the Kennebec Dam, 1834

The following passages are from several petitions that were written in opposition to the proposed building of the Kennebec (later renamed Edwards) Dam on the Kennebec River in Augusta, Maine, in 1834. The dam was supported by local boosters and investors in the city as well as investors from Boston, Massachusetts. While a majority of citizens in the Kennebec River Valley supported the proposed project, a strong minority, 30 to 40 percent, opposed it and articulated their concerns in petitions presented to the state legislature.

1. To what degree do these petitions understand and predict the ecological impact of the proposed dam? What are the economic concerns of the petitioners? What specific problems do they predict?
2. How is the language of the petitions similar to environmental rhetoric today? Where does it differ?

Georgetown Petition

Construction of the dam would injure Georgetown, which has less fertile soil for agriculture and which is dependent on two primary other activities—mills & fishing. The dam would make passage of logs difficult and therefore more expensive and drive Georgetown mills out of business . . . the destruction of fishing privileges would be infinitely more disastrous in its consequences—a large number of our citizens derive their only subsistence from salmon, shad, alewive [sic] and cod fishery . . .

The preservation of the salmon, shad, and alewives in the Kenebec [sic] . . . depends entirely on the free passage afforded them up and down the river. The salmon and shad fishery are a source of considerable profit . . . and create much capital in the state. Many men are employed in taking and curing the fish . . . the alewives are important as a considerable article of trade, as bait for fishermen and as the means of alluring the cod fish to our shores. The cod fishery is perhaps as important as all the above taken together and this fishery depends entirely on the preservation of shad & alewives. On these several fisheries many of our citizens are entirely dependent for the means of livelihood.

Your remonstrants cannot believe that from the great interest your honorable bodies have always manifested for the preservation of their fisheries, as well as the great injury that will be inflicted upon them, that your honorable bodies will not authorize an obstruction, which directly tends to impoverish one part of the community for the possible benefit of another.

Greenfield Petition

Logs in rafts are frequently run from Ticonic bay, and still more frequently from the dead water in Sidney to Hallowell, Gardiner and places below. Under the improved navigation this must cease to be the case, because they cannot be so run over the dam; logs destined for these places must then in future be run singly & caught in tide waters, this we apprehend would subject owners of this species of lumber to a considerable expense.

Ship timber to the amount of from 2500 to 3000 tons is annually run down the river from Sebasticook River to Bath, it is usually run in rafts of about 100 tons, as the timber must be rafted at the dead waters in Winslow & then run to Bath. But the rafts must be inevitably be broken upon passing over the dam & great expense must be incurred in collecting and re-rafting it below; besides it will be subject to considerable loss by some portions being sunk; under the improved navigation we are constrained to say this business must be abandoned.

Woolwich Petition

Where streams of any magnitude have been obstructed by dams it has ever destroyed the fish ascending the streams. When the waters of the Androscoggin were obstructed by dams, it proved the destruction of a vast quantity of salmon which till then, had annually ascended that stream; and should the Kennebec be obstructed by a dam, from shore to shore, we believe the various tribes of fish . . . would be cut at a stroke . . . That the preservation of this species of fishery has been considered by our ancestors, as well as ourselves to have an important bearing in the vast community, not only because it contributes so largely in feeding a vast population, and by causing considerable money to circulate in the country, but that these river fish attract the fish in the sea to approach our shores, and thereby provide employment for all those engaged in the cod & mackerel industries. But as time passes on we find by sad experience that the river fish, if not others, are rapidly diminishing year by year from some cause or other, and the time must soon arrive . . . when the fish will be utterly exterminated.

Document 5.3 "The Wood Thrush," from *Ornithological Biography (Vol. 1)*, *John James Audubon*, 1831

A French-American who immigrated to America in 1803, Audubon had a deep love for and fascination with nature from his childhood on. He is famous for his extraordinary paintings and descriptions of birds, and he identified 25 new bird species as well as numerous sub-species. One of the premier books on birds, his *The Birds of America* is filled with color-plates of American birds in their natural habitat. He also published *Ornithological Biography*, which provides narrative support and explanation for those paintings. This excerpt is taken from a section describing his response to a wood-thrush in the hill-country wetlands of eastern Pennsylvania.

1. How does Audubon describe nature in this section? How are emotions tied to weather and species? How does this compare to the way early explorers and colonists viewed the benefits of nature?
2. What can be assumed about a shifting view of nature based on the level of detail used to describe this bird?

Kind Reader, you now see before you my greatest favourite [sic] of the feathered tribes of our woods. To it I owe much. How often has it revived my drooping spirits, when I have listened to its wild notes in the forest, after passing a restless night in my slender shed, so feebly secured against the violence of the storm, as to shew [sic] me the futility of my best efforts to rekindle my little fire, whose uncertain and vacillating light had gradually died away under the destructive weight of the dense torrents of rain that seemed to involve the heavens and the earth in one mass of fearful murkiness, save when the red streaks of the flashing thunderbolt burst on the dazzled eye, and, glancing along the huge trunk of the stateliest and noblest tree in my immediate neighborhood, were instantly followed by an uproar of crackling, crashing, and deafening sounds, rolling their volumes in tumultuous eddies far and near, as if to silence the very breathings of the unformed thought! How often, after such a night, when far from my dear home, and deprived of the presence of those nearest to my heart, wearied, hungry, drenched, and so lonely and desolate as almost to question myself why I was thus situated, when I have seen the fruits of my labours on the eve of being destroyed, as the water, collecting into a stream, rushed through my little camp, and forced me to stand erect, shivering in a cold fit like that of a severe ague, when I have been obliged to wait with the patience of a martyr for the return of day, trying in vain to destroy the tormenting moschetoes, [sic] silently counting over the years of my youth, doubting perhaps if ever again I should return to my home, and embrace my family!—how often, as the first glimpses of morning gleamed doubtfully amongst the dusky masses of the forest-trees, has there come upon my ear, thrilling along the sensitive cords which connect that organ with the heart, the delightful music of this harbinger of day!—and how fervently, on such occasions, have I blessed the Being who formed the Wood Thrush, and placed it in those solitary forests, as if to console me amidst my privations, to cheer my depressed mind, and to make me feel, as I did, that never ought man to despair, whatever may be his situation, as he can never be certain that aid and deliverance are not at hand.

The Wood Thrush seldom commits a mistake after such a storm as I have attempted to describe; for no sooner are its sweet notes heard than the heavens gradually clear, the bright refracted light rises in gladdening rays from beneath the distant horizon, the effulgent beams increase in their intensity, and the great orb of day at length bursts on the sight. The grey vapour that floats along the ground is quickly dissipated, the world smiles at the happy change, and the woods are soon heard to echo the joyous thanks of their many songsters. At that moment, all fears vanish, giving place to an inspiriting hope . . . Seldom, indeed, have I heard the song of this Thrush, without feeling all tranquility of mind, to which the secluded situation in which it delights is so favourable. The thickest and darkest woods always appear to please it best. The borders of murmuring streamlets, overshadowed by the dense foliage of the lofty trees growing on the gentle declivities, amidst which the sunbeams seldom penetrate, are its favourite resorts. There it is, kind reader, that the musical powers of the his hermit of the woods must be heard, to be fully appreciated and enjoyed.

The song of the Wood Thrush, although composed of but few notes, is so powerful, distinct, clear, and mellow, that it is impossible for any person to hear it without being struck by the effect which it produces on the mind. I do not know to what instrumental sounds I can compare these notes, for I really know none so melodious and harmonical. They gradually rise in strength, and then fall in gentle cadences, becoming at length so low as to be scarcely audible; like the emotions of the lover, who at one moment exults in the hope of possessing the object of his affections, and the next pauses in suspense, doubtful of the result of all his efforts to please.

Notes

1 Martin Bruegel, *Farm, Shop, Landing: The Rise of a Market Society in the Hudson Valley, 1780–1860* (Durham, NC: Duke University Press, 2002), 75.

2 Ibid., 77.

3 Quoted from Ibid., 78.

4 Greenfield, Maine petition against Kennebec Dam, Maine State Archives, 1834.

5 Quoted in Harry Watson, "'The Common Rights of Mankind': Subsistence, Shad, and Commerce in the Early Republican South," in Paul Sutter and Christopher J. Manganiello, editors, *Environmental History and the American South: A Reader* (Athens: The University of Georgia Press, 2009), 131.
6 Henry David Thoreau, *Walking* (Madison, WI: Cricket House Books, 2010), 17.
7 Daniel G. Payne, *Voices in the Wilderness: American Nature Writing and Environmental Politics* (Hanover, NH: University Press of New England, 1996), 33.
8 Donald Worster, *Nature's Economy: A History of Ecological Ideas* (New York: Cambridge University Press, 1994), 108.
9 Henry David Thoreau, *A Week on the Concord and Merrimack Rivers* (New York: Penguin Classics, 1998), 28, 32.
10 Henry David Thoreau, *Walden and Other Writings* (New York: Random House, 2004), 410.
11 Henry David Thoreau, *Walden and Other Writings* (New York: Bantam Books, 1989), 398, 399.
12 Henry David Thoreau, *Walden and Other Writings* (New York: Random House, 2004), 405, 417, 418.

Further Reading

Bernstein, Peter L. *Wedding of the Waters: The Erie Canal and the Making of a Great Nation.* New York: W.W. Norton & Company, 2005.

Bruegel, Martin. *Farm, Shop, Landing: The Rise of a Market Society in the Hudson Valley, 1780–1860.* Durham, NC: Duke University Press, 2002.

Donahue, Brian. *The Great Meadow: Farmers and the Land in Colonial Concord.* New Haven, CT: Yale University Press, 2004.

Egan, Michael and Jeff Crane, eds. *Natural Protest: Essays on the History of American Environmentalism.* New York: Routledge Press, 2008.

Judd, Richard. *Common Lands, Common People: The Origins of Conservation in Northern New England.* Cambridge, MA: Harvard University Press, 2000.

Moore, John Hebron. *The Emergence of the Cotton Kingdom in the Old Southwest: Mississippi, 1770–1860.* Baton Rouge: Louisiana State University Press, 1988.

Steinberg, Theodore. *Nature Incorporated: Industrialization and the Waters of New England.* New York: Cambridge University Press, 1991.

Sutter, Paul S. and Manganiello, Christopher J., eds. *Environmental History and the American South: A Reader.* Athens: The University of Georgia Press, 2009.

Timeline

President Abraham Lincoln's Inauguration	**March 4, 1861**
Confederate Attack on Fort Sumter	**April 12–14, 1861**
The First Battle of Bull Run	**July 21, 1861**
The Capture of Forts Henry and Donelson	**February 1862**
Mississippi River Floods	**1862**
Southern Drought	**Summer 1862**
The Second Battle of Bull Run	**August 28–30, 1862**
The Battle of Antietam	**September 17, 1862**
The Emancipation Proclamation	**January 1, 1863**
The Battle of Gettysburg	**July 1–3, 1863**
The Fall of Vicksburg	**July 4, 1863**
The Fall of Atlanta	**September 1864**
Sherman's March to the Sea	**November–December 1864**
Surrender of the Confederacy at Appomattox Courthouse	**April 9, 1865**
The Assassination of President Lincoln	**April 14, 1865**

Naturally Horrifying 6

Environment in the Civil War

On September 17, 1862, the Army of Northern Virginia under the command of General Robert E. Lee faced off against the Army of the Potomac, led by General George McClellan. The Confederates lined themselves up along Antietam Creek north of Sharpsburg, Maryland, and in cornfields and sunken roads; 60,000 rebel soldiers fought off multiple attacks by a Union army of 100,000 over several hours of terrible fighting. At the end of that bloody day, the Confederates still held the field but were so weak and isolated in the North that they were forced to retreat soon afterwards. This was the bloodiest day in American combat history with 23,000 casualties, approximately 3,600 of which were deaths. A Union surgeon described the Maryland battlefield a week after the fighting.

> But I almost forgot to tell you how the field of battle looked one week afterwards, as the dead were almost wholly unburied, and the stench arising from it was such as to breed a pestilence in the regiment. We were ordered to bury the dead, collect arms, and accoutrements left upon the field and such other acts of cleaning up as is always necessary after an engagement. I have seen, stretched along, in one straight line, ready for interment, at least a thousand blackened, bloated corpses with blood and gas protruding from every orifice, and maggots holding high carnival over their heads. Such sights, such smells and such repulsive feelings as overcome one, are with difficulty described. Then add the scores upon scores of dead horses—sometimes whole batteries lying along side, still adding to the commingling mass of corruption and you have a faint, *very* faint idea of what you see, and can always see after a sanguinary battle. Every house, for miles around, is a hospital and I have seen arms, legs, feet and hands lying in piles rotting in the blazing heat of a Southern sky unburied and uncared for, and still the knife went steadily in its work adding to the putrid mess.[1]

This passage captures the horror of war, the senseless waste of lives, and the destruction of bodies, human and animal. Scenes of this nature, however, were also in and of themselves environmental disasters, transforming the very landscape, turning once fertile fields into reeking morgues that became breeding grounds for invisible pathogens and diseases that neither army understood or could protect themselves against. The costly day at Antietam, arguably the turning point of the war as President Lincoln used that battle to announce his Emancipation Proclamation and turn a war to preserve the Union into one to also end slavery, represents the logic of nature in war. The Confederates were there not

Figure 6.1 The Battle of Antietam, occurring in Sharpsburg, Maryland, on September 17, 1862, was horrifying in its violence and scale of destruction. It remains the bloodiest single day in American combat history with almost 23,000 casualties and more than 3,600 dead. The Union won the battle, but not handily. Victory gave President Abraham Lincoln the ability to announce the Emancipation Proclamation and turn the Civil War into a conflict that was not only about preserving the Union but also to end slavery. Library of Congress.

only to take the war to the North and gain British recognition but also because the soldiers, horses, and other beasts of burden were hungry by September 1862, already running short of food. This invasion cost them dearly.

A nation torn apart by the debate over the expansion of slavery faced off in a war that cost approximately 620,000 lives from a total population of 30 million. Comparing that death toll to the approximately 420,000 deaths from a population of 131 million lost by the United States during World War II is one way to understand the devastation of this conflict. The young Republic faced its greatest challenge and survived—but at great cost, for the South in particular. The destruction of ecosystems on battlefields, the overgrazing of grass by horses and mules, the annihilation and consumption of domestic and game animals across the landscape, the destruction of Southern agriculture and transportation, and the havoc wrought on the bodies of the survivors, all typified the environmental impacts of the war. As the war commenced, the knowledge of landscape and effective use of resources proved integral to success on the battlefield as did the ways the two different economic systems managed and produced resources and tools for their ravenous armies. Both sides were burdened by droughts, heavy rain, and disease, affecting the conduct of the war, undermining or assisting campaigns, battles, and the war effort overall. The American Civil War was an environmental conflict. Fought through the swamps of the Yazoo Delta, the muddy roads of eastern Virginia, and the rich farmland of the Shenandoah, landscape and resources defined success and failure. Similarly, hunger, disease, and abundance had their say in the experience and outcome of the war as well. The armies themselves were vast, consuming and producing ecosystems, as both blue and gray struggled with environmental conditions and changed the landscape as they advanced and retreated over four years of fighting.

When Lee surrendered his sword to Grant in 1865, the Northern landscape and economy produced goods, food, and wealth at a level unprecedented in American history, while the South lay prostrate, a terrain of destroyed fields and infrastructure, torn-down fences, and devastated bodies.

To understand the Civil War comprehensively it is necessary to examine the roles of landscape, pathogens, and bugs in determining the outcomes of battles, campaigns, and the war overall. And while studying the destruction of bodies and environment in war is necessary, of greater historical interest may be comprehending how the land itself changed military leaders and campaigns and even contributed to the growing harshness of the conflict. This had ramifications for the Southern landscape but would extend even into the later western wars against Native Americans.

Understanding the role of environment in the conflict necessarily begins with an assessment of the benefits and detriments of geography, a comparative analysis of food production and distribution, and the role of disease, terrain, weather, and other factors in the conduct of the war, campaigns, and battles. Unlike the North, the South had failed to lay the same infrastructure of roads, canals, and railroads prior to the war and did not experience the same urban and manufacturing growth. Dixie remained a primarily rural society; its men were usually experienced horsemen and hunters and learned to read terrain and weather well. Of course, night riding skill also derived from many poor white men participating in slave patrols. An environmental factor in the South's favor was the fact that it was defending its own homeland, a point the Confederate propagandists announced time and again, drawing specific comparisons to the Americans' fight against Great Britain during the Revolutionary War. Confederates knew the roads, secret paths, mountain passes, and the best river fords. Their knowledge of place certainly played a crucial role in their successes, particularly in the Shenandoah Campaign of 1862 and the Battle of Chancellorsville in 1863. In both cases and many others, local knowledge of terrain proved crucial to Confederate victory.

In the century and a half since the war debates continue over decisions made and not made, battlefield mistakes, and hesitations that lost the day. One criticism of Confederate General Robert E. Lee is that he relied too much on the offensive-defensive, waging a defensive war through use of offensive campaigns, aggressively taking the war to the Union on numerous occasions. Some historians and lost-cause advocates argue that because the Confederacy was 750,000 square miles of mountains, forests, swamps, and rivers, Lee should have conducted fighting retreats such as the Russians did versus Napoleon, drawing the blue-coated soldiers deep into Confederate territory and using local support and raiding to inflict such grievous losses that the Union would abandon its invasion and occupation. This argument ignores a couple of key issues. The essential point is that the Confederacy marshaled limited resources in the war effort. Forage and food for horse and men were limited from the onset of the conflict. Likewise, the Southerners struggled to provide enough medicine, clothing, weapons, and ammunition. Finally, another factor undermining the argument that Lee could merely fight a defensive war and achieve victory through retreat is the fact that Union forces moving through the South would not only free slaves, the much-needed labor force, but slaves would also flee at rumors through the grapevine that Northern soldiers were approaching. This would destroy the slave system and cause the collapse of agricultural production.

While the Confederacy occupied an expansive territory, that 750,000 square miles is actually laced and penetrated with several rivers flowing into, from, or through the heart of the Confederate stronghold. The Mississippi, Tennessee, Cumberland, and others rivers provided routes by which Union gunboats and men could move deeply into Confederate territory. Confederate forts Henry and Donelson in central Tennessee tasked with guarding access to the Tennessee and Cumberland rivers, respectively. General Grant led a combined joint operation between the Union army and navy using infantry and heavily armed and armored gunboats to capture the two forts in February 1862.

These were the first major victories in the western campaign for the Union and provided routes of entry deep into Tennessee via the rivers. The gunboats immediately moved to Nashville and bombed it, and by March 1862 that city was occupied by federal troops. These Union victories early in the war show how river access invalidates the argument of a purely defensive strategy using space, retreat, and attrition of Union forces. The Union also operated a navy that gave it effective control of the seas. For the entire course of the war the Union Navy operated a siege, called Operation Anaconda, with naval ships controlling the flow of cotton out of the South to Europeans and weapons and supplies from the Europeans to the Confederacy. While some smugglers were able to bypass the Union ships, the naval force helped keep the South in a state of constant deprivation throughout

Figure 6.2 General Grant's use of joint forces—naval gunboats and the army—allowed him to use rivers to invade the South and move quickly into its interior by April 1862, when the Union defeated Confederate forces in southern Tennessee at the Battle of Shiloh. Library of Congress.

the conflict. The Confederacy certainly held a strong advantage fighting on its own turf, but winning still required creative strategy, victories, and for other pieces to fall into place.

The Northern agricultural economy provided resources and food far beyond Southern capacity. The development of a market agriculture economy of mixed production of grains, beef, pork, cowhides for leather, and so forth, and a greater infrastructure of roads and railroads meant that Northern farmers produced more food, and that businessmen and the government were better able to consolidate and distribute that food to Union soldiers during the war. Additionally, the Northern states' industrial manufacturing capacity far outstripped that of the South. Of approximately 130,000 industrial businesses in the United States, about 110,000 were in the Northern states. Northern manufacturing factories tended to be better capitalized and larger in size and output than Southern factories. Moreover, Pennsylvania and New York each contained more industrial manufacturing capacity than all Southern states combined. Regardless of Northern and Southern environmental and economic advantages, the war had to be fought to determine a winner. In that contestation, hunger, disease, landscape, and devastation would play crucial roles in making the conflict the worst in American history.

"They May Give Existence to Troops Who Are Idle": Food and Hunger in the Civil War

Many historians like to use the quote of an army traveling on its stomach, but few fully realize the degree to which food for men and animals determined strategic decisions in the Civil War and enabled the Union troops to outlast and finally defeat the Confederacy. Even if the South had enough land set aside for food agriculture and employed the advanced agricultural practices of Northern farmers, a number of problems beset the regions' productive capacity. A hog cholera epidemic wrought destruction on one of the key sources of food for Southerners, civilian and military alike, and a deep freeze in Texas in the winter of 1863–1864 destroyed a substantial portion of the cattle herds in that state. Similarly, flooding along the Mississippi River destroyed crops in 1862 and a drought across the South that same year dramatically reduced the corn harvest. The Confederacy certainly experienced its share of bad luck when it came to weather, disease, and food production, but that should not distract from the fact that their inability to feed themselves also sprang from decisions made in the antebellum era and wartime choices.

Northern farmers raised a variety of crops, including wheat, corn, oats, potatoes, beef, pork, and numerous vegetables and fruit. This diversity of production developed in the prewar decades in response to market demands as well as to the opportunities and limitations of the land. More varied food production over a greater range of ecosystems rendered the Union more resilient to weather threats and, as was the case, to armed rebellion. The South, on the other hand, had committed a large portion of its farmland to the raising of cotton during the antebellum era and, surprisingly, even during the war. King Cotton held such sway that the Southern states were required to import food from Northern states before 1861. An employee of the Confederate War Department wrote, "the most alarming feature of our condition is the failure of the means of subsistence."[2] War meant a dramatic shift in land use and food production for Southerners, one that needed to be accomplished even while white men who farmed and tended cattle marched off to war and slaves were being pulled into wartime duties such as building defensive fortifications or providing service to soldiers in the field. Frederick Douglass noted the fundamental problem with Southern agriculture and dependence on slaves: "The very stomach of this rebellion, is the negro in the condition of a slave. Arrest that hoe in the hands of the negro and you smite rebellion in the very seat of its life."[3] The Southern population in 1861 stood at 9 million versus approximately 21 million in the North. Of those 9 million Southerners, 4 million were slaves—slaves who would flee when given the opportunity and who would work more slowly or commit sabotage as acts of rebellion.

Making matters worse, many cotton planters refused to stop growing the plant. When laws were passed requiring the cultivation of corn rather than cotton, some planters went so far as to plant corn on the edges of fields and cotton in the middle, hidden by the corn. This is a trick marijuana farmers today would recognize. Likewise in Virginia, planters and yeoman farmers continued to grow tobacco even as civilians and soldiers felt the deep bite of hunger and even starvation. One observer noted, "Lynchburg [Virginia] has gone mad—running stark mad—men, women & children—gamblers, doctors, niggers, preachers, lawyers—all tobacco—tobacco—from morning till night from night till morning."[4] Laws restricting the farming of cotton and tobacco as well as capping prices and limiting speculation in salt and food helped increase food production; these laws took time for enactment and implementation and were still resisted. This reality flies in face of the Lost Cause Myth and the belief that all Southerners stood united in defiance to the Union. The limitations of the Southern agricultural economy along with the recalcitrance of some Southern tobacco and cotton planters to support the war meant that the Confederacy always struggled to provide for its soldiers and civilians while the North held the advantage in this area.

Agricultural production in the North flourished and grew for several reasons. First and foremost, a much larger population made it easier to manage farms successfully. When manpower shrank during the war—though never to the degree that the South suffered from loss of workers—farmers turned to increased mechanization. Moreover, food processing potential grew dramatically with the rise of companies producing food to fill military contracts. Both Borden and Armour rose to prominence in their respective condensed milk and meat packing industries on the basis of these contracts. The South had no corresponding food processing industry. The resources of the North, the constantly improving mechanization of production and processing, as well as that region's superior transportation infrastructure meant that Union soldiers generally ate much better than Southern troops. While a well-fed army can accomplish much, fierce Southern resistance and disease would still make this war difficult for the union to win.

Pathogen's Paradise

Many histories and popular celebrations of the Civil War neglect the role pathogens played in this conflict. Not only did disease account for approximately two-thirds of the deaths in the conflict, but outbreaks of various maladies also proved consequential to both battles and campaigns.

At the beginning of the conflict, Southerners assumed the superiority of the Confederate soldier. The classic boast was that one Rebel was worth ten Yankees. The fact that the South was predominantly rural compared to a more urbanized North supposedly made for tougher Southern men growing up hunting, riding horses, and learning how to read the land—all skills that were useful in combat of that era. The South was a more warlike society, deeply invested in the "manly" interests of hunting, horsemanship, dueling, and combat itself. The frequent participation of lower class whites on slave patrols as well as the fact that six of seven service academies were located in the South provided martial training for white men of all classes. But the purportedly less masculine Northern urban men held one key advantage that was impossible to see: immunity to any number of diseases. Training camps and military operations early in the war were struck with epidemics of chicken pox, measles, and influenza. Union soldiers hailing from cities had already acquired immunity to many of these urban diseases.

In that era, the basis of disease was still not clearly understood. Most Americans and some medical professionals still believed in miasma or bad air, much like during the grim days of the bubonic plague. They did associate disease with refuse and garbage and understood the contagious nature of these illnesses without knowing the bacterial or viral basis. Southern soldiers, by virtue of their largely rural backgrounds, suffered disproportionately higher death rates from these epidemics, a loss that could hardly be sustained because the Confederacy was unable to replace lost troops even from the outset of the war. Yellow fever was one exception to this rule as it was predominantly a Southern disease and one-time contact conferred immunity. From the very opening of this conflict the Union's numerical advantage was augmented by the greater resistance to disease of many of its soldiers. Pathogens strengthened the Union cause but increased the misery of soldiers on both sides of the battle line.

Depictions of the Civil War in films like *Glory* or *Gettysburg* rarely capture many base realities of life in the war. While our expectations of the depiction of war in cinema are necessarily low, it is also true that popular and even academic histories set aside some of the more unpleasant details of service in this war, arguing that these conditions are irrelevant to the argument made by an author or to understanding the narrative of the war. Relating the experience of the soldiers themselves is a central mission in any good history, and understanding the lives of both Johnny Reb and Billy Yank is impossible without a consideration of the insects that made the soldiers' lives so miserable and racked up the body count.

Bugs were part of the environment through which soldiers bivouacked, fought, and traveled. The soldiers themselves created whole new ecosystems in which these bugs flourished. Lice, fleas, mosquitoes, and flies all added to the misery of soldiers' experiences and aided in the spread of disease. Lice infested the armies of both sides, driving soldiers to misery and despair. In the time-honored tradition of soldiers, they crafted a new language for the problem. Soldiers called lice graybacks, rebels, zouaves, tigers, and Bragg's body-guard. Soldiers craft a cruder, funnier, and more accurate idiom that more honestly reflects the reality of war than do the letters and speeches of officers and of politicians or the descriptions of the war by the media, politicians, and historians. Soldiers also employed colorful phrases for dealing with graybacks. They referred to killing lice as "fighting under the black flag." Throwing away a lice-infested shirt was called "giving the vermin a parole," and a soldier turning his shirt inside out to obtain temporary relief might comment that he was "executing a flanking movement." A private from Virginia took the humor and frustration to a whole new level when he wrote the following prayer: "Now I lay me down to sleep, While gray-backs oe'r my body creep, If I should die before I wake, I pray the Lord their jaws to break."[5] Lice on the body and in clothing was humiliating for soldiers, and adopting this humor helped them adapt to a reality they were helpless to stop. Lice not only drove them to distraction with itching but also carried typhoid; the bug, therefore, constituted a lethal threat to the health of soldiers. In World War II the U.S. military would distribute DDT in powder form to service members for them to shake on clothing and rub on their body to kill lice and to keep the threat of typhus at bay. There was no such treatment available during the Civil War.

Flies and mosquitoes swarmed soldiers wherever they went, creating their own forms of misery. A Union soldier participating in the siege of Vicksburg wrote his parents,

> last night one of them [mosquito] came into My tent & Attacked me. he grabbed me by the throat & Began to knaw me, I got up Jumpt at my revolver and shot him through the head But did not fase him. So he escaped carrying with him my Boots, hat, & 5,000 doll. In Green Backs.[6]

Soldiers kept smoky fires going to keep bugs down at night, and eventually the Union assigned netting for soldiers. Mosquitoes, particularly bad along the Mississippi River and in the Georgia and Carolina lowlands, carried yellow fever and malaria. Flies transferred dysentery bacteria, cholera, and numerous other diseases.

The Virginia Quick-Step and Tennessee Trots Are Not Dances: Diarrhea and Dysentery in the Civil War

When lecturing on the Battle of Gettysburg and approaching the climactic moment, Pickett's charge, I invariably pause to consider including one last element in the experience of soldiers fighting and losing their lives on this infamous day of American history. Usually I choose not to, focusing instead on the hopes of the soldiers, Lee's tactics, General Meade's expectations, and the slaughter that ensued. Most people do not realize that the majority of Confederate soldiers in that battle suffered from severe diarrhea. The famished, practically starving soldiers in gray consumed too many cherries off the Maryland and Pennsylvania trees as they moved north, igniting their bowels, causing pain and suffering as they faced the greater suffering of battle. Diarrhea and dysentery were constant companions to soldiers on both sides of the conflict throughout the war.

Historically, when forces assembled, disease ensued. Through the centuries of warfare, plagues have accompanied armies and bedeviled camps because of the mixing of a variety of people from different regions, carrying different sicknesses and with a variety of immune systems. Close contact and the poor hygiene of many soldiers, poorly designed sanitation systems (if any at all), malnutrition, and wounds increased the odds of disease. Dysentery and diarrhea in particular plagued soldiers of both sides over the course of the conflict. A Union surgeon described a soldier's death from dysentery:

> Day-before-yesterday [sic] a man of Company "*I*" died of chronic diarrhea, and the doctor must of course see to giving him his last quarters as he failed to give him

> permanent ones here. I have a hole dug in the ground by dint of persuasion and threat, about three feet deep, then put some poles upon the bottom, then deposited the poor fellow with clothes all on, including overcoat and shoes, and drew his blanket around him, then some poles were placed at his side, while cracker box covers formed the ends of his narrow house, then over him other poles were placed—a handful of straw scattered upon the top, and mother earth covered the remains of *Chester G. Alger*—a man beloved by all those who knew him.[7]

Flies helped to spread the misery and death. A Louisiana rebel in Virginia in 1861 described how wretched this insect could make soldiers' lives:

> We find the canvas roofs and walls of our tents black with [flies] . . . it needs no morning reveille then to rouse the soldier from his slumbers. The tickling sensations about the ears, eyes, mouth, nose, etc. caused by the microscopic feet and inquisitive suckers of an army numerous as the sands of the sea shore will awaken a regiment of men from innocent sleep to wide-awake profanity more promptly than the near beat of an alarming drum.[8]

Soldiers often built trenches and cathouses mere yards from their tents or in some cases relieved themselves a few steps from their sleeping areas. These "microscopic feet and inquisitive suckers" carried dysentery, cholera, and other disease bacteria picked up from the feces. Soldiers making camp at night after long hours marching rarely had time to completely inspect the area. A Virginia private wrote in 1863, "on rolling up my bed this morning I found I had been lying in—I won't say what—something though that didn't smell like milk and peaches."[9]

Poor hygiene also contributed to the persistence of these diseases. One North Carolinian wrote about soldiers of the deep South,

> they are like children never away from home before, can't take care of themselves, and need someone to force them to wash themselves and put on clean clothing . . . these are the men which for the most part compose our sick and fill up our Hospitals.[10]

Poor waste disposal in camps, bad hygiene, and a medical corps already struggling to deal with many health and injury problems all contributed to the rise of bugs. Even obtaining enough soap to stay clean was often difficult. General Lee wrote in 1865, "there is great suffering in the Army for want of soap. The neglect of personal cleanliness has occasioned cutaneous diseases to a great extent in many commands."[11] Often, when a soldier became ill with malaria, yellow fever, typhoid, or any other illness, his strength had already been compromised by dysentery or scurvy. Already dehydrated and malnourished, his body was less able to fight off more dangerous diseases.

Dysentery turned at least one battle. The Union Army of the Tennessee following victory at the Battle of Shiloh in April 1862 marched into northern Mississippi to destroy Confederate railroad lines. To stop them, the Confederate army holed up in Corinth, Mississippi. Unfortunately, feces entered the water supply, and the Confederate army was wracked with a devastating dysentery outbreak, littering the town with the corpses of dead soldiers and forcing the ignominious retreat of the Confederates. When Halleck's Union troops occupied Corinth, they, too, suffered through a dysentery epidemic, but they had already achieved victory.

Typhoid, Yellow Fever, Scurvy, and Other Delights

Typhoid killed many men and some women in the war. It is hard to assign death tolls to any of these diseases due to misdiagnosis, soldiers refusing treatment for fear of getting sick at the hospital, and the fact that many died with more than one disease raging through their body. A conservative estimate is that approximately 40–50,000 deaths in the war resulted from typhoid; 30,000 of those on the Union side. The total number constitutes approximately 7–8 percent of the deaths in the conflict. Legions of flies in camps in conjunction with poorly designed sanitation in camps and prison yards, as well as bad personal hygiene, contributed to the spread of the disease since the typhoid bacteria is passed on through feces. Lice also carried the bacteria. Symptoms from typhoid were similar to those of other diseases, making diagnosis and treatment difficult for doctors. Typhoid victims suffered with fever and listlessness and often died from diarrhea and internal hemorrhaging.

Pathogens could be unwitting allies, and Southerners took comfort from the thought that yellow fever might help them win the war. Remembering that contact with this illness conferred lifetime immunity, and by the time of the conflict yellow fever had become a predominantly Southern disease, Southerners hoped that "Yellow Jack" would inflict heavy casualties on Union troops unprepared for the malady. It is true that the presence of the disease in Texas slowed and in some cases prevented military operations and occupations by Northern troops. Union Admiral David Farragut sought to stop the flow of contraband cotton from Brownsville, Texas, into Matamoros, Mexico, but the commander sent to Brownsville refused to put his troops on shore because of fear of the disease. The editor of the *Brownsville Flag* wrote, "Come on Mr. Yankee. Fort Brown is garrisoned at present by Yellow Jack, and if you have a mind to try your strength with him, there will be no particular objections."[12] Yellow Jack kept the Union navy from operating with complete effectiveness, and overall efforts to stop Confederate operations and contraband trade in Texas were limited by the threat of disease early in the war. Historian Andrew McIlwaine Bell writes, "In Texas *Aedes aegypti* mosquitoes created a biological shield that helped thwart the navy's plans to seize Southern ports, a task deemed necessary to shut down the illicit international trade that supplied Confederate armies."[13] Similarly, outbreaks or threats of yellow fever prevented early occupations of Port Lavaca and Galveston, Texas.

Union Troops in the Sea Islands of South Carolina were struck by yellow fever in the summer of 1862. The disease sickened and killed some of the Northern reformers who had traveled south to educate free blacks. Further damaging the war effort, the promising young officer General Ormsby Macknight Mitchel lost his life. In South Carolina the disease interrupted army operations, giving Southern leaders like General Robert E. Lee the confidence that Yellow Jack could be depended upon as a defense, so troops could be pulled from Texas and South Carolina to fight in other combat theaters. New Orleans was a yellow fever petri dish as well as a high value strategic target. Union General Butler and Admiral Farragut did not let the disease stop their invasion and occupation.

Yellow fever was known as the scourge of the South and was an ongoing peril for those making their way into the region. It had existed in the North into the early 1800s, but due to frosts killing the mosquito that carried the fever, quarantines, and the draining of wetlands for agriculture, the disease was largely a memory in that region by the 1850s. In the South, however, it recurred regularly in both the countryside and the cities. Lighter and fewer frosts as well as ongoing trade with Caribbean cargo ships carrying *Aedes aegypti* mosquitoes contributed to the pathogen's persistence in the South. City streets in this region provided a perfect environment for this mosquito. Water troughs, backed-up gutters, and waste-holding water provided breeding sites for mosquitoes. New Orleans suffered a brutal outbreak in 1853 that killed more than 8,000 people in a few weeks. Residents fled the Crescent City in panic, leaving corpses to rot in the sun. In a misguided and desperate effort to dispel the scourge, cannons were fired in the morning and evening in the city and tar burnt on street corners. It was believed this would purify the miasma or dirty air. Yellow fever's far more telling nickname, "Strangers Disease," reflects its disproportionate impact on new arrivals. Survival granted later immunity; many did not survive that first round.

General Butler and Admiral Farragut's rather easy occupation of New Orleans was quickly followed by a tense occupation, at least partially because of Butler's general order that any woman who treated a Union officer or soldier with offense was to be regarded as a prostitute. This meant that their insults and pouring of urine on soldiers from upstairs windows and balconies would earn stern rebukes and punishment. This won him the moniker "Beast Butler." Crescent City residents found comfort in their defeat and occupation with the belief that "stranger disease" would wreak havoc on Union troops and compel a withdrawal from the city. General Butler noted "the rebels are relying largely on yellow fever to clear out northern troops . . . in the churches prayers were put up that the pestilence might come as a divine interposition on behalf of the brethren."[14] Locals talked loudly in hearing range of Union troops about previous outbreaks and rumors of a contract for the construction of 10,000 coffins. This rumor forced Butler to deal with numerous requests for transfer from the city.

The imposition of a quarantine and mitigation of environmental factors for the disease in New Orleans worked wonders. Butler forced citizens to remove rotting animals and debris from the streets, drains and ditches were cleaned and flushed out, and the downtown market scrubbed. Streets were then washed on a regular basis and a schedule for garbage removal established. While General Butler and his medical adviser were unfamiliar with germ theory and the role of the mosquito in carrying yellow fever, their actions were consistent with that understanding, and Union insistence on the removal of standing water proved critical to reducing the disease. The combination of sanitation with the quarantine around the city effectively ended yellow fever in New Orleans for the duration of the war. There were approximately a dozen deaths in the city from the disease during the conflict, compared to more than 3,000 deaths in an outbreak in 1867 only two years after the end of the war. In that respect, at least, "Beast" Butler temporarily improved the lives of the Confederate citizens of the Crescent City.

As if typhoid, dysentery, and yellow fever were not enough, scurvy also caused plentiful problems for Northern and Southern troops; an in-depth analysis of disease in the war might better determine whether this illness or others determined the outcome of battles or campaigns.

Symptoms of scurvy, a disease that arises from several months without Vitamin C, are listlessness, unwillingness to perform assigned duties, depression, pain, aches, bleeding in the joints, heavy bleeding from light wounds, and gums so soft that teeth become loose and wobbly. Early in the conflict, in 1861 and 1862, General George McClellan, commander of the Union Army of the Potomac, led 100,000 troops in the Peninsula Campaign. The intent of the operation was to capture Richmond, the Confederate Capitol, and force surrender. Setting aside McClellan's strategic misunderstanding of the war and his mistakes in this campaign, the Union soldiers during this operation suffered from scurvy as well as other ailments. Is it possible that the listlessness engendered by this sickness made soldiers slower to respond to orders, to move into position, to even fight effectively in battle? What makes such an analysis difficult is the fact that the Confederates likely suffered from scurvy at the same time as well as many other ailments, and it is difficult to ascertain first the numbers of sick and then draw a connection to preparatory and battlefield actions. But, if we simply assume the hearty healthiness of soldiers in these battles, then our understanding of Civil War history is no better than that of the well-fed re-enactors who line up to play at war. Also worth considering is that during the Seven Days Battles, when Lee's soldiers put an end to the Peninsula Campaign, General Stonewall Jackson, one of the heroes and great generals of the Confederacy, bordered on dereliction of duty, appearing late numerous times and unaccountably falling asleep during the day multiple times over the course of these battles. Might he have suffered from scurvy as well? Or was this merely remnant fatigue from the brilliant Shenandoah Valley Campaign he had just completed?

Military leaders and surgeons on both sides recognized the dangers of scurvy and sought anti-scorbutics such as onions, potatoes, and other fresh produce. Foods like this were a premium in wartime. As a result, foraging was common on both sides, both at the behest of commanders and at the initiative of soldiers. Blackberries were particularly popular, and General Sherman commented, "I have known the entire skirmish line, without orders, to fight a respectable battle for the possession of some old fields full of blackberries."[15] A comparison of command responses to the threat of scurvy reveals much about the resource wealth and transportation infrastructure of the opposing sides. In a March 1863 letter, Lee wrote,

> The troops of this portion of the army have for some time been confined to reduced rations . . . I do not think it [the food available] enough to continue them in health and vigor, and I fear they will be unable to endure the hardships of the approaching campaign. Symptoms of scurvy are appearing among them and, to supply the place of vegetables, each regiment is directed to send a daily detail to gather sassafras buds, wild onions, garlic, lamb's quarter, and poke sprouts; but for so large an army the supply obtained is very small.[16]

In stark contrast stand the efforts conducted in support of Union soldiers by women's aid societies working in cooperation with the U.S. Sanitary Commission. They organized food drives, requesting donations of onions and potatoes as well as other foods and goods. Talent shows were organized with the cost of admission being a potato or onion. A slogan coined during the war urged "don't send your sweetheart a love letter. Send him an onion." A Sanitary Fair organized in Chicago by women's aid societies drew huge support with the city offices closing for the day so employees could attend. A parade of 100 wagons with potatoes, cabbage, onions, and barrels of cider and beer passed by. For the following two weeks food continued to flood in—in one case six young girls donated five barrels of potatoes they had grown themselves. These goods were shipped south via the extensive network of rail and steamboats to keep the Union soldiers healthy. The North's superior ability to provide greater amounts of food, and anti-scurvy scorbutics, specifically contributed greatly to their victory over the Confederacy.

Laudable Pus: The Fight against Infection in the Civil War

Combat inflicted great damage on human bodies. Rifle bullets, bayonet and saber wounds, and explosive artillery ordnance tore holes in bodies and removed limbs. Grape and canister added to the maiming. Canister were metal balls packed into shells and fired from cannon like giant shotgun blasts; grape was balls connected like the fruit it is named for. While grape was used more commonly in naval warfare because of its effectiveness in tearing rigging and sails and clearing decks, it saw use on Civil War battlefields as well. Surgeons had all they could handle when the wounded, horribly disfigured, and dying were brought to field hospitals or barns, tents, and houses for triage and treatment. Those with stomach wounds were usually set aside to die because surgeons of that era did not possess the skill to repair abdominal wounds. Other injuries were treated as quickly as possible to stop bleeding first and infection second. Once finished, surgeons watched the wounds carefully, hoping for what was called "laudable pus." This did not have a bad odor, was yellow in color, and attained a creamy appearance; that pus drew a sigh of relief. Ichorous or malignant pus smelled bad, was thinner, and typically contained blood. This indicated an infection, that if allowed to spread, could become pyemia or blood poisoning. Once established, pyemia was almost impossible to eradicate. Amputations and, often, second more severe amputations were one way to stop the spread of infection. Another way was through debridement, the cutting away of rotting or dead tissue around the infected wound to stop pyemia's spread. Once the infection traveled beyond control, surgeons could accurately predict how quickly death would arrive for the doomed.

One reason for the high rate of infection arose from the ignorance of the bacteria streptococci and staphylococci, the key sources of infection. An important American surgeon of the late 19th century who served at Antietam as a medical cadet described the consequences of their ignorance of germ theory:

> We operated in old blood-stained and often pus-stained coats, the veterans of a hundred fights . . . We used undisinfected instruments from undisinfected plush-lined cases, and still worse, used marine sponges which had been used in prior pus cases and had been only washed in tap water.[17]

He explained that medical instruments and sponges dropped on the ground were washed with water and then used again.

> Our silk to tie blood vessels was undisinfected . . . The silk with which we sewed up all wounds was undisinfected. If there was any difficulty in threading a needle we

> moistened it with . . . bacteria-laden saliva, and rolled it between bacteria-infected fingers. We dressed the wounds with clean but undisinfected sheets, shirts, tablecloths, or other old soft linen rescued from the family ragbag. We had no sterilized gauze dressing, no gauze sponges.[18]

The lack of sterile technique in a germ-rich environment added to the Civil War death toll.

While lack of sterile technique and filthy field hospitals certainly increased the number of deaths, the nature of Civil War weaponry greatly increased the odds of infection. Wounds were worsened and the risk of infection increased by the low velocity and tumbling movement of minnie balls fired by both sides. They ripped flesh, bounced around inside the body, and destroyed bones, carrying dirty cloth into the wound and rarely exiting the body. The use of canister and grape did great damage to the body and carried dirty, strep and staph infected cloth from uniforms deep into wounds. Hospital conditions did not help matters, and in addition to infection, lethal forms of gangrene developed in that environment.[19]

Conditions were occasionally so bad observers were able to make odd observations. Flies swarmed hospitals and therefore maggots were common in the wounds of soldiers. The wiggling white worms, consuming the dead flesh of open and covered wounds, disgusted observers but caused the patients no pain. In fact, some surgeons noted a beneficial result. According to a Confederate surgeon, "a gangreneous wound which had been thoroughly cleaned by maggots healed more rapidly than if it had been left by itself."[20] It was also noted that rats would carefully eat dead flesh without hurting the live flesh and consequently aided healing.

Environment and Combat

There are several arguments for the invasion of the North that culminated in the Battle of Antietam on September 17, 1862. This bloodiest day of combat in American history was the result of General Lee convincing President Davis that an invasion was necessary for ongoing successful negotiations with Great Britain; a powerful alliance rested in the balance of conflict. Southerners had been belligerent in the debates over the expansion of slavery in the decades prior to the war partially because of their confidence in "King Cotton." This belief in cotton and its power played at least an indirect role in the rise of war and the invasion of Maryland in 1862.

King Cotton in the pre-war era meant that the Southerners believed that the North's dependence on Southern cotton for its textile mills would prevent it from cutting its own economic throat by risking war with the South over slavery. Two-thirds of Southern cotton went to British textile mills, comprising almost the entirety of British cotton imports. If Northerners were foolish enough to go to war, Southern leaders believed that economic logic and dependence on the commodity so integral to the British industrial economy that it would compel that empire to recognize the Confederacy in a war against the North and then lend aid. There was significant support in Britain and in Parliament to recognize the sovereignty of the Confederate government and to provide economic and military assistance beyond the small amounts that slipped through the Union naval siege on blockade runners.

At one point an apparent majority in Parliament supporting recognition was lobbying for a vote but were stopped by Prime Minister Palmerston in July 1862. If Britain had provided recognition, France and Russia would have quickly followed. British, French, and Russian support of the Confederacy would have ensured their successful separation from the United States. It is worth remembering that an agricultural commodity drove these negotiations of foreign policy over all other matters. In this way, too, an organism can affect history.

Lee understood the necessity of victory for British recognition. Moderate leaders such as Palmerston sought solid proof that the Confederacy could win a key battle before Great Britain committed to war with the United States. So the dice-rolling general sought that opportunity when he took his troops north across the Mason-Dixon line. Other reasons for the invasion are often cited. The opportunity to win a great victory on Northern soil immediately before the Union's mid-term elections might result in more Peace Democrats being elected. If this were accomplished, Lincoln's ability to manage a war-time government would be greatly weakened. Confederates were also tempted by the possibility that the border state of Maryland might join them. Recent victories at the Seven Days Battles and the Second Battle of Bull Run had left Lee and other officers confident of victory and their troops swollen with pride and high morale. In contradiction, Union troops were dismayed

Figure 6.3 Antietam is the bloodiest day of combat in American history and some of the fiercest fighting occurred in the center of the line. With the Confederate troops taking cover in a sunken road, Union soldiers advanced directly into rebel fire and almost broke the line in the center. If General McClellan, commander of the Union Army of the Potomac, had sent the 20,000 soldiers waiting in reserve into combat as the center of the line broke, the Confederate Army of Northern Virginia might have been destroyed or captured on that day in September 1862. The corn field and sunken road were filled with the bodies of the fallen. The sunken road became known as "bloody lane," and observers attested that it was filled with bodies two to three deep and one could walk the length of the road without ever setting foot on soil. As it was, Lee's army was able to retreat and fight for almost three more years. Library of Congress.

and depressed by their string of losses at the hands of Southern troops. Lee assumed that this state of affairs would persist. He was wrong.

While his troops and horses were not suffering in the summer of 1862 in the way they would the following summer, the prospect of feeding his horses and troops Maryland grass and farm products provided Lee with additional incentive for the invasion. In spite of McClellan's hesitations and mistakes on the battlefield, the Army of Northern Virginia was defeated on September 17, 1862, in the bloodiest day in American combat history.

The central irony of this battle relates back to King Cotton and a Southern culture, economy, and ideology built around slavery and a commodity: cotton. Union victory at Antietam made it possible for President Lincoln to announce the Emancipation Proclamation, thus making the Civil War no longer a war simply to preserve the Union but also a war to end slavery. The Confederacy's lack of resources forced them to roll the dice and take a big gamble. Their consequent defeat changed the very nature of the war to one designed to ensure the freedom of African-American slaves and resulted in a dramatic expansion of democracy as African-American men gained the vote and were protected politically by the Thirteenth and Fourteenth Amendments to the Constitution.

Fighting for Food: Gettysburg

The second invasion of the North culminated in the brutal three-day Battle of Gettysburg. In this case, the political and strategic goals were muted in comparison to the dire needs of the Army of Northern Virginia. This rebel army used hundreds of thousands of horses and oxen, and by the summer of 1863 the landscape of Virginia was stripped bare. Moreover, feeding the Confederate armies was a persistent problem through the war, made particularly bad in 1863 because of Mississippi River flooding and drought throughout the Southern states the summer before, severely reducing the corn harvest. Lee wrote in the spring of 1863,

> their ration consists of 1/2 pound of bacon, 18 ounces of flour, and 10 lbs of rice to each 100 men, every third day, with a few peas and a small amount of dried fruit occasionally as could be obtained. They may give existence to troops who are idle but certainly will cause them to break down when called upon for severe exertion.[21]

These were not soldiers "who are idle." When in camp they were required to perform a variety of tasks—wood gathering, foraging, weapons maintenance, guard duty, and so forth. When on the move, they could average a dozen miles a day or more of fast marching; hunger was their boon companion. Therefore, heading north not only offered potential strategic and political victories, but that relatively untouched landscape promised a larder of grass, beef, milk, grain, and fruit to feed both the horses and men of the hungry, gray army. A reading of Lee's letters in the months of April through June 1863 reveal his persistent anxiety about feeding soldiers. He referred to troops under General Longstreet in North Carolina foraging for provisions to bring back to the Army of Northern Virginia, asserting to the Secretary of War and President Davis the fact that his army was not intact and barely ready to fight because of the need for forage and food and also pointing out that moving north would relieve the hunger of soldiers and animals. In letters to subordinate commanders during the invasion north, he pointed out the benefits of travel routes because of forage and food available and reminded them to gather information on sources of provisions and to collect as much food as possible. In one case he ordered a general to gather every wagon on hand and make sure to appropriate all the grain in local grist mills. While Lee's communications in the month of June asserted the strategic and political opportunities of such an invasion, they were foregrounded by months of concern about food and the health of his army. In fact, the need for food for men and beasts of burden may be the most important factor in the decision to invade the North that culminated in the Battle of Gettysburg.

Once they crossed into Union territory, according to historian Mark Fiege,

> the Army of Northern Virginia systematically stripped the terrain. Rebel soldiers diligently rounded up horses, mules, cattle, and sheep, even searching out animals that farmers had hidden in remote corners of their properties or in nearby mountains . . . When marching through the countryside, the voracious men and animals emptied barns and granaries, ransacked larders and stores, devoured crops, and literally drank wells dry. In each town, officers summoned the citizenry and demanded food, clothing, saddles, harness, horseshoes, and other resources that the South's impoverished environment could no longer furnish in sufficient quantities.[22]

The Confederates never really had any chance of defeating the Union at Gettysburg. Not only did the Northern troops occupy the high ground from which they could deliver heavy rifle and cannon fire from behind fences and other defenses on Confederate troops advancing uphill across rough terrain and open fields, but they also outnumbered the Southerners almost two to one. While an environmental analysis of this battle might emphasize the high ground, or the rocks of the Devils Den or the forests of Little Round Top, the crucial environmental factor was food and hunger. That was the primary reason that Lee and his army were in Pennsylvania, and put it them in the position to suffer a terrible defeat. After this devastating loss, the Army of Northern Virginia was never able to fully recover. From driving north to forage for desperately needed food, the most important Confederate army barely escaped intact, so weakened that it never approached its pre-Gettysburg strength again.

The Landscape Makes the Man: Vicksburg and General Ulysses S. Grant

One project of environmental history is to map out how nature affects culture. It is important to understand how environment affects not only the conduct of a battle or a campaign but also how it impacts individuals. In the case of General Ulysses S. Grant, the swamps, Mississippi River, high bluffs, and other environmental conditions and terrain features surrounding Vicksburg transformed him into a more creative and aggressive general, creating a corresponding impact on the remainder of the war. Cashiered from the army in 1854 for alcoholism, Grant had done little to distinguish himself by the time the nation mobilized for war in spring 1861. He was able to reenter federal military service via a political appointment at the rank of captain. In the ensuing campaigns along the Cumberland, Tennessee,

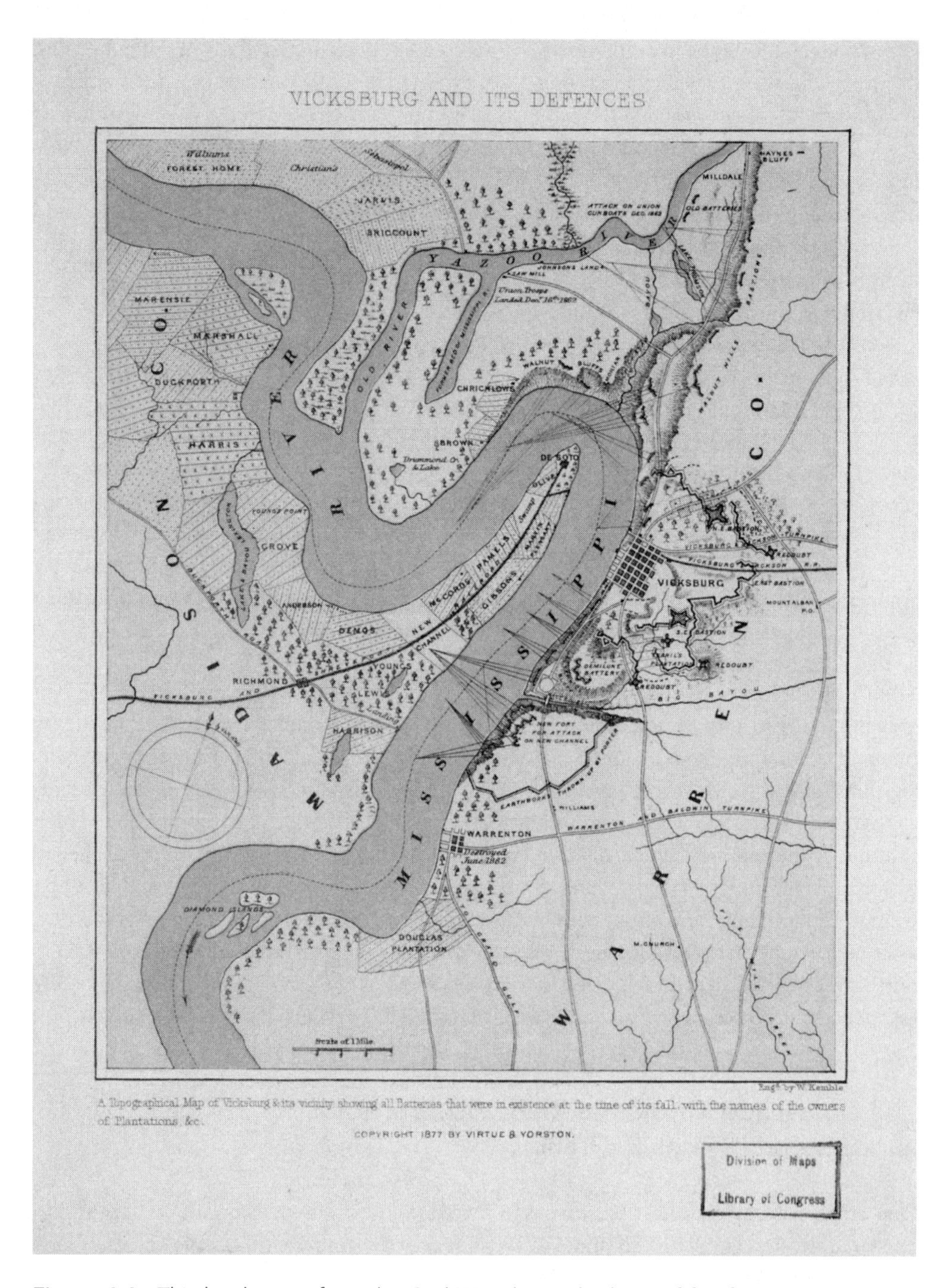

Figure 6.4 This battle map from the Civil War shows the lines of fire for the artillery batteries defending Vicksburg and some of the terrain features that made attacking the city so difficult. Note the steep bluffs to the north of the city and the long exposure of gunboats to artillery fire while approaching the bend to the north and then running straight south past the city. National Archives.

and Mississippi rivers, in the western theater, Grant quickly showed his abilities as a military leader, rising rapidly through the ranks.

After victories at Forts Henry and Donelson, opening the Cumberland and Tennessee rivers for the invasion of the upper South by gunboats and troops, and after barely achieving victory at the Battle of Shiloh in April 1862, Grant moved quickly to capture the Confederacy's prize military possession, the so-called Gibraltar of the Confederacy, Vicksburg. It earned this moniker because of its ideal location and strength. Sitting high on steep bluffs directly over the river, just north of the city, batteries of artillery offered a strong defense from any potential river-based attack such as that employed by Grant at Forts Henry and Donelson. The wetlands north of the city made movement of troops practically impossible, as did swamps west of the river. The combination of ecosystems, terrain, and deployment of military assets seemingly rendered Vicksburg invulnerable. This strength was important because both the Confederacy and the Union believed that control of the Mississippi River was key to victory in the war. Troops, equipment, and food supplies were transported on the river, and Union victory at Vicksburg would enable the Union to stop the movement of Confederate war materiel and also allow Union transport of troops and supplies from the agricultural and industrial production centers in the North all the way to New Orleans and points in between. Additionally, taking this mighty river would also cut off Texas, Arkansas, and Louisiana from the rest of the Confederacy.

To accomplish this daunting task Grant set off cross-country over Mississippi, basing himself in Oxford and sending troops forward to the Yazoo River. This campaign was conducted by the rules of traditional warfare. His long supply line from Memphis moved men and equipment further and further south into Mississippi. His hope was to extend this line all the way to the northeast of Vicksburg and from solid ground to attack and lay siege to the city. Unfortunately, the Confederates deployed too many forces in the Magnolia State, and the long supply line was vulnerable to attack. Stopping the blue-coated soldiers at the Yazoo River, the rebels used cavalry raids to strike the supply line, most famously at Holly Springs, and thereby weakening the Union invasion.

General Earl Van Dorn's successful raid on the Union supply depot in Holly Springs in December 1862 convinced Grant to initiate previously rejected measures. Previously, he had ordered his soldiers to treat civilians with courtesy and not seize their food and domestic animals. This was consistent with the efforts by many military leaders and politicians to prevent the war from becoming too "hard." However, facing the open derision of the best of Oxford, Mississippi, society, where he was headquartered, Grant ordered soldiers into the countryside. Grant wrote of the moment,

> They came with broad smiles on their faces, indicating intense joy, to ask what I was going to do now without anything for my soldiers to eat. I told them that I was not disturbed; that I had already sent troops and wagons to collect all the food and forage they could find for fifteen miles on each side of the road. Countenances soon changed, and so did the inquiry. The next was, "What are *we* to do?" My response was that we had endeavored to feed ourselves from our own northern resources while visiting them; but their friends in gray had been uncivil enough to destroy what we had brought along, and it could not be expected that men, with arms in their hands, would starve in the midst of plenty.[23]

The wagons returning to town stuffed to the brim with hams, vegetables, and milk convinced Grant that his soldiers could live off the Mississippi landscape if required.

Forced to call off the overland invasion, Grant retreated and launched yet another attempt from Memphis down the Mississippi River to north of Vicksburg. The goal was to use the river in support of a siege, but he found himself stymied by the landscape of river, swamps, and bluffs. So, the West Point–trained engineer tried to alter the environment to assist him in taking the city, continuing an earlier project to try and change the flow of the Mississippi River by digging a channel to direct the river away from Vicksburg's guns. This failed. Grant also put soldiers to work digging a channel through the swamps on the western side of the river in Louisiana to be used by gunboats and barges to slip past the city. This project was abandoned. In yet another attempt to alter the environment in support of traditional military tactics, Union troops breached an upstream dam to release reservoir waters to flood the Yazoo River Delta north of Vicksburg. Grant sought to float his gunships through this maze of swamp, forest, and stream within bombarding range of the city. The size of the gunboats combined with the twisting waterways and heavy forest growth rendered navigation nearly impossible. Some sailors had the unenviable task of sweeping snakes, bobcats, and raccoons off the decks of the boats as these animals were shaken loose from trees struck by the gunboats. Cable and pulley systems helped turn the large vessels through tight corners.

Confederates dropped trees in the boats' paths to slow them down and easily picked off Union troops with rifle fire from cover. This effort was also called off. In the end, the natural environment was too much for Grant to overcome and could not be changed to serve Grant's strategic interests.

Having failed to bend the landscape to his purposes, Grant became a bolder, more aggressive strategist and military leader; the environment changed him. When the defenders of Vicksburg observed the retreat of Grant's troops and gunships, they celebrated their Gibraltar holding strong. One night in April 1863, while dancing waltzes to celebrate Grant's failure, the city's elite civilians and officers were interrupted by the sound of heavy cannon fire. Dashing to the porch and looking to the river below, they could see by the light of riverbank bonfires the outlines and shapes of Union gunboats dashing downstream past the city defenses. The bluff-top artillery blasted away but failed to sink a single vessel. What was Grant's bigger plan? He had confounded not only General Pemberton (the Confederate commander of the forces in Vicksburg) but even some of his own officers, particularly General William Tecumseh Sherman who complained up the chain of command against his commanding general's outlandish strategy.

In a campaign comparable to General Stonewall Jackson's Shenandoah Valley Campaign of 1862, Grant ferried his small force of 30,000 men across the Mississippi River south of Vicksburg and then marched inland toward Jackson, the state capital. Like Jackson, Grant's forces faced superior enemy numbers. Also, like Jackson, the Union troops defeated the Confederates multiple times using speed, diversionary tactics, and aggressive attacks. The difference between Jackson's much ballyhooed and still studied Shenandoah campaign and Grant's relatively unknown Mississippi campaign? Grant led this operation deep in enemy territory, while Jackson fought on home ground. Grant operated without a supply line and little support,[24] while Jackson enjoyed the assistance of Virginia civilians. In conducting this operation Grant abandoned his traditional use of

Figure 6.5 175 feet long and 51 feet at their widest point, these gunboats had a relatively shallow draft of 6 feet, allowing them to operate in bayous and tributary rivers. Armor was 2.5 inches thick in most places, and the ships generally carried 13 guns, a mix of 32 and 42 pounders. The U.S.S. *Cairo* saw action in fighting on the Yazoo River north of the city but was sunk by Confederate underwater mines in December 1862. Civil War historian Edwin C. Bearss used maps to locate the water and silt-covered wreck. After years of work the gunboat was salvaged and restored. It is now open for viewing at the Vicksburg National Military Park. Library of Congress.

slow, steady pressure and advance supported by a long tail of supply lines to the north, instead leading his small army deep into enemy territory, foraging to feed themselves and using the element of surprise to defeat Confederate soldiers. In so doing, Grant defied critics and unequivocally defeated Confederate forces in Mississippi multiple times. Having secured his rear, he laid siege to Vicksburg from solid ground. During the operations in Mississippi, Union soldiers not only abandoned the old restraints against foraging and looting, but they also burned most of Jackson to the ground. The old ideas about avoiding hard war were clearly no longer tenable.

Unable to overcome the terrain of Vicksburg that made attack so difficult, Grant became a daring and even more aggressive general, launching himself and his troops into one of the most brilliant campaigns of the war. When Vicksburg fell after months of starvation for the Confederate civilians and soldiers in Vicksburg and disease on both sides, Grant's fame and reputation were sealed. President Lincoln had found the aggressive and competent general he desperately had sought, and Grant's continued victories finally put him in a position to directly oppose General Lee and the Army of Northern Virginia in the East.

Destroying the Countryside: Shenandoah and Georgia

Early in the war, in November of 1862, a surgeon serving in the Army of the Potomac scribed his observations and concerns as the force moved into Virginia:

> Yesterday we marched about 17 miles and encamped upon the belt of a piece of timber near a small water course, where were it not for *guards,* forage for the horses might easily be attained: but, as it is, our poor horses *look* and *starve,* while wheat, corn, and oats are within a stone's throw of them, I am sick and tired of guarding the property of rebels; and why our men should be compelled to guard and respect anything belonging to them, is a mystery hard to solve by those who have left comfortable homes and happy firesides to endure the hardships of a soldier's life.[25]

Dr. Holt then reflected on the soldier who offered apples gained from a "friend" and noted that soldiers were already slipping from camp to steal food from Confederate farms. Early in the war, officers on both sides attempted to prevent pillage on enemy territory. General Grant endeavored to maintain strict discipline among his troops on this issue early in the Vicksburg campaign. General Lee insisted on payment with Confederate currency for any goods seized in his two invasions of the North. The increasing violence and duration of the war, changes in tactics, and desperate need would make the concerns stated by Dr. Holt in November 1862 as alien and strange as South Carolina Senator James Chestnut's prediction at the beginning of the war that he would drink the blood spilled in the war from a thimble. The Civil War would become a war of utter destruction and devastation—a conflict bringing wrack and ruin to homes, towns, industry, bodies, and the very land itself.

Famed Confederate General Stonewall Jackson covered himself with glory in his remarkable campaign in the Shenandoah Valley in spring of 1862, defeating multiple Union armies using geographical advantage, speed, and aggressiveness. The valley was a natural thoroughfare almost to Washington, D.C., itself, and because of its fertility and productive farming communities, provided forage and food for Confederate horses and men. It was a tightly contested stretch of land with the town of Winchester, Virginia, changing hands 72 times, by one count. With command of the Union army in 1864, General Ulysses S. Grant decided to bring the valley under complete control once and for all and devastate its economy and agriculture so the area could no longer provision the rebels. After failures with other generals, Grant appointed the fierce cavalry general Philip Sheridan to command the Army of the Shenandoah.

Unleashing "the fiery little terror of a man,"[26] Grant ordered him to wage the kind of war he had developed in the long, difficult Vicksburg campaign. Unlike Grant's own approach early in the war and in the Mississippi campaign, there would now be almost no restraints on Union soldiers. Rather, the opposite was true—they were ordered to ravage the very economy and landscape of the valley. The famous quote from Grant clearly reveals his intent: "Eat out Virginia clear and clean as far as they go, so that crows flying over it for the balance of the season will have to carry their provender with them."[27] The landscape would no longer be a collateral victim; it was the direct target of war as the Union sought to bring this long, bloody war to an end.

Sheridan defended this campaign and his style of war in his memoir:

> I do not hold war to mean simply the lines of men shall engage each other in battle, and material interests be ignored. This is but a duel, in which one combatant seeks the other's life; war means much more, and is far worse than this . . . the loss of property weighs heavy with the most of mankind; heavier often than the sacrifices made on the field of battle. Death is popularly considered the maximum punishment in war, but it is not; the reduction to poverty brings prayers for peace more surely and more quickly than does the destruction of human life, as the selfishness of man has demonstrated in more than one great conflict.[28]

Sheridan and his soldiers wrought havoc on the valley. According to historian Lisa M. Brady the amount consumed, appropriated, or destroyed came to "435,802 bushels of wheat, 77,176 bushels of corn, 20,397 tons of hay, 10,918 beef cattle, 12,000 sheep and 15,000 hogs."[29] Union troops also burnt 1,200 barns and stole or killed more than 4,000 mules and horses. As Brady points out, not only did the strategy of the Army of the Shenandoah cause short-term deprivation in the valley, rendering it unusable for Confederate forces, but it also broke asunder the relationship between residents and the land, limiting their productive capacity for several years after the war. Some historians question the degree to which the agricultural economy was destroyed as crops began to recover quickly in peacetime, but the key point is that the landscape itself became a military target and suffered as a result.

Figure 6.6 Popular painter Felix Octavius Carr Darley's depiction of General Sherman's march to the sea shows his troops destroying railroads and telegraph lines and buildings in flame. Library of Congress.

With the fall of Atlanta in the summer of 1864, President Lincoln's reelection was ensured. But the Confederates under Robert E. Lee in Virginia and General Hood in Georgia fought on tenaciously. General Sherman proposed a destructive march to the coast of Georgia to wreak havoc on one of the few remaining agriculturally productive regions in the South untouched by battle and thus bring the war home to civilians. In so doing, he desired to shorten the war and incorporated the model of fighting he learned under Grant in Mississippi and employed by Sheridan in the Shenandoah. But there was anger, too, in these measures and Sherman's defense of them:

> These people made war upon us, defied and dared us to come south to their country, where they boasted they would kill us and do all manner of horrible things . . . We accepted their challenge, and now for them to whine and complain of the natural and necessary results is beneath contempt.[30]

It is true that the broader war had grown uglier and more destructive since the campaign in Mississippi in 1863, but recall that Sherman had strongly advised against Grant operating deep in enemy territory. When they had reached Vicksburg after the campaign and five battlefield victories in Mississippi, turning to Grant, General Sherman acknowledged the genius of the plan. Now, in the fall of 1864, Sherman sought to repeat the same process in Georgia, albeit against a weaker enemy. Covering an approximately 40-mile wide swath while marching 200 miles to the coast, Sherman's troops "ate out" the countryside, consuming beef, pork, chickens, and farm products to their hearts' content. After arriving at Savannah, Sherman wrote to his wife,

> we came right along living on turkeys, chickens, pigs &C. bringing along our wagons to be loaded as we started with bread &c. I suppose Jeff Davis will now have to feed the People of Georgia, instead of collecting provisions of them to feed his army.[31]

No concerns with malnutrition or scurvy troubled the mind of Sherman on this gambol through the country in Georgia. Little resistance stood before them, and they set to their work with diligence, destroying the farm animals they could not consume, appropriating horses and oxen for their own use, and killing the surplus. They burned the plantations of active Confederate sympathizers, government and military facilities, harvested cotton and destroyed railroads tracks. The rails were melted and twisted into shapes called "Sherman's neckties," rendering them unusable. As Brady argues, the troops in Georgia, as in the Shenandoah Valley, not only destroyed and consumed the agricultural goods of the landscape but sought to destroy the very underpinnings of the agricultural society created in these two regions.

The war ended with the slaves freed, the Northern economy humming along as the Northern states expanded economically, and the nation continuing west throughout the conflict. For the South, however, the ending was much darker. Fences were torn down across the Southern landscape by Northern and Southern soldiers collecting firewood fuel and wood for temporary cabins. The hogs that survived the war now ranged widely, consuming crops. The key labor force of the Southern economy had been freed from bondage, and their future, along with the future of labor in the South, remained unclear. The loss of horses, mules, and cattle weakened the Southern economy; the bodies of Southern men, wracked by disease or torn apart by bullets and shrapnel, limited the ability of Southerners to rebuild their ravaged landscape. Burnt bridges, artillery-and-rifle-blasted forests and fields, and burned buildings all spoke to the devastating economic and environmental impact of the Civil War on the South.

Document 6.1 From *Personal Memoirs of U.S. Grant,* 1885

General Ulysses S. Grant rose from the rank of captain at the beginning of the Civil War to the commander of the Union Army by 1864. Early in the conflict he rose to prominence in the western theater, leading soldiers in combat along the Cumberland, Tennessee, and Mississippi Rivers. Employing a bold, attacking strategy, he used a joint force of armored, shallow draft gun boats and soldiers to attack and capture Forts Henry and Donelson. These victories brought him fame and increased the morale of a Union concerned at the string of losses in the eastern theater. After near defeat at Shiloh, Tennessee, in April, 1862 and the occupation of Corinth, Mississippi, General Grant decided to capture Vicksburg, the "Gibraltar" of the Confederacy. Following his success there, he rose through the ranks until he commanded the entire Union Army. Grant directed the final campaigns that forced the surrender of the Confederacy. His victory at Vicksburg is one reason for his rise to power in the Union Army. In his memoir he sought to explain the necessity of the Union military actions against the Confederacy and told, in a clear, precise fashion, the story of his own military experiences and decisions. He was dying of cancer as he wrote this book, and it was published by Mark Twain.

1. How does Grant describe the physical challenges and efforts to overcome them? Do you believe his assertions that these were ploys or ways to kill time, or were these serious efforts?
2. Evaluate Grant's understanding of terrain and leadership. To what degree does this understanding influence a commander's success on the battlefield?

The real work of the campaign and siege of Vicksburg now began. The problem was to secure footing upon dry ground on the east side of the river from which the troops could operate against Vicksburg . . . [this city] is on the first high land coming to the river's edge, below that on which Memphis stands. The bluff, or high land, follows the left bank of the Yazoo for some distance and continues in a southerly direction to the Mississippi River, thence it runs along the Mississippi to Warrenton, six miles below. The Yazoo River leaves the high land a short distance below Haines' Bluff and empties into the Mississippi nine miles above Vicksburg. Vicksburg is built on this high land where the Mississippi washes the base of the hill. Haines' Bluff, eleven miles from Vicksburg, on the Yazoo River was strongly fortified. The whole distance from there to Vicksburg and thence to Warrenton was also intrenched, with batteries at suitable distances and rifle-pits connecting them.

From Young's Point the Mississippi turns in a northeasterly direction to a point just above the city, when it again turns and runs south-westerly, leaving vessels, which might attempt to run the blockade, exposed to the fire of the batteries six miles below the city before they were in range of the upper batteries. Since then the river had made a cut-off, leaving what was peninsula in front of the city, an island. North of the Yazoo was all a marsh, heavily timbered, cut up with bayous, and much overflowed. A front attack was therefore impossible, and was never contemplated; certainly not by me. The problem then became, how to secure a landing on high ground east of the Mississippi without an apparent retreat. Then commenced a series of experiments to consume time, and to divert the attention of the enemy, of my troops and of the public generally. I, myself, never felt great confidence that any of the experiments resorted to would prove successful. Nevertheless I was always prepared to take advantage of them in case they did.

In 1862 General Thomas Williams had come up from New Orleans and cut a ditch ten or twelve feet wide and about as deep, straight across from Young's Point to the river below. The distance across was a little over a mile. It was Williams' expectation that when the river rose it would cut a navigable channel through; but the canal started in an eddy from both ends, and, of course, it only filled up with water on the rise without doing any execution in the way of cutting. Mr. Lincoln had navigated the Mississippi in his younger days and understood well its tendency to change its channel, in places, from time to time. He set much store accordingly by this canal. General McLernand had been, therefore, directed before I went to Young's Point to push the work of widening and deepening the canal. After my arrival the work was diligently pushed with about 4,000 men—as many as could be used to advantage—until interrupted by a sudden rise in the river that broke a dam at the upper end, which had been put there to keep the water out until the excavation was completed. This was on the 8th of March.

. . . On the 30th of January, the day after my arrival at the front, I ordered General McPherson, stationed with his corps at Lake Providence, to cut the levee at that point. If successful in opening a channel for navigation by this route, it would carry us to the Mississippi River through the mouth of the Red River, just above Port Hudson and four hundred miles below Vicksburg by the river.

Lake Providence is a part of the old bed of the Mississippi, about a mile from the present channel. It is six miles long and has its outlet through Bayou Baxter, Bayou Macon, and the Tensas, Washita and Red Rivers. The last three are navigable streams at all seasons. Bayous Baxter and Macon are narrow and tortuous, and the banks are covered with dense forests overhanging the channel. They were also filled with fallen timber, the accumulation of years. The land along the Mississippi River, from Memphis down, is in all instances highest next to the river, except where the river washes the bluffs which form the boundary of the valley through which it winds. Bayou Baxter, as it reaches lower land, begins to spread out and disappears entirely in a cypress swamp before it reaches the Macon. There was about two feet of water in the swamp at this time. To get through it, even with vessels of the lightest draft, it was necessary to clear off a belt of heavy timber wide enough to make a passage way. As the trees would have to be cut close to the bottom—under water—it was an undertaking of great magnitude.

On the 4th of February I visited General McPherson, and remained with him for several days. The work had not progressed so far as to admit the water from the river into the lake, but the troops had succeeded in drawing a small steamer, of probably not over thirty tons' capacity, from the river into the lake. With this we were able to explore the lake and bayou as far as cleared. I saw then that there was scarcely a chance of this ever becoming a practicable route for moving troops through an enemy's country. The distance from Lake Providence to the point where vessels going by that route would enter the Mississippi again, is about four hundred and seventy miles by the main river. The distance would probably be greater by the tortuous bayous through which this new route would carry us. The enemy held Port Hudson, below where the Red River debouches, and all the Mississippi above to Vicksburg. The Red River, Washita and Tensas were, as has been said, all navigable streams, on which the enemy could throw small bodies of men to obstruct our passage and pick off our troops with their sharpshooters. I let the work go on, believing employment was better than idleness for the men. Then, too, it served as a cover for other efforts which gave a better prospect of success. This work was abandoned after the canal proved a failure.

Lieutenant-Colonel Wilson of my staff was sent to Helena, Arkansas, to examine and open a way through Moon Lake and the Yazoo Pass if possible. Formerly there was a route by way of an inlet from the Mississippi River into Moon Lake, a mile east of the river, thence east through Yazoo Pass to Coldwater, along the latter to the Tallahatchie, which joins the Yallabusha about two hundred and fifty miles below Moon Lake and forms the Yazoo River. These were formerly navigated by steamers trading with the rich plantations along their banks; but the state of Mississippi had built a levee across the inlet some years before, leaving the only entrance for vessels into this rich region the one by way of the mouth of the Yazoo several hundreds of miles below.

On the 2d of February this dam, or levee, was cut. The river being high the rush of water through the cut was so great that in a very short time the entire obstruction was washed away. The bayous were soon filled and much of the country was overflowed. This pass leaves the Mississippi River but a few miles below Helena. On the 24th General Ross, with his brigade of about 4,500 men on transports, moved into this new water-way. The rebels had obstructed the navigation of Yazoo Pass and the Coldwater by felling trees into them. Much of the timber in this region being of greater specific gravity than water, and being of great size, their removal was a matter of great labor; but it was finally accomplished, and on the 11th of March Ross found himself, accompanied by two gunboats, under the command of Lieutenant-Commander Watson Smith, confronting a fortification at Greenwood, where the Tallahatchie and Yallabusha unite and the Yazoo begins. The bends of the rivers are such at this point as to almost form an island, scarcely above water at that stage of the river. This island was fortified and manned. It was named Fort Pemberton after the commander at Vicksburg. No land approach was accessible. The troops, therefore, could render no assistance towards an assault further than to establish a battery on a little piece of ground which was discovered above water. The gunboats, however, attacked on the 11th and again on the 13th of March. Both efforts were failures and were not renewed. One gunboat was disabled and we lost six men killed and twenty-five wounded. The loss of the enemy was less.

Document 6.2

The following image (Figure 6.7) was the cover of *Harpers Weekly,* a popular magazine based in the North. It is a print of Mosby's Raiders plundering a Union supply train. For much of the war this group operated with impunity in the Shenandoah Valley, until finally caught during General Sheridan's campaign in 1864.

1. How are the rebel guerilla fighters depicted in this image? How does this conform to stereotypes about the Confederates?
2. What will a careful analysis of this picture, informed by an understanding of Confederate scarcity, tell the viewer about the importance of raiding for Confederate troops?

HARPER'S WEEKLY.
A JOURNAL OF CIVILIZATION

Vol. VII.—No. 349.] NEW YORK, SATURDAY, SEPTEMBER 5, 1863.

Figure 6.7 John Mosby commanded a unit of Virginia Cavalry and was known as the "Gray Ghost" for his ability to elude Union pursuit. His raiding and victories made him a popular figure in the South and helped build Southern morale during the conflict. Library of Congress.

Document 6.3

The following images (Figures 6.8, facing, and 6.9, overleaf) are taken from the Battle of the Wilderness (1864) and the Battle of Gettysburg (1863). In Figure 6.8 from the Wilderness, damage to the forest and Confederate entrenchments are clearly seen. In Figure 6.9 of Culp's Hill at Gettysburg, destruction of trees by artillery and rifle fire are clearly evident.

1. Discuss the obvious impacts of the war on habitat and wildlife. To what degree is this discussed in Civil War histories, and how would the scale of this for the entire war be calculated?
2. Should the destructiveness of war in regards to habitat loss, the killing of game, and disruption of ecosystems be calculated in the costs of war and given greater weight in discussions about entering war?

Figure 6.8 The Battle of the Wilderness was fought in Virginia on May 3–4, 1863. It was fought over the same ground as the Battle of Chancellorsville a year earlier. During this ferocious battle, soldiers of both sides fought amid the dug up bones and skulls of those who died in the earlier battle. Union victory under General Grant's leadership forced General Lee to begin a series of fighting retreats to attempt to extend the war as long as possible. Library of Congress.

Figure 6.9 Culp's Hill was the site of ferocious fighting during the Battle of Gettysburg, July 1–3, 1863. Gettysburg is the most famous of the Civil War battles. Fought in the town of the same name in Pennsylvania, over three days of fighting both sides suffered a total of 50,000 casualties. The Army of Northern Virginia was never able to recover from its losses in this battle. Library of Congress.

Notes

1 James M. Greiner, Janet L. Coryell, and James R. Smither, editors, *A Surgeon's Civil War: The Letters and Diary of Daniel M. Holt, M.D.* (Kent, OH: The Kent State University Press, 1994), 27, 28.
2 Quoted in William Blair, *Virginia's Private War: Feeding Body and Soul in the Confederacy, 1861–1865* (New York: Oxford University Press, 1998), 71.
3 Quoted in Joe A. Mobley, *Weary of War: Life on the Confederate Home Front* (Westport, CT: Praeger Press, 2008), 23.
4 Quoted in Blair, *Virginia's Private War,* 73.
5 Bell I. Wiley, *The Life of Johnny Reb and The Life of Billy Yank* (Baton Rouge: Louisiana State University Press, 1994), 251.
6 Quoted in Michael B. Ballard, *Vicksburg: The Campaign That Opened the Mississippi* (Chapel Hill: The University of North Carolina Press, 2004), 375.
7 Greiner, Coryell, and Smither, editors, *A Surgeon's Civil War,* 58
8 Quoted in Wiley, *The Life of Johnny Reb,* 248, 249.
9 Ibid., 248.

10 Ibid., 246.
11 Ibid., 248.
12 Andrew McIlwaine Bell, *Mosquito Soldiers: Malaria, Yellow Fever, and the Course of the American Civil War* (Baton Rouge: Louisiana State University Press, 2010), 39, 40.
13 Ibid., 54.
14 Alfred Jay Bollet, *Civil War Medicine: Challenges and Triumphs* (Tucson, AZ: Galen Press, 2002), 298.
15 Quoted in Ibid., 352.
16 Ibid., 352, 353.
17 Major W. W. Keen, M.D., "Military Surgery in 1861 and 1918," *The Annals of the American Academy of Political and Social Science,* Volume LXXX, November 1918, 11–22.
18 Ibid., 91.
19 One episode of hospital gangrene was so bad that a nurse who scratched some bug bites became infected through those scratches and died.
20 Quoted in Bollet, *Civil War Medicine,* 212.
21 Ibid., 343.
22 Mark Fiege, "Gettysburg and the Organic Nature of the American Civil War," in *Natural Enemy, Natural Ally,* edited by Richard P. Tucker and Edmund Russell (Corvallis: Oregon State University Press, 2004), 101.
23 Ulysses S. Grant, *Memoirs and Selected Letters: Personal Memoirs of U.S. Grant, Selected Letters, 1839–1865.* 2 vols in 1 (New York: Library of America, 1990), 291.
24 Grant's forces did receive some useful information from local slaves. This occurred throughout the South and was one way the home-field advantage was undercut by the institution of slavery.
25 Greiner, Coryell, and Smither, eds., *A Surgeon's Civil,* 42.
26 Lisa M. Brady, *War Upon the Land: Military Strategy and the Transformation of Southern Landscapes During the American Civil War* (Athens: The University of Georgia Press, 2012), 79.
27 Quoted in Ibid., 90.
28 Ibid., 90.
29 Ibid., 86.
30 Brooks D. Simpson and Jean V. Berlin, eds., *Sherman's Civil War: Selected Correspondence of William T. Sherman, 1860–1865* (Chapel Hill: The University of North Carolina Press, 1999), 760.
31 Ibid., 767.

Further Reading

Ballard, Michael B. *Vicksburg: The Campaign That Opened the Mississippi.* Chapel Hill: The University of North Carolina Press, 2004.

Bell, Andrew McIlwaine. *Mosquito Soldiers: Malaria, Yellow Fever, and the Course of the American Civil War.* Baton Rouge: Louisiana State University Press, 2010.

Bollet, Alfred Jay, M.D. *Civil War Medicine: Challenges and Triumphs.* Tucson, AZ: Galen Press, 2002.

Brady, Lisa M. *War Upon the Land: Military Strategy and the Transformation of Southern Landscapes During the American Civil War.* Athens: The University of Georgia Press, 2012.

Grant, Ulysses S. *Memoirs and Selected Letters: Personal Memoirs of U.S. Grant, Selected Letters, 1839–1865.* 2 vols in 1. New York: Library of America, 1990.

McPherson, James. *Battle Cry of Freedom: The Civil War Era.* New York: Oxford University Press, 1988.

Mobley, Joe A. *Weary of War: Life on the Confederate Home Front.* Westport, CT: Praeger Press, 1998.

Timeline

California Gold Rush	1849
Beginning of Hydraulic Mining in the West	1853
Homestead Act	1862
Creation of Yosemite National Park (The First National Park)	1864
The End of the Civil War	1865
Bozeman Road War	1864–1867
First Columbia River Salmon Cannery	1866
Completion of first Transcontinental Railroad	1869
Bison Hide Chemical Tanning Process	1870
Creation of Yellowstone National Park	1872
Attempted National Bison Legislation	1874
The Battle of Little Bighorn	1876
The End of Bison Market Hunting	1883
The Johnson County Range War	1892
Completion of the Great Northern Railway	1893
Frederick Weyerhauser Pacific Northwest Land purchase	1900

7
Western Lands of Wealth and Violence

Susan Magoffin traveled the Santa Fe Trail through the southern Great Plains to Santa Fe, New Mexico, in the summer of 1846 with her husband, Samuel Magoffin, a businessmen involved in the Santa Fe trade. This was the beginning of the Mexican-American War, and they followed the army of Colonel Stephen Kearney, whose force invaded and occupied Santa Fe and then traveled south into Mexico. Magoffin records many details of life on the trail, natural elements and, later, her observations of Mexican society, Indian pueblos, and the roles of women in these societies. The sections of her journal from the journey on the trail show a fascination with this new western landscape, the abundance of game and profusion of wild fruit, as well as the beauty of the land along the Arkansas River. Like so many other scribes of western travels, her writing also reveals fear and anxiety about weather, predators, and nuisance insects. These passages serve as a reminder of the challenges faced by those traveling west and their fascination with what was to them a new world.[1]

In one section of her journal she describes a night of wolf song:

> Last night I had a wolfish kind of serenade! May Pan preserve me from the like tonight. Just as I had fixed myself for sleep after faning [sic] off to some other quarters the muquitoes [sic], the delightful music began. Bak! ba! gnow, gnow, in such quick succession, it was almost impossible to distinguish one from the other. It a mixture of cat, dog, sheep, wolf . . . It was enough to frighten off my sleep and everything else.[2]

She describes her dog running out of the tent to drive off the wolves, causing one to consider whether they were actually coyotes. In another part of the journal Magoffin describes the threat so many western travelers were forced to deal with at one time or another. Miles without water left the oxen in dire condition:

> some of the most fatigued absolutely crept under the wagons for shade, and did not move till they were driven up in the evening. One poor thing fell in the road and we almost gave him up for lost. His driver though, rather a tender hearted lad I presume, went with a bucket to a *mud hole* and brought the *wet mud* which was a little cool, and *plastered his body over with it.* He then got all the water from the water kegs after the men had drank, which was not more than two or three tin cups full; he took this and opening the ox's mouth poured it down his throat. He then made a covering over him with the ox yokes standing up and blankets spread over.[3]

The oxen finally made it to his feet but wandered off for water, and they were unable to find it before moving on. The extreme effort made by the driver is typical of the adaptations necessary for travelers to successfully complete their journey. From terrifying thunderstorms and hail to swollen rivers and deep canyons, the westward migration required courage and tenacity. But for those who set out to the California Gold fields, the farmlands of the Willamette Valley, or into the desert by the Great Salt Lake, the potential rewards eased their worries. They were drawn by great natural wealth, including fertile soil, gold-laced mountains, salmon-rich streams, and deep stands of forest. Embarking on their various pilgrimages they did not question the sacrosanct idea of Manifest Destiny, nor did they wonder at the future of the Indians who occupied the very lands Americans desired. This was their birthright, and as God intended this land for Americans (as they believed), they forged ahead.

Before and after the suppression and consolidation of Indian populations on reservations, a flood of settlers, miners, businessmen, and capital migrated to salmon rivers, gold fields, grasslands, and ancient forests to convert the western storehouse of natural capital to monetary wealth. This process, a repeat of earlier experiences in American history, occurred much more rapidly in the late 19th century as the machinery of harvest and distribution made extraction and transformation of the landscape a more efficient and devastating process while Americans still lacked a corresponding land ethic or philosophy of stewardship. As the bison were hunted to near extinction, the banks and streambeds of gold streams washed away, and the forests laid low, the devastation of the American Western landscape helped close three centuries of unrelenting and unrestrained resource extraction in American history and serve as the bridge to a conservationist era.

Three Forms of Western Agriculture: New Mexico, Oregon, Utah

Oregon's Willamette Valley

America is a nation of immigration and migration, and central to its celebratory narrative of exceptionalism are stories such as the Lewis and Clark Expedition and westward migration on the Oregon Trail. These celebratory narratives neglect the negative impact on native peoples as well as marginalize other great stories of migration, such as the steady flow of Spanish, Mexican, and Hispanic settlers north into and across America. Regardless, westward migration on the Santa Fe Trail, the Oregon Trail, and the Mormon Trail constituted a key moment in the process of westward expansion and conquest and settlement of the continental United States. Hundreds of thousands of settlers traveled west to find their opportunities and pursue social mobility in the fertile soils and temperate climate of Oregon's Willamette Valley and spread from there into California, Washington, and other areas.

The Willamette Valley lays on the western side of the Cascades and east of the coastal range in Oregon. Much of the soil is of volcanic origin and was highly fertile. The flow of warm and moist Pacific air makes for a conducive agricultural climate of long seasons and well-watered crops. Middle-class farmers from the Upper South constituted the vast majority of settlers to that region in the 1840s and 1850s. They shared in the popular American belief in Manifest Destiny as a reason and justification for westward movement. Simply

Figure 7.1 This famous mural, "Westward the Course of Empire Takes Its Way," was painted by German born artist Emanuel Luetze in the U.S. Capitol building in 1861 and 1862. While such a bold painting of American expansion and solidarity was probably a much-needed message during the dark days of the Civil War, he captures American themes and patterns of settlement recognizable from the nation's colonial period to the late 19th century. Standing on the Continental Divide toward the West and the Pacific, arms eagerly pointed west, the golden light suggests God's ordination and new beginnings for those expanding onto land owned by others. The frontiersmen, the horses, cattle, and wagon depicted in the mural capture the process of westward migration in American history while providing a celebratory narrative of manifest destiny. Courtesy of the Smithsonian American Art Museum, Bequest of Sara Carr Upton.

put, Americans believed their nation was blessed by God with a distinct mission to spread Protestant Christianity, democracy, and free-market capitalism to other peoples and to settle and conquer western lands, making them part of America. Also, these intrepid immigrants believed they were required to subdue nature and convert it from "wilderness" to agricultural production. Other less ideological issues factored into decisions to head west, for instance, the desire to escape malaria and yellow fever. Of no small import was the desire to escape the seemingly expanding power and reach of plantation planters in the South. Yeoman farmers from the Upper South feared the ability of plantation farmers to buy up land and gain disproportionate power in state government. The Oregon Territory seemed to offer an opportunity to reorganize society on terms more conducive to middle-class farmers. This is one reason they banned African-Americans from the territory in 1850, to prevent the spread of slave labor–based agriculture.

Western travel brought its own set of challenges and difficulties. A 5-month journey by trail, crossing multiple rivers, gorges, and mountain ranges as well as long stretches of desert, required capital and careful planning. A wagon with a team of oxen or mules was a large investment, particularly because prices increased the nearer the migrants got to the jumping-off point. The need for food for the entire family, grain for the beasts of labor, and supplies to build a cabin or lean-to and start their crops upon arrival required a significant outlay of cash before even hitting the trail. This was not an endeavor for the poor; these migrants were largely middle-class farmers and merchants seeking to improve their lot in life. However, status and preparation provided no buffer against the greatest threat on the trail, disease. Thousands of Americans died from disease while traveling west on the Oregon Trail, as well as to the gold fields of California. Cholera, a bacteria that is passed in feces and causes severe dehydration via diarrhea and vomiting, was the most dangerous pathogen for travelers. Trails were marked not just by the bleached white bones of abandoned oxen and cattle but also with the simple graves of loved ones lost along the way.

Once established, the farmers brought their land into use employing a mix of traditional methods dating from the colonial period and mixed land-use systems developed during the 18th century. The earliest settlers in the Willamette Valley established their homesteads on the tree-line between forest above and grassland below. Because of Indian burning, extensive open grasslands already existed for cattle. Pasture, farmland, and woodlot were all easily available from those early choice locations. Access to forests also enabled a greater reliance on game populations before being lost to overhunting and habitat destruction. Spanish longhorns brought from California and cattle driven west on the Oregon Trail were the foundation of the Willamette Valley cattle industry and were initially set free to fend for themselves. Swine also roamed the land, tearing up the camas so important in the Indian food economy and so delicious to the pigs. Livestock caused other forms of ecological damage like what occurred before in other places and earlier times in American history. Many native grass and plant species of the Willamette Valley and other agricultural areas in the Northwest were quickly displaced by European weeds and grasses; some went extinct.

Settlers brought the land into production relatively quickly and almost as quickly oriented themselves toward external markets. The impetus for that was the California gold rush and a rapidly growing San Francisco–based economy suddenly hungry for grain, beef, and other resources. The growth of California's population from the gold rush and related commercial activities as well as the rise of coastal logging camps in Oregon and Washington's Puget Sound caused a great demand for wheat, driving the price skyward. Willamette farmers responded by specializing agricultural production to benefit from the boom. While farmers were predominantly Southerners from the border states, the vast majority of merchants were Northerners, well tutored in the realities of the market economy and arriving west from a society transformed by the Market Revolution. Like farmers from the very beginning of American history, they looked for business opportunities locally and abroad; by the early 1850s, they produced great amounts of grain and beef for a Pacific coast economy. The consequence of the gold rush demands for wheat, cattle, and lumber was that the stages of economic development from subsistence to market agriculture in the Willamette Valley were compressed to a few years.

The Mormon Utopia in Utah

Whereas Willamette pioneers and California gold miners sought fertile lands and gold strewn streams, Mormon westerners seemingly sought desolation. Having been repeatedly attacked and driven from Missouri and Illinois, leaders of the Church of Jesus Christ of Latter-day Saints believed that their best chance to survive and create a thriving, theocratic society required gaining separation from the Protestant Americans who bedeviled them. The interior West was largely unsettled by Anglo Americans in 1847 when the first Mormons arrived weary and foot-worn in the Salt Lake Valley and immediately began putting the land to work. Although very dry, the valley had one particular attribute that helped the Mormons transform the desert into a land of milk and honey: the rushing streams that drained the Wasatch Mountains standing sentinel above them. Mormon traders had observed Hispanic irrigation in the Southwest and, like them, had little capital to invest in their first efforts. On City Creek, the site of Salt Lake City, in July 1847, the first farmers planted corn, beans, and potatoes, dammed the creek, and diverted water to this first field. Expanding the operation, the Mormons bolted two planks into an A shape and used this to cut canals across the land, employing a pan of water as an ad hoc level. In the beginning, they piled rocks and brush into rudimentary dams. The expansion of irrigated agriculture in Mormon society was centrally controlled, with the Church directing efforts, providing some funding and expertise, requiring labor service, and specifying locations for new colonies and irrigation. Within three years of arriving, more than 16,000 acres of land were irrigated. By 1890 that number was 263,473 acres. In 1850 they produced 44,000 bushels of potatoes and almost 108,000 bushels of wheat. Four decades later the potato harvest had increased by more than 1,000 percent and the wheat yield by even more. Regardless of religious perspective, all Americans understood the foundation of their success to be productive, agricultural land.

Bringing the land into production yielded cultural outcomes as well. Historian Donald Worster argues that irrigation

> provided a unified scheme of development and—within the limits of the available technology—a maximization of resource exploitation, and it freed the communities from individuals squabbling over water rights. It allowed the amassing of capital to undertake new projects and provided a cushion of security when projects failed. And, most importantly, it claimed to speak with the voice of God.[4]

The transformation of the desert to flourishing fields underpinned the exponential growth of Mormon society and caught the attention of the U.S. government. Moving to take over control of Utah, the church leadership decentralized the management of the irrigation system to local communities in order to prevent federal government control and preserve ownership of the water, so essential to success in the West, within the church community. The abundance and prosperity created by irrigated agriculture served as the basis of the expansion of the Mormon church. Irrigation is not simply a byproduct of their success—it is one reason for their remarkable growth and expansion. The control of nature enabled greater control over church members and a consolidation of wealth and resources that was used to support the rise of the church.

Hispanic Agriculture in Northern New Mexico

Even though forced to become American citizens following Mexico's defeat in the Mexican-American War (1846–1848) and the Treaty of Guadalupe Hidalgo of 1848, the Hispanic citizens of the Southwest continued to follow land-use practices developed over centuries. In the treaty, Mexico attempted to protect its former citizens by compelling the U.S. government to recognize land and water ownership of Hispanic Americans as well as traditional communal resource use. This protection would be sorely tested over the course of the 19th century. The failure of the U.S. government to effectively protect these citizens forced regular changes in agricultural practices and land use in their effort to remain in homes and communities they and their ancestors had occupied as far back as the 1600s.

Land ownership and distribution followed a long Hispanic tradition that differed from the Anglo-American system. Each family owned a plot of land directly by their home where a kitchen garden provided a majority of the food consumed by the family. Corn, beans, chilies, melons, squashes, tomatoes, and other produce was grown in this garden by the women of the family. Each family also used a large section of land outside of town where they grew corn or ran sheep; most of this production was generally for sale in local and external markets, and this sphere was controlled by the men. The land outside of town was not owned by the family but rather was allocated by a town board that also managed and distributed water resources, forest use, and the maintenance duties and labor on land and irrigation systems required of members of the community. This system could persist as long as Anglo-American capitalists stayed outside of the region and did not pursue control of the land and water and import capital to invest in the sheep industry. This old tradition in northern New Mexico continued to function well until the arrival of the railroad in the 1870s; then everything changed.

With the arrival of the Atchison, Topeka, and Santa Fe Railroad in 1879, northern New Mexico cattle herds increased exponentially from approximately 137,000 head in 1880 to around a million by 1885, and a few years later reaching almost 1.3 million. Spending winters in valleys and plains, the cattle inflicted extensive damage as they were pushed high into the Sangre de Cristo and other mountain ranges for fresh grass, even as the snow was still melting. The need for feed was so severe that cattle grazed the high mountain meadows before the grasses were grown and before they had time to drop seeds for new grass. Sheep perpetuated the damage and worsened it, as they were typically brought into grazing lands after it was too poor for continued cattle use.

Organized in a partido system controlled by a handful of Anglo-American and a few rich Hispanic land owners and businessmen, sheepherders worked in a sharecropping system from which they gleaned a portion of the lambs produced by the herd they tended as compensation. The impoverished Hispanic sheepherders, desperate to feed their families, drove sheep into every valley, cirque, draw, and saddle of the high mountains, shearing the soil and gravel of its cover. The hoofed locusts increased from 619,000 in 1870 to approximately 5 million before 1890. The overgrazing of the land by cattle and sheep occurred not only in northern New Mexico but on the Navajo reservation in Arizona and western New Mexico, on the short grass prairie of Wyoming, and in southern Idaho. In all these cases and others, erosion, degradation of ecosystems, and decline in native flora and fauna resulted.

William de Buys writes of the ecological double whammy produced by grazing cattle and sheep in these mountains:

> It was to the good fortune of the meat and wool growers and the great misfortune of the land that cattle and sheep favored different kinds of mountain rangelands. Sheep liked the open spaces, short grasses, and tangy forbs of the high divides and found the deep bunchgrasses of the lower parks relatively unpalatable. Cattle, on the other hand, relished the bunchgrasses and avoided the windy, unprotected alpine slopes. As a result, every square foot of the mountain rangeland was put to full use.[5]

Environment and Western Indian Wars

Even as the United States struggled to suppress the Confederacy, President Lincoln's Republican administration promoted expansion and settlement in the West and deployed military force against resisting Indians. The Homestead Act, passed in 1862, provided up to 160 acres of land to settlers seeking opportunities in the West. Similarly, the Pacific Railroad Acts of 1862 and 1864 easily passed in a Republican-controlled, wartime government, providing generous land subsidies to support the construction of a transcontinental railroad. To encourage westward expansion and protect those willing to risk the ire of Indians by settling on their land, the American government assigned federal troops to garrisons scattered throughout the region. The positioning of military garrisons often exacerbated tensions and engendered conflict with Indians. Elliott West writes,

> usually the forts' locations are explained in terms of potential flashpoints. The army built near trading posts and stations and at spots especially important for emigrants, where Indians and travelers were most likely to come to blows . . . Roadside businesses and emigrant camps had not been placed arbitrarily. Merchants, traders, and travelers chose a particular setting because it offered something that other spots did not—precious environmental assets that everyone needed. The soldiers simply followed suit.[6]

By the 1860s western tribes were already in serious decline, having suffered onslaughts of cholera, small pox, measles, and other diseases. Their perilous situation was exacerbated by their own overhunting of bison. This, joined with much more destructive market hunting of bison and the displacement of wildlife by wagon trains and increased settlement, pushed these tribes to the very brink. Troops were positioned exactly where the Indians needed to be, competing with them for critical resources. The crucial locations for survival on the Great Plains were the river bottoms. These provided wood for fires and fort construction as well as water for horses and humans. Game populations were higher along waterways, and during the winter, when temperatures could drop to 40 below with wind chills of 70 below, bottoms provided refuge from bitter northern winds. Having used these sites for generations, Plains Indians were seized with fear and anger at the sight of a fort being raised where they traditionally set up camp, grazed their horses, and hunted for and gathered food. Moreover, the forts represented a firm government commitment to securing land for settlers. By the 1860s, Indians on the Plains understood that forts meant ranches, farms, roads, roadhouses, stagecoaches, railroads, and everything that came with that; the loss of their way of life.

In the 1860s some native peoples conducted a fierce campaign to stop this process from even occurring. The construction of the Bozeman Road, running from Fort Laramie, Wyoming, to Bozeman, Montana, through the very heart of bison grasslands in the Powder River country, convinced the Sioux and Northern Cheyenne to rise in defense of their way of life. Bison were still abundant in this area. Numerous streams and rivers draining the Big Horn Mountains provided a steady supply of water year round, and the snowfall and rain that watered the eastern slope of the mountains rendered this area relatively lush for the region and profuse with game such as deer, antelope, and elk. This was not the first contestation of this land; in the 1820s the Sioux arrived from the east and drove the Crow to the western side of the Big Horn Mountains. Now, the Sioux and Northern Cheyenne, who had migrated north from Colorado, were determined to defend themselves, understanding that the penetration of their territory by a road and the construction of forts would lead inevitably to the destruction of their land and culture.

Chief Red Cloud and Crazy Horse emerged as the key leaders in the Bozeman Road War. While moderate and accomodationist chiefs argued for peaceful resolution and continued negotiation with the American government, the war faction decided that Crazy Woman Creek would be their Rubicon; if American troops crossed that line, they would fight. Refugees from the Sand Creek Massacre streaming into their camps strengthened their resolve to fight. When the U.S. Army built three forts in the Powder River country, the Indians launched their defense by attacking woodcutters, firing on the forts, and killing travelers on the Bozeman Road. Sioux and Northern Cheyenne held a natural advantage in the rolling terrain of hills shouldering up against the mountains cut with many winding ribbons of streams and rivers (see Document 7.2). The draws, valleys, and streams with dense brush and trees in their bottoms made surreptitious movement possible, as Captain Fetterman and his soldiers discovered in December 1866.

When a small band of Indians attacked a group of woodcutters in December 1866, Fettermen was dispatched from Fort Phil Kearny with approximately 80 men and orders to rescue the woodcutters but to not pursue the Indians. Crazy Horse and a handful of carefully selected young men played the role of bait. When Fetterman and his troops arrived, the young warriors fired some shots and then fled to a valley of confusing draws, ridges, and hiding places.

The aggressive commander took the bait, and as his troopers caught up with the young Indians, approximately 800–1,000 Sioux and Cheyenne swarmed from their hiding places and fell upon the soldiers. They killed the woodcutters and wiped out Fetterman's command. This humiliating defeat and ongoing fighting wore down the will of the U.S. government to continue the campaign, culminating in an agreement to abandon the Bozeman Road and the three forts. Once they were abandoned and destroyed by his warriors, Chief Red Cloud traveled to Fort Laramie to negotiate and sign the Fort Laramie Treaty of 1868. It guaranteed the Sioux and Northern Cheyenne the western half of South Dakota, including the game-rich and spiritually significant Black Hills. Also stipulated in the treaty was the right of the Indians to continue to use and hunt in the land of northeastern Wyoming they had fought so fiercely to protect.

Figure 7.2 The ridgeline under the perspective of the photographer west toward the ravines below is the site of the December 21, 1866, Fetterman Massacre. In this battle during the Bozeman Road War, Sioux and Northern Cheyenne Indians tricked Captain Fetterman into ordering a pursuit of several decoys, including a very young Crazy Horse. They then sprung the trap and thousands of Indians rode and ran out of the ravines and draws, wiping out Fetterman's command. The Sioux and Northern Cheyenne eventually won this war, and the Bozeman Road and three U.S. forts were abandoned. The wooded and well watered land with its profusion of game that was so essential to the persistence of Indian culture is clearly observed in this photo. Stockmen would contest this same terrain with small ranchers and cowboys in the Johnson County Range War of 1892. Courtesy of the author.

Across the West, the federal government pursued a middle position with Indians for several years, offering limited annuities (yearly payments of food supplies) to gain access to land. Unfortunately, the resources provided by the government under treaty agreements were generally so limited that Indians could not replace lost game and other traditional food sources with the flour, corn, and cattle provided by the government. They left the reservation to fend off starvation. The government also used force in response to uprisings, resistance, and the panic of western settlers demanding government action. The appointment of General Philip Sheridan as commander of the Department of the Missouri marks a shift to increased use of force and even total war against the western Indians. Sheridan's approach to Indian campaigning reflected the methods of war he had developed during the Civil War in the Western campaign and in the Shenandoah Valley. General William Tecumseh Sherman, his former commander during the Civil War and now the Commander of the Army, supported this strategy. During the last conflict, both Sherman and Sheridan had targeted food production in an effort to weaken the Confederacy. Now, they launched a similar strategy against western Indians. As long as bison existed, some Indians would leave the reservation to hunt them out of necessity, to experience temporary freedom, and to sustain the sacred patterns of their old lives. Upon leaving the reservation they became an audience and potential recruits for non-reservation leaders such as Sitting Bull and Crazy Horse preaching resistance. For Sheridan and Sherman it was a simple formula: no bison, no resistance. Army garrisons provided free bullets to bison hunters and offered them protection from Indian raids where needed. When the Texas legislature debated a bill to preserve the remnants of the southern plains bison herd, Sheridan testified against it. Similarly, federal legislation to check the destruction of the bison in 1874 was vetoed by President Grant.

Gold was discovered in the Black Hills by a government expedition led by Colonel Custer in 1874. Miners, merchants, and thousands of others flooded into the sacred land of the Sioux and immediately cut into this key resource base. Instead of enforcing the federal treaty agreed upon by the U.S. government and the Sioux and Cheyenne in 1868, the Grant administration decided to reduce the reservation by removing the Black Hills and other

Figure 7.3 While critical of the Bureau of Indian Affairs' management of reservations, Colonel George Custer also firmly believed that native peoples in the West needed to be restrained on reservations and that western settlement would speed that process. To that end and to support American growth, he hoped to discover gold and led an expedition into the Black Hills of South Dakota in 1874. Sacred land to the Sioux, these mountains were part of the Great Sioux Reservation at the time of the military expedition. Custer's foray did find gold and triggered a rush of miners and merchants onto the reservation. These events inflamed American and Indian conflict and contributed to the decision to reduce the reservation and force resistant Sioux and Northern Cheyenne onto the reservation, leading to Custer's own demise at the Battle of the Little Bighorn in 1876. National Archives.

territory while also compelling resistant Indians living off-reservation in the Powder River and Tongue River country to return to government supervision. These actions by miners, settlers, the U.S. government, and the military increased the militant mood of a large part of the Sioux population, stiffening the resolve of leaders such as Sitting Bull and Crazy Horse to step up their opposition to the U.S. government, military, and reservation.

The army's campaign, led by General Terry, was supposed to be a three-pronged attack into the Powder River and Tongue River country in the heart of winter. Political in-fighting and slow preparation delayed the campaign until summer, drastically changing the environmental milieu for fighting with the advantage shifting to the Sioux and Northern Cheyenne. Army horses were bigger, stronger, and faster and could travel further in a day than smaller Indian horses. However, they required a steady diet of grain. In the summer, native peoples held the advantage because their horses performed well on a diet of plains grasses. In the winter, however, the U.S. cavalry gained the upper hand because grass was buried under snow or was dead, and the Indians did not have grain. Therefore, their horses were weaker and could only cover short distances. The U.S. cavalry gained the speed and mobility advantage. However, the army failed to capitalize on this tactical edge. By the time the campaign began, the snow was gone, and the grass of early summer was green and lush; the environment supported Indian warfare. The Indians' determination to protect the grasslands and their way of life combined with the failure of the military to move expeditiously when it had the advantage in winter set in motion the defeat of Custer at the Greasy Grass.

The Battle of Little Bighorn and the slaughter of Custer's command is the greatest failure of American troops in the West and the greatest success of Indians seeking to stop the destruction of their way of life. Of course, in the months following this battle, American forces ran the resistant Indians to the ground, killing leaders and numerous Indians in several battles and forcing the rest onto reservations. This was but one of many battles fought by Indians to protect their way of life. In the Southwest the Apaches struggled tenaciously to maintain their independence through the 1880s. A year after the embarrassing defeat of U.S. troops in Montana, Chief Joseph and the Nez Perce conducting a stunning fighting retreat over a period of four months. Protecting approximately 500 women, children, and elders, 250 Nez Perce warriors defeated U.S. troops on numerous occasions as they traveled 1,200 miles from northeastern Oregon through Idaho and into northern Montana. This fighting retreat, ending in surrender and relocation to reservations for the Nez Perce, was also the result of the violation of a federal treaty by American settlers and the U.S. government and the desire of Indians to remain on their own, beloved land.

The End of the Bison

Indians of the southern and northern Plains continued their commercial harvest of bison to the bitter end, desperate for the powder, bullets, guns, and other goods they needed to protect their control of the land and access to the very animals they were helping drive into such precipitous decline. Vast horse herds hastened this collapse as they consumed the grass the bison also needed.

The primary responsibility for the near extinction of this iconic western animal is borne by the market hunters and the commercial market. Buffalo Bill Cody, for example, built his first significant income on an edifice of bison flesh, harvesting the animal by the thousands to feed soldiers, railroad workers, and settlers. The plains was scattered with numerous other Buffalo Bills, Bobs, and Mikes, earning their monikers through the prodigious slaughter of bison. While many of these men hunted for military posts, reservations (ironically), and railroad construction crews, the majority of market hunters harvested hides for eastern markets, replicating, albeit much more quickly, the patterns of the deerskin industry over a century earlier. The perfecting of a chemical tanning process in Philadelphia in 1870 and the arrival of railroads on the plains at the same time greatly reduced the labor involved in harvesting hides. No longer needing to spend hours stripping meat and fat from the hides, tanning them, and lightening them for transport by horse or wagon, hunters could now harvest substantially more bison in a shorter amount of time.

Market hunting was a model of simplicity. The shooter mounted his rifle on a tripod then carefully picked off one animal after another, staying downwind to keep his scent from the herd and hoping to not spook them and cause a stampede. If the herd didn't panic, a market hunter could easily drop dozens or more per day. In one case a man killed at least 112 bison from one spot in approximately 45 minutes. Some of these hunters went deaf from the loud and constant report of their rifles. Two men handled the skinning while

Figure 7.4 This illustration from *Harpers Weekly* shows a bison hide and bone processing center. Native peoples had traditionally relied on the meat of the bison for survival, but the flesh of this animal was largely wasted in the market hunting economy. Although many bison hunters made their living killing the animal to supply meat to military posts and Indian reservations, the vast majority of bison were slaughtered for their hides and the meat left to feed wolves, eagles, ravens, and to rot. The market economy's need was hides for industrial uses and bones to be ground and used as phosphorus fertilizer for worn out soils in the Eastern and Midwestern agricultural economy. Library of Congress.

another cooked and stretched skins back in camp. The meat was invariably left to rot. An army officer described the landscape of slaughter.

> Where there were myriads of buffalo the year before, there were now myriads of carcasses. The air was foul with a sickening stench, and the vast plain, which only a short twelve months before teemed with animal life, was a dead, solitary, putrid desert.[7]

An English traveler, viewing the cunning of killing methods along the Arkansas River wrote,

> for some thirty or forty miles along the north bank of the Arkansas River . . . there was a continual line of putrescent carcasses, so that the air was rendered pestilential and offensive to the last degree. The hunters had formed a line of camps along the bank of the river and had shot down the buffalo, night and morning, as they came to drink.[8]

With this rate of killing, it was inevitable that hides would go to waste. Approximately 3–5 hides were left to rot on the carcasses for each one that actually reached market. The southern herd was wiped out by 1878 and the northern herd by 1883. In the summer of that year, many hunters journeyed onto the plains for another round of slaughter, finding nothing but grass, antelope, and bleached bones stripped clean by wolves, ravens, and the scouring wind. The last remnants of the bison were finally used as bones were collected and stacked in massive piles by railroad tracks, shipped east, and then ground up to serve as phosphate rich fertilizer for depleted soils in the Midwest and East. Even as the once abundant bison herds of millions were reduced to a handful of survivors, the bones were used to try and restore farmlands.

With the extinction of the bison drawing near, Congress passed legislation in 1874 banning the killing of female bison by non-Indians and also prohibiting squandering the animals' flesh. President Grant's Secretary of the Interior stridently opposed this legislation, writing,

> I would not seriously regret the total disappearance of the buffalo from our western prairies, in its effect upon the Indians. I would regard it rather as a means of hastening their sense of dependence upon the products of the soil and their own labors.[9]

This belief that the extermination of the bison would help confine Indians to the reservation and force their cultural and economic assimilation was shared by Generals Sherman and Sheridan, most members of the Grant administration, and most importantly by President Grant himself. He used a pocket veto to block what would have been revolutionary federal legislation; the extermination continued.

Range Wars and Environment

Movies like *The Virginian* and *Shane* as well as western novels by Zane Gray, Owen Wister, and Louis L'Amour imprinted in Americans' minds a romantic view of range wars with culminating quick draw gunfights as the climax of the narrative. In these depictions of violence in the contestation of land between varying groups, popular culture makes no reference to the ecological dimensions. One of the clearest examples of range war as environmental conflict is the Johnson County Range War of 1892 in the Powder River region of Wyoming. There was also the Tonto Basin Range War in Arizona as well as the lynching of numerous Hispanic sheepherders across the West. This conflict was similar to Bacon's Rebellion in some respects. It pitted well-funded stockmen of the upper class against small ranchers, cowboys, and homesteaders of the lower class. Like the colonial revolt, the poorer citizens sought access to land to improve their own social and economic status and found this stymied by the powerful Wyoming stockmen and their supporters. The major difference here is that no one was seeking the destruction of the Indians for their land; they had already been defeated and removed. Second, the lower class did not revolt but found themselves targeted by ranching elites seeking to retain control of public resources.

Once the Sioux and Northern Cheyenne were removed from this region by force, the near extermination of the bison left a rich grass resource where cattle operations could flourish. Well-financed stockmen quickly moved to set up massive ranches. Capital for most of these operations migrated from Great Britain, Scotland, and from the Northeast and Chicago. In the cattle industry literature of that era, authors sometimes promised profits up to 35 percent a year. One promoter wrote,

> since it was very well known that the cattle business is very safe, and that the larger capital therein the greater the ratio of profit, as the expenses do not increase at the same rate with the capital, the cattle-raiser has found no difficulty in borrowing an amount of money equal to his investment; and, though he might pay from 10 to 15 per cent annum for the use of it, yet he might fairly expect to realize therefrom 25 to 35 per cent.[10]

The author, not a rancher himself but a Prussian who loved the plains grasslands, predicted that $250,000 invested in a ranch, following his plan, would double in value within five years. The central irony was that even as he wrote this book, profits were plummeting, and its publication in 1885 corresponded with a collapse in beef prices. This was a consequence of overstocking and overgrazing of the grasslands that would also contribute to the massive die-off of cattle in the brutal winter of 1886–1887. In short, the bubble had already burst.

The large operations used the public lands as if it belonged to them while claiming or purchasing (both legally and fraudulently) the water essential to ranching in an arid environment. This effort to grab land illegally and the overuse of public lands is somewhat understandable because under the terms of the Homestead Act the maximum an individual with a family could obtain was 160 acres (up to 320 acres with the passage of the Timber Culture Act in 1873). This was not nearly enough for a small ranching operation in this region, much less a grandiose operation of thousands of head. The millions of acres of public land used by ranches throughout the West was considered "accustomed range," meaning that prior use denoted de facto ownership. Attempting to own public land, these stockmen and ranchers therefore demonstrated little tolerance for other small ranchers, cowboys, or the homesteaders who began moving into the area in droves in the 1880s.

With the arrival of increasing numbers of poor cowboys and small ranchers, tensions over the uses of public land in Wyoming grew quickly. The passing of a maverick law in 1884 to restrict the branding of weaned calves (known as mavericks) by cowboys or cattlemen not recognized by the Wyoming Stock Growers Association (WSGA) created simmering tensions and escalating conflict for the following eight years. The reduction of cowboys' wages, blacklisting of hands who ran their own cattle, and charging for meals for out-of-work cowboys drove the wedge further between stockmen and small operators. Stockmen also built enmity by ordering their hands to drive cattle across homesteaders' legally owned land, knocking down their fences, consuming needed grass, and churning their crops and gardens into the ground.

The brutal winter of 1886–1887 is often cited as the reason for the end of open-range ranching, but it was the practices of the stockmen that exacerbated losses. Years of overloading the range in pursuit of maximum profits while overgrazing the grass meant that the cattle facing that harsh winter entered that season already weak, malnourished, and barely ready for the devastating cold and lack of food. The cold and snow that winter was unrelenting with arctic fronts dropping the temperature to 46 degrees below zero in places. Unlike bison, cattle walk away from storms, prolonging their exposure to the elements. Cattle piled up on fences and froze standing up. Ravines and draws filled with cattle piled several feet deep, with the horror of hundreds or thousands of dead cattle not revealed until the Chinooks melted the blanket of snow concealing them. Many Wyoming and Montana residents approached insanity listening to the lowing of desperate cattle round the clock, milling about towns, eating frozen garbage, and the bark from log cabins as well as the tarpaper from the sides of shacks. When the warm weather finally arrived and the snow washed away, the landscape was littered with hundreds of thousands of dead cattle. Observers as far south as St. Louis saw thousands of dead cattle, day after day, bobbing downstream on the Missouri River. Some ranches suffered 100 percent losses, and overall, the approximate loss of cattle on the northern plains was 30 percent. It is difficult to provide precise numbers because of the failure of many ranches to keep close counts of their wide-ranging herds and due to the fact that some large operations concealed earlier losses and sales from investors by including them in this horrific winter kill.

With the diminishment of the massive herds, more public land appeared open to small ranchers and homesteaders, with a larger influx of those groups into the Powder River region after the winter of 1886–1887. That competition joined with ongoing financial difficulties for the stockmen as well as continuing and increasing tensions from the abuses by the WSGA led to a series of lynchings organized by the wealthy cattle operations and finally an extermination campaign known as the Johnson County Range War, occurring in April 1892. In this event, a group of 55 men, including some powerful stockmen and their foremen, 7 U.S. Marshal deputies, and 22 hired gunmen from East Texas, invaded the Powder River country. Each of the invaders was armed with new pistols, a rifle, and 50 rounds for each. A support wagon carried surgeon's supplies and two trunks of dynamite. Their mission was to execute 70 people named on a "death list." This list included the Mayor of Buffalo, the sheriff, the county commissioners, the editor of the town newspaper, the leading merchant in town, and several other legitimate small ranchers, cowboys, and homesteaders who were both economic and political opponents of the powerful stockmen. After killing two men, neither of whom was a rustler, this invading army was surrounded by local defenders numbering close to 400 men, rising up to defend themselves against elites' efforts to preserve their illegal use of public lands. This invading army might have been exterminated if not for the intervention of the U.S. military. The governor of Wyoming, the two U.S. senators, and leading businessmen helped organize and fund this invasion and got the vigilantes off the hook and acquitted once they were defeated.

The consequence of these events was essentially twofold. The days of the sprawling ranch of hundreds of thousands of acres largely came to an end, and smaller cattle operations prevailed. Small ranchers put more time into haying to provide fodder for the winter, building corrals and barns, and making other improvements to protect their herds. They learned to run smaller herds over the land and fenced private land and grass. Moreover, many of the ranchers diversified their economic activities to be more resilient. The other consequence was the introduction of sheep. The massive kill-off of cattle created an opening for sheep who could find food on an overgrazed landscape where cattle might starve. Where the "hooved locusts" scoured the terrain, the deterioration of the land continued.

It is of no small significance that battle over battle ensued over this landscape. The Sioux and Northern Cheyenne fought to protect their access to this very ecosystem and lost; the stockmen tried to preserve their control while small ranchers, cowboys, and homesteaders fought back fiercely to access land guaranteed them by a federal government that had removed the Indians. The fierceness of these conflicts reminds us that land and resources are the foundation upon which cultures and societies are built, and something as simple as access to grass can trigger the bloodiest of conflicts.

The Gold Age

Gold attracted the first Virginia colonists and their investors, and although they failed to find the lucrative mineral, efforts to find and profit from it figure prominently through American history. The story best known to and mythologized by Americans is the California Gold Rush of 1849. Fortunately for Americans, the nation is blessed with numerous, rich gold deposits. In the West, the ore would be found in the Front Range of Colorado, the mountains of Nevada, the Bitteroots of Idaho, the Rockies of Montana, the Yukon, and many other places. The gold fields, mines, and gold rush towns caused radical environmental changes in the surrounding countryside and downstream and triggered economic, cultural, and environmental changes in nearby and distant communities. These gold economies served as the basis of local development and helped finance the rise of the American economy as currency production and economic growth were tied to the ability to find and control even more quantities of gold. Extracting gold and building communities to profit from that process also further diminished native peoples and removed even more land from their control.

Tens of thousands of Americans traveled by trails or by ship to get to the gold fields as quickly as possible. Chileans, Mexicans, Chinese, and others also migrated in the boom year of 1849 and after. By the mid-1850s every nation on earth except Russia and Japan had representatives working the sluices or mines, cleaning hotels, running restaurants, offering sexual services, and trying to make their way in this economy where one might achieve great wealth but rarely actually accomplished that feat. Chinese referred to California as "Gold Mountain" and migrated with the hope of making money to take back to China to pay taxes or purchase land. Free blacks migrated to California for economic opportunities and refuge from persecution in other places; slaves were brought to work in the gold fields as well.

The gold rush devastated the native peoples of California. Indians participated in the gold economy in the very early stages but were poorly paid and then pushed aside. Although the state banned black slavery, making it a free state, the California government at the same time passed a law making it legal to enslave Indians. A thriving trade in the kidnapping and sale of Indians quickly developed. Indians were forced off their rivers and away from their villages as river ecosystems were destroyed, forests cut down, and game hunted by miners and hunters selling meat to miners. In some extreme cases, Indians were hunted and killed by newly arrived Americans. In addition to these stressors, the influx of different peoples and pathogens from around the world launched multiple epidemics of disease among California Indians. The population of natives in California in the mid-1840s has been estimated at 150,000. By the late 1850s their numbers were reduced to approximately 16,000. This destruction, like the impact of gold mining on the environment, is just one of the many costs of gold mining not accounted for in the stories America tells about itself.

Indians were not the only ones targeted for removal. When the easy profits were gone, American born miners' resentment grew at the non-citizens working "their" gold fields. They lobbied for and got a foreigner miners' tax of $20 a month to limit foreign participation in this economy. While some left, as did many Americans once the easy gold was gone, many others persisted. The Chinese managed to pay these unfair taxes and make profits while working already mined and abandoned gold claims.

The iconic image of the gold miner is of a man stooped over a pan, standing knee-deep in a cold, mountain stream. Scooping up some gravel and sand and using water to swirl it in the pan, shaking out the debris, the miner hoped to find heavier "color," flakes or nuggets in the bottom of the pan. The typical miner put in 10 to 12 hours a day of this exhausting, numbing work. Typically, miners worked the rivers and riverbeds quite heavily, and panning was quickly abandoned as too slow. While rockers were an improvement over panning, the sluicing system represented a substantial increase in both resources used and the amount of gravel that could be worked in a day. Originally constructed with hollowed out logs then with planks, the sluices were typically 12 feet long, but it became common to string them together in a row to make sluices hundreds of feet in length. Miners shoveled gravel, dirt, and sand into the sluice and ran water through it. Riffles at the end of the sluice would catch most of the gold while the other material washed out. This water intensive system required a gamut of tools from simple troughs and pumps to diversions of creeks and miles-long aqueducts. The sluice system targeted not only the river bed itself, but because rivers and creeks meander over time, miners flushed out dry riverbed and banks as well. In some instances the miners moved rivers entirely to access the riverbed. Building a dam upstream they would channel the water a different direction then work the now dry riverbed for its gold.

The environmental impacts of this type of mining were considerable. The panning, digging, and sluicing of riverbeds disturbed and destroyed river habitats for insects, fish, birds, amphibians, and others while extending damage up the food chain to species dependent on those lost in destroyed and degraded rivers. Fish beds were destroyed, killing eggs and young fish. Water turbidity increased substantially, making it unhealthy for salmon and trout because of both the dirtiness of the water and increased water temperatures. Also, the digging of these gravels and their sluicing created an increased flow of sediments downstream from the goldfields, extending the damage sometimes hundreds of miles.

Sluices, troughs, and aqueducts for transporting water all required a heavy use of timber. That, and the consumption of trees for firewood and the construction of cabins in mountain environments, led to a rapid and complete deforestation of the areas surrounding gold fields. Naturally, this removal of trees led to even greater erosion, higher stream turbidity, and warmer water temperatures—both from turbidity and the removal of tree shading of streams and irregular water flow into streams and rivers. The removal of forests also eradicated habitat for game. Hunting to feed the miners generally depleted deer, squirrel, rabbit, bear, and elk populations as quickly as the trees could be removed. Market hunters worked distant forests and meadows to provide meat for miners, thereby extending the environmental impacts of the goldfields far beyond their own immediate areas.

Moving Mountains for Gold

While panning and sluicing created great wealth in the California gold fields and other areas, hydraulic mining was the most economically productive and environmentally destructive of gold extraction processes. It required a large initial outlay of capital for equipment such as pipes, nozzles, the construction of aqueducts, and much more. Hydraulic mining reduced the amount of labor required while giving the operation owners greater control over their environment and work regimen. By consolidating control of water and its consistent distribution to the hydraulic mining sites, they reduced their dependence on weather and stream hydrology and, as historian Andrew Isenberg argues, effectively systematized the removal of gold.[11]

The cost of controlling water was so exorbitant that external capital was required. However, businessmen hesitated to invest without infrastructure in place, so local mining companies worked fervently to store and move water. The projects were oftentimes rickety but also immense for that time and development of the western economy. One consisted of a ditch 18 miles long with a 1,000-foot long aqueduct carrying water across a deep ravine. On the Stanislaus River a company built a system that included four large dams, the largest reservoir with a capacity up to 200,000 cubic feet of water. As Chinese laborers became available in the 1860s and 1870s, grander projects were constructed in the gold fields. The Natoma Water Project contained 40 miles of ditches, and the Eureka Lake and Yuba Canal Company built a system that cost more than a million dollars with a carrying capacity of 1.13 million cubic feet of water. These improvements increased the infrastructure necessary for expanded hydraulic mining operations and elicited outside investment with 4 million pounds sterling in British investments alone between 1870–1873. This increased flow of capital into the region hastened the transition to full-scale hydraulic mining. The environmental impacts were nothing short of catastrophic.

The infrastructure of hydraulic mining was similar to placer mining but even more cost and resource intensive. Timber was needed for flumes, aqueducts, bridges, and other structures. This resulted in massive and rapid deforestation while forests were also lost to the rising waters of the reservoirs. The washing away of gravel, rock, and dirt flushed material downstream in stunning amounts. By one estimate, 885 million cubic yards was sluiced away in the California mountains from the mid-1850s to 1885. Several rivers were negatively effected, including the American, Bear, Feather, and others, not to mention countless streams and tributaries. An 1872 newspaper description of the Bear River captures the scale of downstream sediment deposits: "Tall pine trees, formerly far above the stream, have been gradually engulfed season after season, until now only the top branches appear above the current."[12] These heavy sediment flows raised river beds several feet, forcing water up over the banks onto farms and into estuaries in a series of unprecedented floods.

Once sparkling streams, fed and cooled by mountain snowmelt, were degraded into channels of slow-moving, muddy, warmer water. Flooding increased in intensity and occurrence with the debris filling up rivers and the fluctuations in flow made more extreme because of upstream deforestation. River waters escaped their banks and dumped a mixture of sand and gravel that lacked nitrogen and phosphorus and was also highly alkaline. Croplands and

Figure 7.5 Hydraulic mining replaced panning for gold and sluice mining. Requiring large amounts of capital for the construction of waterworks and the purchase of pumping and spraying equipment, hydraulic mining made the extraction of gold more efficient. As evidenced in this photo, it also caused enormous erosion with negative consequences for habitat and downstream farmers and fishermen. National Archives.

pasture were destroyed as a result—productive land lost to the unregulated and relentless pursuit of gold. Extracting an abundance of gold created a corresponding decline in fertile soil. As much as 40,000 acres of land valued at more than $2,500,000 was destroyed by this sediment on the Feather, Bear, and Yuba Rivers. In desperation, downstream communities built levees to protect themselves from these transformed rivers. According to Isenberg,

> By 1878, stretches of the Bear River's bed were actually higher than the surrounding countryside. Only the walls of the levee kept the waters of the river from overflowing. Levees, however, were expensive. According to one estimate, the complete system of levees on the Sacramento and its tributaries cost over $2 million, nearly 80 percent of the value of the farmland destroyed by tailings. Levees created new problems. Water seeped through them, and, with nowhere to drain, pooled alongside them. Salts leached to the surface of the saturated soil, salinating the topsoil and rendering it unfit for cultivation.[13]

Salmon were destroyed by hydraulic mining as well. California Fish Commissioners in 1878 estimated that half the salmon habitat in the state had been destroyed by this practice. The sediments covered river bottoms and salmon redds, suffocating the salmon inside and reducing the quality of streambeds for future reproduction. Also, the conversion of the rivers from clear and cold to slow and muddy meant increased turbidity, warming the water to levels unhealthy for salmon, and also causing the gills of salmon to be coated with dirt and mud, asphyxiating them. As if this weren't enough, the flow of sediments downstream also filled in many estuaries, which function as important spawning and smolt habitat. It is difficult if not impossible to calculate the economic benefits of the enterprise of hydraulic mining versus the external costs inflicted on the landscape and the other economies of that region. But it is clear that any discussion of wealth produced must be balanced with an analysis of the costs incurred both ecologically and on other economic activities. Gold is precious for any number of reasons, but in that era it was precious to Americans because it backed currency, and without more gold, economic growth was limited. The growth of the American economy based on destructive gold extraction, irrational as it may have been, was predicated once again on the wealth of nature.

With the spread of hydraulic mining to Idaho, Oregon, and other areas, the negative environmental impacts rippled across the West. With the loss of habitat and destruction of existing salmon (along with overharvest), the Sacramento Valley commercial salmon industry was essentially destroyed. Salmon cannery owners like the Hume brothers fled to the Columbia River in pursuit of other fish. Hydraulic mining caused severe economic losses in the northern California salmon industry and, in so doing, caused earlier, aggressive harvest of Columbia River and Puget Sound salmon.

Removing Western Forests

The gold rush triggered a logging rush, and great swathes of forests were laid low to provide lumber for mines, smelters, aqueducts, cabins, and town buildings as well as heat for those buildings. While mills sprang up all over northern California to serve these needs, the swift liquidation of forests drove prices for lumber in San Francisco skyward. This could be as much as $1,000 per thousand board feet in 1850. The price for pine at some sawmills in 2014 is still $1,000 per thousand board feet; in today's dollars the 1850 lumber was worth at least $30,000. No wonder then that lumbermen's gaze turned to Washington and Oregon for their untapped forests. Mead Simpson migrated from the Maine logging economy like many other early California and Pacific Northwest lumbermen, and although he originally sought gold in California, his experienced eye saw the potential logging profits. By early 1851 he was already exploring the northern reaches of the U.S. Pacific coast, and he promptly established mills in Astoria and Coos Bay, Oregon, as well as South Bend, Washington, by 1858. Operations at the mouth of the Umpqua River in Oregon and increased mill production kept his California lumberyards bustling. The Coos Bay mill alone, by 1868, required a fleet of 16 ships to haul lumber to the San Francisco yard. Historian William Robbins notes, "these isolated areas [became] extractive tributaries for the urban markets in California."[14] As the forests of the Atlantic shore appeared to Europeans from shorn over lands and contributed to the growth of the Spanish, French, and English states, so too California businessmen coveted the northwestern trees. This lumber supported the growth of the West coast economy and was a prized commodity in the growing Pacific trade system.

The great northwestern swale of green and the vast reservoirs of lumber convinced these experienced lumbermen and capitalists to seek out other markets. The Port Gamble mill in Washington began shipping lumber to Hawaii by 1856, and the Puget Mill Company quickly supplied not only California and Hawaii but Australia and South America, too. Resentments arose against Californian control of the economy. One editor referred to the California lumber magnates as "evil birds of omen,"[15] remonstrating that they controlled industry and ripped off workers to enrich themselves. Noting these concerns, historian Thomas R. Cox argues in response that the arrival of capital, jobs, and workers in these isolated logging communities modernized the economy more quickly than would have been possible with the agricultural economy of the region at that time.

The northwestern logging industry jumped in size with the arrival of the Northern Pacific transcontinental railroad and the purchase of 900,000 acres of land from James J. Hill by Frederick Weyerhauser in 1893. Titans of the modern railroad and logging industry as well as neighbors in the upscale Summit Avenue neighborhood overlooking St. Paul, Minnesota, Hill and the lumbermen negotiated one of the most grandiose private land deals in American history. Weyerhauser had risen to great wealth logging the forests of Minnesota and Wisconsin. Having contributed to the destruction of those vast swathes of trees, he needed new forests. The 900,000 acre purchase at 6 dollars an acre was a great value. While some lumbermen were originally incredulous at the huge purchase, they realized the logic of migration into the Northwest and swiftly followed Weyerhauser's lead. There was then, in the early 20th century, a rush of capital and established logging operations from other regions into Pacific Northwest forests. Consequently, the rate of harvest escalated dramatically. In the period from 1898–1914 timber harvest and milling increased threefold. Of course Weyerhauser was the grandee among them, owning 26 percent of the all the timberlands of Washington and almost 20 percent of Oregon's.

The removal of the forests, while providing short-term wealth and jobs, resulted in a wide variety of environmental problems—some of them short-term, others persistent. Early logging occurred along the shores of Puget Sound, Washington, and Coos Bay,

Oregon, but as the forests within easy reach of water fell and technology improved, loggers were able to extend both their reach into the mountains and the speed of tree removal. The immediate impact was the destruction of forest habitat for a variety of species, including grizzly bear, black bear, wolves, cougars, owls, bald eagles, ravens, and numerous other animals and birds, triggering a dramatic decline in their numbers. Of course, this very process had occurred on the Atlantic seaboard in the colonial period and then across the nation over the course of three centuries. What is amazing is how little changed regarding the liquidation of nature. The annihilation of massive stands of forest regardless of environmental consequences remained standard business policy with little dissent. In fact, an American ideology of abundance and constant economic growth required it. The only real change over three centuries was the speed and efficiency with which the forests' were harvested. The continued abundance of natural resources made available as new lands were settled and exploited meant that the discussion of stewardship and conservation could continue to be ignored as nature continued to undergird the growth of American wealth.

As forests were cut, trees were dropped directly into salmon-rich streams, either for transport or from simple thoughtlessness. No efforts were made to preserve trees alongside streams as has become standard practice since the late 20th century. This meant that salmon streams were subject to increased erosion with no tree roots to hold the soil or branches to break the fall of rain on the land. The flow of soils and sands into the water meant the suffocation of salmon redds, increased turbidity, and changes to the hydrological cycles of the rivers, which was detrimental to fish who had evolved to these stream patterns over millenia. Mudslides closed off rivers and tributary streams as well. Once logging moved from the shorelines, corduroy roads and wooden chutes enabled loggers to transport lumber to the water where it would be hauled to sawmills. Corduroy roads were simply smaller trees cut and laid parallel to each other to create a road of sorts that oxen and donkeys could drag logs down. In some cases, dogfish were pressed for their oil, which was used as a lubricant on these "skid roads." Flume systems also consumed a great deal of wood. Splash dams were built on streams in the foothills and mountains to store water in a small reservoir. When enough logs were cut and assembled at the flumes for transport, the dam was breached to release the stored water and float the lumber downhill.

Still, the lumbermen moved further inland from the shore. As they moved up foothills and mountains and the gradients steepened as deep river canyons cut the landscape, narrow gauge railroads proved instrumental to continued lumber extraction. With their shorter distance between rails and the ability to use smaller locomotives, the narrow gauge rails could be laid and traveled at steeper inclines and better follow the twists and turns of ridges, hills, and canyons then could standard gauge rails. What had been inaccessible to corduroy roads, oxen and mules, flumes, and rivers were now opened to loggers, and the forests fell up the sides of the mountains.

Engines, in fact, industrialized logging as the steam donkey and high-lead logging were developed alongside the rails. The steam donkey, a steam-powered machine, replaced animal labor in moving logs to shipping sites. Long cables ran from the machine down steep slopes to creek and river bottoms and up hills and slopes. A chokesetter wrapped the cable around the log, a whistlepunk blew a whistle to signal to pull it up, and sliding, bucking, and churning and tearing up the landscape in troughs and tears, the downed tree was pulled to

Figure 7.6 Because of mild winter temperatures and rugged terrain, rivers were used regularly in the West for moving logs downstream to mills and or join rafts to be towed to mills by boat. In order to have enough water to float logs, splash dams were built on streams to store water. When enough logs were ready for market, water would be released from behind the splash dams to rush downstream and carry the lumber with it. An efficient response to the challenges of terrain, this system was also ecologically devastating. Courtesy of the Oregon Historical Society.

the steam donkey and railroad. The use of high-lead logging extended the reach of steam donkeys further as tall trees were turned into spars in the tops of which were placed pulleys to run cables through.

This efficient extraction of lumber wealth led to ecological chaos, costs that were not captured in the price of the commodity but absorbed and born by nature and some communities. Logs tore ruts and troughs into the ground, smashed underbrush and grass, and flattened smaller trees. The exposed soil eroded much more quickly when the rains came. The loss of soil was bad for the forest ecosystem and also damaged salmon and trout populations with warmer streams, increased turbidity, choked nests, and even blocked streams and rivers. The impact on fish rippled out to other species dependent on salmon such as bear, wolves, raccoons, ravens, eagles, and even orca. The dependence on distant and foreign markets, fluctuating prices, and the heavy debt and fixed costs of many operations created a cut at all costs mentality. The need to service debt, make payroll, and feed employees meant that stands of trees were leveled even as prices plummeted, driving prices down further by contributing to the glut while destroying forest ecosystems and removing the resource from the local community.

Figure 7.7 Logging operations used steam donkeys to haul logs to railroad tracks for shipment to mills. The steam donkey can be seen in the back of this picture. High-lead logging incorporates spars (former tall trees used as poles) with pulleys on top to run long cables through. It extended the reach of logging operations down and up steep slopes. Use of extraction systems liked this quickened the pace of clearing a forest, and dragging the logs up and down slopes created a great deal of damage to the environment. A close observation of the photograph reveals the nature of the ecological deterioration. Courtesy of the University of Washington Libraries, Special Collections.

Rivers of Wealth

The Hume brothers left the destroyed Kennebec River fisheries and joined the frenzy of salmon harvesting on California's Sacramento River in the 1850s. While managing their cannery in California, they experimented with methods for preserving and canning salmon, dramatically improving a process that created a revolution in the salmon industry, making it possible to ship the protein and calorie rich fish around the world while also preserving better tasting, less salty salmon; this western commodity quickly became an internationally popular food. The overharvest of these fish, the conversion of wetlands to farmlands, and the ongoing downstream flow of rocks, gravel, and sand from hydraulic and other forms of mining in the Sierra Mountains drove salmon to the edge of collapse within a few short years on the rivers of northern California. With this diminishment of the salmon resource, the Hume brothers led the salmon merchants' migration north to the sustainably used Pacific Northwest salmon runs. In 1866 they opened a small cannery on the lower Columbia River and in that first year produced 6,000 cases, each holding 48 1-pound cans of Chinook salmon, for a total value of $64,000. Canned salmon of the Columbia River and other rivers of the Pacific Northwest and, later, Alaska was shipped worldwide, becoming a popular source of cheap protein in industrial cities on the East Coast, in Australia, and in Europe and becoming standard rations for British troops in India. Like lumber, canned salmon was immediately shipped and sold through a vast, global economy. The abundance of Northwest rivers, sustained by native peoples for thousands of years, now fed factory workers in the dingy cities of Industrial England and British troops on the subcontinent. Cans of fatty, protein- and calorie-rich salmon subsidized British colonialism and the industrial revolution and provided great wealth for those who built the canneries and jobs for numerous Swedish, Norwegian, Japanese, and Chinese immigrant laborers.

A boisterous and thriving salmon economy developed with hundreds of canneries sprawling throughout the Northwest—50 on the banks of the Columbia and its tributaries alone within a few years. The increased salmon harvests on the Columbia alone was astounding. The Hume Brothers' facility canned 272,000 pounds of salmon in 1866. A few years later the number had increased to five canneries processing 10 million pounds, and by 1875 14 canneries had multiplied the harvest to an impressive 25 million pounds of canned salmon. The rich and easily accessed resource drew in more businessmen and capital so that by 1884 there were 37 canneries canning 42 million pounds of salmon on the Columbia. Busy fishermen and cannery workers had no way to know this would mark the peak of the salmon harvest on this river, with dramatic declines in following years to half that harvest in 1889. Concerned by the precipitous drop off, legislators in Oregon and the Washington Territorial Government sought to pass management laws stipulating closures and creating no harvest days, but with no staff to enforce them, these laws were ignored. As historian Carlos Schwantes has noted,

> cutthroat competition and unsound practices, like stringing traps and nets across river mouths so that few adult fish survived to spawn, characterized commercial fishing almost from the beginning. At times such an armada of fishing craft jammed the lower Columbia that it seemed a person could walk across the river on their decks.[16]

The same free-market economic practices with no stewardship or restraint that devastated forests and farmlands, both from logging and the mining economy, now wrought havoc on Pacific Northwest salmon runs.

There were a number of ways to harvest salmon. The most common method was drift gill nets, with 60 percent of the harvest caught by this means. Seine netting was popular as well, with some nets as long as 2,400 feet and requiring horses to pull in the weight of thousands of pounds of squirming fish.

In the search for ever more efficient ways to harvest resources and reduce labor costs, salmon harvesters put the first fish wheel into use in 1879. A boat with a wheel on the back, typically, it would be anchored in a narrow or fast section of a river or along a bank where fish swim. The wheel would spin with the flow of the current, scooping up fish and dumping them into the boat to be collected later. This automatic system could be quite effective,

Figure 7.8 In the effort to harvest as many salmon as possible on the Columbia River and in Washington's Puget Sound, salmon wheels were invented. A model of simple genius, simply scooping up fish for tenders to haul to canneries, they also generated great waste and were eventually banned. Courtesy of the University of Washington Libraries, Special Collections.

with one fish wheel in 1906 gathering up 417,855 pounds of salmon. By 1899 there were 76 fish wheels on the Columbia River. Lackadaisical observation and unloading of the haul meant a great deal of waste as fish died, rotted, and slipped off the sides of growing piles on these contrivances. At the behest of gillnet fishermen and also because of concern at the profligate waste, fish wheels were finally banned, first in Oregon in 1926 and then in Washington in 1934.

Waste was not a problem peculiar to fish wheels; it was symptomatic of the entire industry. If a fisherman came in with a fresh load of salmon but the cannery floors were already stacked with fish, he would have no choice but to dump the fish and then immediately head out to harvest more salmon. Waste accounted for 30 to 40 percent of the harvest at some canneries. Approximately 7 million pounds of salmon were discarded by Columbia River canneries in 1895 alone. It was essential to always maintain a ready supply of fish in order to keep the 200 to 400 workers in the facility occupied. Cannery owners had fixed costs to serve, the pay and provision of the laborers living on site in most cases, the debt for machinery and building construction, and of course the expectations of investors. Therefore, it was better to get whatever value you could from fish, even if systematically harvesting them without limitations led to gluts that drove prices down, sometimes as low as two or three cents a pound.

The problem driving this waste and destruction was the fact that there was no reward for restraint. This fact drives what ecologist Garett Hardin termed the "tragedy of the commons."[17] Because salmon were a public resource, not constrained by political borders or contained within the boundaries of private land, management of the resource was quite difficult. Moreover, because the salmon were not owned by any one individual or business,

Figure 7.9 Because native-born Anglo Americans did not want to work fish canneries, businessmen were compelled to hire Indian women and increasingly Asian labor. Chinese men were recruited to work in canneries from China. Once Chinese immigration was banned with the Alien Exclusion Act of 1882 and its reapproval in 1892, cannery owners hired Japanese workers. With the Gentlemen's Agreement of 1907 stopping the flow of Japanese labor, cannery owners then turned to Filipino laborers. Like with the Norwegians and Scandinavians working the fishing boats, the salmon industry led to an increasingly diverse northwestern population as people of different ethnicities and nationalities immigrated for economic opportunities. Asian cannery workers labored under difficult conditions, slicing up fish, gutting them, and canning them while maintaining their balance standing on a thick layer of blood, guts, and fish scales. Courtesy of the Museum of History and Industry, Seattle, Washington.

like cattle or private forests, there was no incentive to harvest less and no benefit to practicing any form of stewardship. If one person or business moderated their exploitation of the resource, someone else would harvest it instead. Similarly, if the salmon were depleted in one location, fishermen could simply go further and deeper in pursuit of the valuable fish. The abundance of the resource suggested an endless wealth of fish, but many of the owners of canneries and even fishermen had participated in earlier fishery economies in Maine and California and had contributed to and seen their rapid destruction. They knew full well there was no such thing as an infinite resource and how quickly the days of plenty could come to a screeching halt. These businessmen did not overharvest in innocent bliss of the consequences of their actions. The tragedy of the commons, joined with lack of regulation and combined with fixed costs, led to overharvests that created gluts and drove prices down further, leading to even more harvests as businessmen struggled to avoid bankruptcy. The result was the rapid decline of this once abundant resource in a short three decades.

By the end of the 19th century, the American West had been exploited, transformed, and degraded in numerous ways. Having fought to protect their land and culture, western Indians were contained on numerous barren reservations while the bison teetered on the edge of extinction. Much of the mountains had been washed to the farmlands below, and as the mountains were mined for gold, so too were the rivers for their salmon. Forests were laid low for the wealth contained in that wood. The aggressive pursuit of these valuable commodities drove local economies and related industries while also playing havoc on widespread ecosystems. Unquestionably these economies helped drive the expansion of American capitalism and contributed to general but inconsistent prosperity. Finally, as the 19th century drew to a close and the nation fully occupied and settled, the question of conservation began to gain greater credence as many Americans viewed the devastated landscape and considered the ecological and economic costs of their actions.

Document 7.1 Powell Report to Congress, *John Wesley Powell*, 1878

John Wesley Powell was a geologist and professor, but he is most famous for the three-month expedition in 1869 that he led down the Green and Colorado Rivers, mapping the rivers and learning about the western landscape along the way. He perceived the inherent limits of the arid West and proposed a new model of land use that offered a higher potential for success and preservation of land in the hands of the middle class than what actually occurred. In his *Report on the Lands of the Arid Region of the United States* (1878) to Congress, he argued that west of the 100th meridian the 160-acre homestead model should be scrapped in favor of large sections of land per farm family and land use and community organization policies reflecting Spanish and French practices in North America.

1. What problems does Powell anticipate in the West if the Homestead Act continues to be the basis of land distribution without careful planning and dispensation of resources to settlers? What steps does he recommend to avoid environmental and agricultural crises?
2. Does his model of land use and resource allocation represent a major departure from traditional practices to that point? In what ways and from what cultures does he borrow ideas?

Irrigable Lands

Within the Arid Region only a small portion of the country is irrigable. These irrigable tracts are lowlands lying along the streams. On the mountains and high plateaus forests are found at elevations so great that frequent summer frosts forbid the cultivation of the soil. Here are the natural timber lands of the Arid Region—an upper region set apart by nature for the growth of timber necessary to the mining, manufacturing, and agricultural industries of the country. Between the low irrigable lands and the elevated forested lands there are valleys, mesas, hills, and mountain slopes bearing grasses of greater or lesser value for pasturage purposes . . . In discussing the lands of the Arid Region, three great classes are recognized—the irrigable lands below, the forest lands above, and the pasturage lands between.

Advantages of Irrigation. There are two considerations that make irrigation attractive to the agriculturist. Crops thus cultivated are not subject to the vicissitudes of rainfall; the farmer fears no droughts; his labors are seldom interrupted and his crops rarely injured by storms. This immunity from drought and storm renders agricultural operations much more certain than in regions of greater humidity. Again, the water comes down from the mountains and plateaus freighted with fertilizing materials derived from the decaying vegetation and soils of the upper regions, which are spread by the flowing water over the cultivated lands . . . It may be anticipated that all the lands redeemed by irrigation in the Arid Region will be highly cultivated and abundantly productive, and agriculture will be but slightly subject to the vicissitudes of scant and excessive rainfall.

Cooperative Labor or Capital Necessary for the Development of Irrigation. Small streams can be taken out and distributed by individual enterprise, but cooperative labor or aggregated capital must be employed in taking out the larger streams.

The diversion of a large stream from its channel into a system of canals demands a large outlay of labor and material. To repay this all the waters so taken out must be used, and large tracts of land thus become dependent upon a single canal. It is manifest that a farmer depending upon his own labor cannot undertake this task . . . When farming is dependent upon larger streams such men are barred from these enterprises until cooperative labor can be organized or capital induced to assist . . .

In Utah Territory cooperative labor, under ecclesiastical organization, has been very successful. Outside of Utah there are but few instances where it has been tried; but at Greeley, in the State of Colorado, this system has been. . . .

The Farm Unit for Pasturage Lands. The grass is so scanty that the herdsman must have a large area for the support of his stock. In general a quarter section of land alone is of no value to him; the pasturage it affords is entirely inadequate to the wants of a herd that the poorest man needs for his support.

Four square miles may be considered as the minimum amount necessary for a pasturage farm, and a still greater amount is necessary for the larger part of the lands; that is, pasturage farms, to be of any practicable value, must be of at least 2,560 acres, and in many districts they must be much larger.

Farm Residences Should Be Grouped. These lands will maintain but a scanty population. The homes must necessarily be widely scattered from the fact that the farm unit must be large. That the inhabitants of these districts may have the benefits of the local social organizations of civilization—as schools, churches, etc., and the benefits of cooperation in the construction of roads, bridges, and other local improvements, it is essential that the residences should be grouped to the greatest possible extent. This may be practically accomplished by making the pasturage farms conform to topographic features in such manner as to give the greatest possible number of water fronts.

The great areas over which stock must roam to obtain subsistence usually prevents the practicability of fencing the lands. It will not pay to fence the pasturage fields, hence in many cases the lands must be occupied by herds roaming in common; for poor men cooperative pasturage is necessary, or communal regulations for the occupancy of the ground and for the division of the increase of the herds. Such communal regulations have already been devised in many parts of the country.

Document 7.2 Description of Powder River Country, *Captain Benjamin Franklin Rockafellow,* 1864

The following passage was written by Captain Rockafellow. He was a veteran of the Civil War, having been wounded at the Battle of the Wilderness in 1864 and served under General Patrick Connor in a campaign against the Sioux and Northern Cheyenne in the Powder River region. This military expedition was a failure. Rockafellow went on to serve in the Colorado legislature.

1. In reading this description of the landscape, what observations can you make about the native species and abundance of the region?
2. Thinking like an officer on the ground, what tactical problems do you see arising from the nature of the landscape described here?

We came to several fine flocks of Antelope. One flock of eight I flanked as they went around a hill. Though I dismounted and popped away at them with my Revolver, which the Genl [sic] is like p—g against the wind, did not get one. Just as I got through came up to three Buffalo Bulls. Genl killed one and I put three shots into another which Genl afterwards gave the finishing shot. My mare Julia [was] very much frightened by them at first. After this fine sport [we] followed down [the] dividing ridge on Maj. Bridger's old wagon trail made ten years ago. Genl decides to adopt route as his wagon road and we struck off to third ravine west of [Pumpkin] Buttes in search of water. Then [we] got separated, old Maj [Bridger] going up it while we went down. This was about noon and last we saw of him that day. When we found water [it] was late in [the] afternoon below forks and in main channel of Powder River which with exception of two water holes was as dry as a powder horn. This forenoon [I] saw [the] first snow I ever saw in Aug. Was on peaks of Big Horn Mountain far beyond Powder River range. [Pumpkin] Buttes are four mountains which lay N & S [north and south], are level apparently & the same height. Made me thing of telegraph letter a long mountain mark, a short one, one still longer and then one shorter still, the longest about sixty rods on top, shortest about 25 rods. Elevation above plain more than 200 feet. At [a] point below forks where we discovered water along steep bank on north side which is a ledge of coal situated much as below we discovered a herd of 11 Buffaloes 5 or 6 Bulls, 4 cows & two calves which were laying off to bluff southward from us. [Janise] went off to [the] left of them and got a shot at a calf. This started [the] herd and away we went after them over the bluff, my mare jumping water course after water course which were such sharp courses that [I] believe Julia jumped one fully 16 feet wide. We followed Buffalo close along [the] edge of [a] high bluff. Genl Connor went so much faster than we could that he flanked them and kept them there while we popped into them. Finally they made a break down the bluff and the Genl after them went clear around [the] point out of sight while we took after one we got detached. To escape he climbed an almost perpendicular bluff, receiving one shot from Lieut. Jewett which bled him freely. We, Lt. Jewett, Guide and myself went up [a] narrow watercourse which under other circumstances [I] would not thought of climbing. By this time the Buffalo got about sixty rods start on of us and we after him urging our horses over rather level course at [the] height of their speed. I got the lead and leaning [forward], was pressing my mare to utmost speed when she over reached and turning somer sault as quick as thought came down on my breast or chest. I thought it crushed in and was sure I could not live. Was rather pleased when found I could speak. I could not hardly move. My breast bone and right chest were badly bruised and my back seemed much jellified. They stopped the chase and Genl said [we] would get along to [a] spring so [he] helped me on my horse. I had to get off but trying it again got to the bottom. Unsaddled and Genl kindly furnished whisky from his flask to rub my body with and bade me take some inwardly. I could not lie down and [Janise] fixed me up well against a tree. Lt. Jewett got water and took care of Julia. We remained until [the] moon came up when [we] marched about miles up channel and camped in a nice grove or rather a cluster of cottonwood trees. [It] was about midnight before I could get in [a] horizontal position. Buffaloes [were] bellowing and tramping near camp and Genl ordered guard to keep them from stampeding our few horses. Were at least 25 miles from camp.

Document 7.3 Chinese and Salmon from *American Notes, Rudyard Kipling,* 1891

British writer and advocate of European Imperialism Rudyard Kipling traveled through the United States in the 1880s collecting observations and using these to write and publish a series of essays on his experience in *American Notes.* Written in a sometimes sarcastic style, with plenty of criticism for American society, there is still a clear appreciation and often awe for the abundance of nature and the beautiful landscape.

1. Evaluate Kipling's description of salmon in the various locations. To what degree is abundance central to his narrative? What does he say regarding the strength and beauty of salmon and its place in nature?
2. Compare the description of Chinese workers in the cannery and Kipling and "California's" own pursuit of salmon. Keep in mind that Kipling was a fervent imperialist and Anglo-Saxonist. Also, he was one of those concerned with what was called the "Yellow Peril," i.e. the rise of Asian populations and civilization and the supposed threat they presented to western societies. Which environment is more degrading? Which is more manly? How does use of the salmon resource denote or contribute to these characteristics?

The steamer halted at a rude wooden warehouse built on piles in a lonely reach of the river, and sent in the fish. I followed them up a scale-strewn, fishy incline that led to the cannery. The crazy building was quivering with the machinery on its floors, and a glittering bank of tin scraps twenty feet high showed where the waste was thrown after the cans had been punched.

Only Chinamen were employed on the work, and they looked like blood-besmeared yellow devils as they crossed the rifts of sunlight that lay upon the floor. When our consignment arrived, the rough wooden boxes broke of themselves as they were dumped down under a jet of water, and the salmon burst out in a stream of quicksilver. A Chinaman jerked up a twenty-pounder, beheaded and detailed it with two swift strokes of a knife, flicked out its internal arrangements with a third, and cast it into a blood-dyed tank. The headless fish leaped from under his hands as though they were facing a rapid. Other Chinamen pulled them from the vat and thrust them under a thing like a chaff-cutter, which, descending, hewed them into unseemly red gobbets fit for the can.

More Chinamen, with yellow, crooked fingers, jammed the stuff into the cans, which slid down some marvellous [sic] machine forthwith, soldering their own tops as they passed. Each can was hastily tested for flaws, and then sunk with a hundred companions into a vat of boiling water, there to be half cooked for a few minutes. The cans bulged slightly after the operation, and were therefore slidden along by the trolleyful to men with needles and soldering-irons who vented them and soldered the aperture. Except for the label, the "Finest Columbia Salmon" was ready for the market. I was impressed not so much with the speed of the manufacture as the character of the factory. Inside, on a floor ninety by forty, the most civilized and murderous of machinery. Outside, three footsteps, the thick-growing pines and the immense solitude of the hills. Our steamer only stayed twenty minutes at that place, but I counted two hundred and forty finished cans made from the catch of the previous night ere I left the slippery, blood-stained, scale-spangled, oily floors and the offal-smeared Chinamen.

. . . The next cast—ah, the pride of it, the regal splendor of it! the thrill that ran down from finger-tip to toe! Then the water boiled. He broke for the fly and got it. There remained enough sense in me to give him all he wanted when he jumped not once, but twenty times, before the up-stream flight that ran my line out to the last half-dozen turns, and I saw the nickelled reel-bar glitter under the thinning green coils. My thumb was burned deep when I strove to stopper the line.

I did not feel it till later, for my soul was out in the dancing weir, praying for him to turn ere he took my tackle away. And the prayer was heard. As I bowed back, the butt of the rod on my left hip-bone and the top joint dipping like unto a weeping willow, he turned and accepted each inch of slack that I could by any means get in as a favor from on high. There lie several sorts of success in this world that taste well in the moment of enjoyment, but I question whether the stealthy theft of line from an able-bodied salmon who knows exactly what you are doing and why you are doing it is not sweeter than any other victory within human scope. Like California's fish, he ran at me head on, and leaped against the line, but the Lord gave me two hundred and fifty pairs of fingers in that hour. The banks and the pine-trees danced dizzily round me, but I only reeled—reeled as for life—reeled for hours, and at the end of the reeling continued to give him the butt while he sulked in a pool . . .

A wild scutter in the water, a plunge, and a break for the head-waters of the Clackamas was my reward, and the weary toil of reeling in with one eye under the water and the other on the top joint of the rod was renewed. Worst of all, I was blocking California's path to the little landing bay aforesaid, and he had to halt and tire his prize where he was.

"The father of all the salmon!" he shouted. "For the love of Heaven, get your trout to bank, Johnny Bull!"

But I could do no more. Even the insult failed to move me. The rest of the game was with the salmon. He suffered himself to be drawn, skipping with pretended delight at getting to the haven where I would fain bring him. Yet no sooner did he feel shoal water under his ponderous belly than he backed like a torpedo-boat, and the snarl of the reel told me that my labor was in vain. A dozen times, at least, this happened ere the line hinted he had given up the battle and would be towed in. He was towed. The landing-net was useless for one of his size, and I would not have him gaffed. I stepped into the shallows and heaved him out with a respectful hand under the gill, for which kindness he battered me about the legs with his tail, and I felt the strength of him and was proud. California had taken my place in the shallows, his fish hard held. I was up the bank lying full length on the sweet-scented grass and gaspiong in company with my first salmon caught, played and landed on an eight-ounce rod. My hands were cut and bleeding, I was dripping with sweat, spangled like a harlequin with scales, water from my waist down, nose peeled by the sun, but utterly, supremely, and consummately happy . . .

Very solemnly and thankfully we put up our rods—it was glory enough for all time—and returned weeping in each other's arms, weeping tears of pure joy . . .

Notes

1 Susan Shelby Magoffin, *Down the Santa Fe Trail and into Mexico: The Diary of Susan Shelby Magoffin* (Lincoln: University of Nebraska Press, 1962).
2 Ibid., 13.
3 Ibid., 32.
4 Donald Worster, *Rivers of Empire: Water, Aridity, and the Growth of the American West* (New York: Oxford University Press, 1985), 77, 78.
5 William deBuys, *Enchantment and Exploitation: The Life and Hard Times of a New Mexico Mountain Range* (Albuquerque: University of New Mexico Press, 1985), 221.
6 Elliott West, *The Contested Plains: Indians, Goldseekers, and the Rush to Colorado* (Lawrence: University Press of Kansas, 1998), 275.
7 Quoted in William Cronon, *Nature's Metropolis: Chicago and the Great West* (New York: W.W. Norton & Company, 1991), 217.
8 Quoted in Richard White, *It's Your Misfortune and None of My Own: A New History of the American West* (Norman: University of Oklahoma Press, 1991), 219.
9 Quoted in Michael J. Robinson, *Predator Bureaucracy: The Extermination of Wolves and the Transformation of the West* (Boulder: The University Press of Colorado, 2005), 25.
10 Walter Baron von Richthofen, *Cattle-Raising on the Plains of North America* (1885) (Norman: University of Oklahoma Press, 1964), 81, 82.
11 Andrew Isenberg, *Mining California: An Ecological History* (New York: Hill and Wang, 2005).

12 Ibid., 43.
13 Ibid., 45, 46.
14 William Robbins, *Hard Times in Paradise: Coos Bay, Oregon* (Seattle: University of Washington Press, 2006), 14.
15 Ibid., 20.
16 Carlos Schwantes, *The Pacific Northwest: An Interpretive History* (Lincoln: University of Nebraska Press, 1996), 203.
17 Garrett Hardin, "The Tragedy of the Commons," *Science* 162 (3859): 1243–1248, 1968.

Further Reading

Bray, Kingsley M. *Crazy Horse: A Lakota Life.* Norman: University of Oklahoma Press, 2006.
Cox, Thomas R. *The Lumberman's Frontier: Three Centuries of Land Use, Society, and Change in America's Forests.* Corvallis: Oregon State University Press, 2010.
Davis, John. *Wyoming Range War: The Infamous Invasion of Johnson County.* Norman: University of Oklahoma Press, 2010.
deBuys, William. *Enchantment and Exploitation: The Life and Hard Times of a New Mexico Mountain Range.* Albuquerque: University of New Mexico Press, 1985.
Donovan, James. *A Terrible Glory, Custer and the Little Bighorn: The Last Great Battle of the American West.* New York: Back Bay Books, 2009.
Hyde, Anne F. *Empires, Nations, and Families: A New History of the American West, 1800–1860.* Lincoln: University of Nebraska Press, 2011.
Isenberg, Andrew C. *The Destruction of the Bison: An Environmental History 1750–1920.* New York: Cambridge University Press, 2000.
———. *Mining California: An Ecological History.* New York: Hill and Wang, 2005.
Kelman, Ari. *A Misplaced Massacre: Struggling Over the Memory of Sand Creek.* Cambridge, MA: Harvard University Press, 2013.
Monnett, John H. *Where a Hundred Soldiers Were Killed: The Struggle for the Powder River Country in 1866 and the Making of the Fetterman Myth.* Albuquerque: University of New Mexico Press, 2008.
Robbins, William. *Hard Times in Paradise: Coos Bay, Oregon.* Seattle: University of Washington Press, 2006.
———. *Landscapes of Promise: The Oregon Story, 1800–1940.* Seattle: University of Washington Press, 1997.
West, Elliott. *The Contested Plains: Indians, Goldseekers, and the Rush to Colorado.* Lawrence: University Press of Kansas, 1998.
White, Richard. *Railroaded: The Transcontinentals and the Making of Modern America.* New York: W.W. Norton & Company, 2012.
———. *It's Your Misfortune and None of My Own: A New History of the American West.* Norman: University of Oklahoma Press, 1991.
Worster, Donald. *A River Running West: The Life of John Wesley Powell.* New York: Oxford University Press, 2001.

Timeline

Creation of the Audubon Society	1886
Creation of the Boone and Crockett Club	1887
Founding of Hull House	1889
Creation of the Forest Service System	1891
Adirondack State Park Established	1892
Creation of the Sierra Club	1892
Creation of the Forestry Division in the General Federation of Women's Clubs	1902
Creation of the Forest Service	1905
Banning of the Use of White Phosphorus in Matchstick Production	1911
Publication of John Muir's *My First Summer in the Sierra*	1911
Hetch Hetchy Fight	1903–1913
Creation of the Chicago City Waste Commission	1913
Publication of John Muir's *Travels in Alaska*	1915
Creation of the National Park Service	1916

Conserving Resources, Saving Sacred Spaces, and Cleaning the Cities

8

America in the Conservation Era

The officer who arrested a poacher in Yellowstone National Park was humble in his account of events to a reporter, attributing his success to luck and favor. He had hit the trail early in the morning and quickly found a tepee and a cache of buffalo heads. Soon after he heard six shots and discovered five dead bison. This occurred in 1894 when bison hovered on the edge of extinction. Spotting the poacher with a rifle, the army officer knew he was outgunned with his .38 caliber revolver, and he had to cover 400 yards of open space to get to the poacher without being shot.

> Howell's rifle was leaning against a dead buffalo, about 15 ft. away from him. His hat was sort of flapped down over his eyes, and his head was toward me. He was leaning over, skinning on the head of one of the buffalo. His dog, though I didn't know it at first, was curled up under the hindleg of the dead buffalo. The wind was so the dog didn't smell me, or that would have settled it. That was lucky, wasn't it? Howell was going to kill the dog, after I took him, because the dog didn't bark at me and warn him. I wouldn't let him kill it.[1]

The officer ran fast across the snow in his snowshoes and even jumped a 10-foot wide ditch and managed to catch the poacher off guard.

The reporter's admiring tone shifted when he described the poacher:

> we found, a most picturesquely ragged, dirty and unkempt looking citizen. His beard had been scissored off. His hair hung low on his neck, curling up like a drake's tail. His eye was blue, his complexion florid. In height he seemed about 5 ft. 10 in. His shoulders were broad, but sloping. His neck stooped forward. His carriage was slouchy, loose-jointed and stooping, but he seemed a powerful fellow. Thick, protruding lips and large teeth completed the unfavorable cast of an exterior by no means prepossessing.[2]

The author of the article also commented on his dirty, greasy clothing and the fact that he had no shoes and only one pair of thin socks. While the description of his physical appearance and bearing are negative, the language shifts again in the description of the poacher's woodcraft:

> His snowshoes (*skis*) were a curiosity. They were 12 ft. long, narrow, made of pine (or spruce), Howell himself being the builder of them. The front of one had its curve

> supplemented by a bit of board, wired on. All sorts of curves existed in the bottoms of the shoes. He had them heavily covered with resin to keep the snow from sticking to them. To cap the climax he had broken one shoe while in the Park—a mishap often very serious indeed, as one must have two shoes to walk with, and elsewise cannot walk at all. With the ready resources of a perfect woodsman, Howell took his axe, went to a fir tree, hewed out a three-cornered splice about 5 ft. long, nailed it fast to the bottom of his broken shoe, picked out some pieces of resin, coated the shoe well with it, and went on his way as well as ever.[3]

The reporter concluded by commenting admirably on the man's physical strength, evidenced by his ability to haul a load of 180 pounds on a toboggan over several miles of rough terrain and snow. He also described the man's cheery attitude as he consumed 24 pancakes, joked with army officers, and explained that his profits exceeded any fines the government would assess him.

The article about the poacher and the army officer that captured him was published in an 1894 issue of *Forest and Stream,* a magazine edited by George Bird Grinnell, an important early conservationist who helped found the Boone and Crockett Club, strongly influenced Theodore Roosevelt, and fought for game laws to protect wildlife. This piece, like many others, publicized the poaching crisis and helped generate support for 1894 legislation to strengthen national parks' ability to protect game. The piece reveals the tension between admiration for the frontiersmen and the need for the state to protect diminishing resources. Also, class attitude shows itself; the character of a man who poaches in a time when upper-class men were promoting sporting hunting seemingly reveals itself in his primitive physical appearance and seeming poverty. Most of the men who promoted conservation of land and resources came from the upper class, and their attitude toward the lower class was less than admirable. At the same time, animals that once seemed abundant struggled to survive, and only strong laws and new boundaries could protect them. In this time of ecological crisis in the late 19th and early 20th century, American activists and leaders struggled to craft creative, new solutions to seemingly entrenched problems. There would inevitably be casualties.

After centuries of ecosystem transformation and environmental destruction with limited and unsuccessful efforts to manage the worst damage, federal and state governments entered into the debate over land use and economic growth with a variety of strategies ranging widely in their success. The conservation agenda was not simply a monolithic movement imposed on Americans by their government but originated from multiple sources with diverse goals. The result was a complex moment and movement in American history that defies simple analysis or categorization. To frame this movement correctly it is essential to note the work of urban women reformers as they fought for cleaner cities, playgrounds, and the removal of dangerous chemicals from the workplace. Other conservationists, typically male, sought to use science to increase yields of natural resources and also began restricting and regulating harvest and use of resources. In opposition to these utilitarians stood advocates for the preservation of parks. The various actions and beliefs that constituted the conservation movement sprung from a deep-seated concern over the destruction of nature and the widening gap between Americans and the natural world and the negative impact this may have on nature and on American citizens themselves. The abundance of resources and availability of land were clearly threatened across the nation by the late 19th century. Conservationists sought to preserve and restore abundance while preservationists saw a need to create a system of restraint and protection for places of great beauty and of sacred importance.

Protecting Natural Resources

The traditional understanding of the conservationists aligns them with the Progressive era, rising to power and accomplishing a great deal of reform, regulation, and protection in roughly the time period of the 1890s to 1917. Of course, the conservation impulse can be located earlier in American history. Conservation as a government strategy to control the loss of resources emerged in New England and Maine in the 1860s and 1870s. Maine Fish Commissioners and state fisheries managers in the New England region were stymied by the power of the logging and milling industries and their own inability to save the remaining populations of fish in a state that had once flourished with tens of millions of striped bass, shad, sturgeon, Atlantic salmon, and numerous other species. Dams built without fish passageways, debris from log drives, erosion, and numerous other problems drove many species to extinction and others to the very cusp. These early conservationists sought to limit both poaching and overharvest as well as force dam owners to allow fish passage. They failed to achieve the last goal, and as a result New England fisheries managers were the first in the United States to use hatcheries to boost river populations. While they were tasked with the responsibility of being state stewards of the fisheries, they lacked the regulatory power to truly protect the resource. Therefore, they began to rely on science years before the traditionally accepted timeframe for the conservation era. Fisheries managers in New England attempted strategies that would be recognized as classic conservation in the early 20th century. If they could not preserve abundance through regulatory power, then they would restore it and hopefully sustain it through hatcheries.

Conservationists sought to efficiently and rationally manage resources through the use of harvest regulation, science, and technology to reduce waste and provide a steady supply of commodities for present and future use. In so doing, they did not oppose capitalism but actually sought to make it more predictable and orderly. Ranchers, logging companies, cannery owners, and others fulminated that the rise of regulation and enforcement undermined their own economic rights. Conservationists working for the government at the federal and state level sought to use science and regulation to both reduce waste and increase yield, with forester Gifford Pinchot famously comparing forest management to agriculture.

It is hard to imagine the current landscape of parks, refuges, and national forests existing without the bold actions of President Theodore Roosevelt. He laid the foundations of the great public lands system the United States has today by creating 16 national monuments, including Grand Canyon and Chaco Canyon, and 5 national parks, such as Crater Lake and Mesa Verde, as well as establishing over 50 wildlife reserves. Western legislators concerned with the pace of his setting aside or "locking up" of federal land from private use pushed through a bill in 1907 stipulating legislative approval for the creation of any more reserves in six western states. With the bill awaiting his signature, Roosevelt plowed ahead and created and enlarged 32 reserves encompassing 75 million acres. When Roosevelt took office,

Figure 8.1 Theodore Roosevelt loved nature his entire life and possessed an encyclopedic knowledge of flora and fauna. His love of nature included a deep passion for hunting, and he slew thousands of animals in America, Africa, and Brazil. He is photographed here astride his horse following a bear hunt in Glenwood Springs, Colorado, in 1905. Library of Congress.

40 reserves of a little more than 50 million acres existed. Upon leaving the White House, the nation could boast of 159 national forests across 194 million acres. This accomplishment greatly enlarged the federal presence in American life, particularly in the West, provided significant protections to important ecosystems and recreational landscape, and firmly embedded a tradition of the federal government setting aside land for protection that has grown and expanded over the course of the 20th century. The failure of Americans to develop a philosophy of stewardship and lack of government regulation led Roosevelt and supporters to the conclusion that borders protecting resources, archeological sites, and monumental landscapes were necessary.

Young Theodore's life was suffused with nature. Suffering from asthma as a child, nature books caught his fancy, and as a young boy he spent countless hours playing alone outside, observing wildlife. By age 7 he was collecting specimens and endeavored to write natural history by the wizened age of 10. These interests continued through college, where he kept wild animals in his dorm room at Harvard. His fierce love of hunting resulted in many victims—on one trip he killed 203 animals. But he also knew bird song, animal habits, and behavior and through his life acted on his strong love for nature. Following the tragic deaths of his young wife and mother within 24 hours of each other in 1884, he fled west to rebuild himself in the rugged landscape of western North Dakota. Through hunting, ranching, and labor Theodore learned much about the western landscape and land use by western ranchers and hunters. Likewise, the young man gained a firsthand knowledge of the collapse of the bison population, saw the evidence of overgrazing on the northern plains, and experienced and heard about the abuses of public lands and homesteaders by the stockmen. In 1886 Roosevelt expressed his views of this damage in clear terms:

> The buffalo are already gone; a few straggling individuals, and perhaps here and there a herd so small that it can hardly be called more than a squad, are all that remain. Over four-fifths of their former range the same fate has befallen the elk. . . . The shrinkage among deer and antelope has been relatively nearly as serious. There are but few places left now where it is profitable for a man to take to hunting as a profession; the brutal skin-hunters and greasy Nimrods are now themselves sharing the fate of the game that has disappeared from before their rifles.[4]

The last comment neglects the role of market forces and military policy in the destruction of the bison in favor of class disparagement. But, his western experience not only reinforced Roosevelt's belief that rural landscapes needed to be preserved in order to give men a place to test and strengthen themselves away from the supposedly softening or feminizing influences of urban society and middle-class life but also to maintain access to resources for the nation's economic health. In short, western landscapes needed to be preserved or conserved for both economic and environmental reasons and to preserve masculinity.

This love of nature and for what he perceived as the remnant of the "wild frontier" also drove his and others' work for nature. After returning from North Dakota in 1887, he and famous conservationist George Bird Grinnell founded the Boone and Crockett Club, an organization of well-heeled hunters and fishermen that launched a powerful tradition of sportsmen supporting conservation activism. Hinting at the future of conservation, after Gifford Pinchot joined the club, he quickly changed the group's policy of setting aside land for preservation to one that emphasized use. For some sportsmen of that era and today, hunting and killing wild animals did not negate their love for nature—rather it was inherent in that passion. Theodore Roosevelt once wrote, "when I hear of the destruction of a species, I feel just as if all the works of some great writer had perished."[5] Roosevelt's scientific knowledge of nature continued to surprise acquaintances and colleagues and helped build support for his conservationist agenda. While president he often visited the home of Dr. C. Hart Merriam, the Head of the Biological Survey, to view his in-home museum of skins and bones. According to Merriam,

> one evening at my house (where I then had in the neighborhood of five thousand skulls of North American mammals) he [Roosevelt] astonished every one—including

> several eminent naturalists—by picking up skull after skull and mentioning the scientific name of the genus to which it belonged.[6]

Appropriately enough, it was while finishing a hike that Roosevelt was informed that President McKinley was dying from an assassin's bullet. His assumption of the office of president would launch one of the great conservationist careers in American history, and in his First Annual Message to Congress he expressed his intent to promote conservation, create forest reserves, and protect wildlife: "The preservation of our forests is an imperative business necessity . . . we have come to see clearly that whatever destroys the forests, except to make way for agriculture, threatens our own well being."[7]

One reason for Roosevelt's success was Gifford Pinchot, one of the more controversial figures of conservation and environmental history. While he did not create utilitarian conservation, Pinchot does receive much credit for this direction as well as criticism for its mistakes, the multiple-use mandate, and a pragmatic philosophy that many believe was too sympathetic to business. Because there was no forestry education in the United States, the young man pursued his forestry studies in France with the strong encouragement of his father, who, having reaped great profits as a lumberman, came to question the practice that left behind devastated landscapes. That training and observation of German forestry practices led the young Pinchot to his belief that trees should be grown like crops of corn or wheat in forests of single species trees like white or ponderosa pine. It is worth noting that deterioration of forest habitat led the Germans to abandon that model in the 1920s.

Much of his influence and career success derived from his relationship with Teddy Roosevelt and the favor and support of TR before and during his presidency. Roosevelt provided Pinchot entry into the Boone and Crockett Club and also supported and expanded Pinchot's reach during his presidency. Attributing Pinchot's success to the president's support alone understates his skills and ambition, however. His abilities, self-promotion, and political shrewdness and, maybe most importantly, refashioning of conservation into a business-friendly model, also propelled him forward to great success. When he took over forestry under President McKinley, it was housed in the Department of the Interior. To expand his influence and control over the forest reserves, Pinchot lobbied for the transfer of forestry to the Department of Agriculture in 1905, crafting compromises with business and gaining the support of groups like the Sierra Club. As important as his political acumen was his skillful promotion of the agency and his self. The Forest Service budget increased a hundredfold under his leadership, and this is at least partially due to his use of mailings promoting trees that the Forest Service sent to magazines and newspapers. Managing a mailing campaign targeting 800,000 Americans, Pinchot seems to have mastered mass marketing before its 1920s and 1930s heyday.

He would have been nothing without the legions of eager conservationists that desperately sought solutions for resource loss and ecosystem destruction. Historian Nancy Langston's description of forester H. D. Langille's efforts serves as a reminder of what they were up against and hoped to achieve. Working for the Bureau of Forestry in the Blue Mountains of Oregon, he wrote inspection reports in the 1890s in a frantic race to catalog and protect timber lands against loggers, corrupt Oregon politicians, and land swindlers scrambling in ways both legal and illegal to grab up and clear as much forest and land as they could with no pretense stewardship or for collective posterity. As she writes,

> After Langille had ridden five nights alone on a tired horse in an unsurveyed forest, he was regarded with suspicion by those who saw him. His head ached and he was cold inside his heavy, wet leather coat, and his horse stumbled and the water dripped off his hat in cold rivulets down his neck, and the forest seemed dark and alien around him. Every few hours, he wrote his notes, crouched over the saddle horn, his fingers cramped with the cold. To keep his spirits up, he reminded himself that this was an almost sacred mission. He was on a quest to save the heart of the country: to preserve the life-giving springs from those who would destroy their protective forests.[8]

This man was deeply aware of the economic processes and destruction that had previously occurred and would be repeated if the forest was not protected.

Under Pinchot's careful guidance, the Forest Service trained and launched a generation of young foresters onto the public lands. Prepared first in the Yale forestry school started by Pinchot and other schools as the field grew, they applied principles of scientific management and production to complex forest ecosystems. But their efforts to increase forest yield were stymied by their lack of understanding of forest ecosystems. Foresters saw old growth forest as chaotic and wasteful. For them, rotting trees did not represent habitat for birds, bugs, and chipmunks or nutrients reintroduced into the soil for the next generation of plants and trees. Rather, this represented poor management, inefficiency, and the loss of valuable economic resources. Similarly, the diversity of tree species in a forest that today are understood as an indication of ecosystem health supportive of multiple species of insects and animals, to them represented an unproductive landscape filled with undesirable trees. "Garbage trees" such as larch or hemlock occupied space better employed for the production of marketable trees such as ponderosa pine. Therefore, in national forests across the West, foresters promoted massive cutting to sell as much lumber as possible in order to clear a mosaic ecosystem that was ecologically healthy but economically wasteful, and replace it with heavily managed and marketable tree crops. The forests of conservationists' vision led to monolithic woodlots of few tree species and reduced biodiversity and natural abundance.

To protect these valuable tree crops the Forest Service initiated policies that resulted in severe and unanticipated ramifications. Seeking to remove all threats to productivity, the federal agency and, increasingly, state offices and private foresters emulating the federal agency targeted pests, disease, and fire. These efforts to create uniform crops engendered a series of problems that have plagued the Forest Service through most of the 20th century. First and foremost, fire suppression actually made fires worse and more destructive while hindering the growth of certain tree species. The control of fire seemed logical on the surface because burning forests appeared wasteful. But in suppressing fires completely, the Forest Service allowed debris to accumulate on forest floors, making later fires more likely and, with more tinder, also stronger, hotter, and more destructive. Certain species, such as Ponderosa Pine, need fire for their seeds to germinate. Fire suppression then meant the accidental and unanticipated loss of even desired species. Native use of fire created a mosaic landscape of patches of different varieties of trees and clearings in the forest. Not only had this resulted in a much more diverse ecosystem supporting a wider range and higher number of species than a forest of only one or two tree species, but it also rendered the forests more resilient to disease and insect infestations. With the simplification of the forest ecosystem and the suppression of fire, the national forests became much more vulnerable to damage from fires, insect infestations, and disease.

Better Rivers through Science

In the Pacific Northwest, salmon runs were collapsing under the combined impact of overharvest, dam construction, and destruction of habitat from logging, mining, grazing, and other economic activities. By the beginning of the 20th century numerous salmon runs were driven extinct with a strong general decline throughout the region. For Northwest conservationists it was clear that the only way to bring the losses under control was through a combination of regulation, increased enforcement, and the use of science to increase runs. Worsening the situation for the salmon were the numerous small dams built without fish passageways. Boosters across the West sought to reap the benefits of hydroelectricity to drive economic growth. The first power plant producing hydroelectricity constructed in the United States was built in 1889 at Willamette Falls in Oregon City, Oregon. That company then dammed Oregon's Clackamas River in 1907. The Puget Sound Power Company was organized in 1902 to construct a 25,500-kilowatt dam and plant on Washington's Puyallup River, and several dams were built in quick order on the eastern side of the state on the Spokane River and in northern Idaho. Back in western Washington, the White River Dam and power plant was completed in 1911, and two years later the Condit Dam was built on the White Salmon River of Washington. Dams were constructed on the Missouri River in Montana in 1890, 1910, 1915, and 1918, with the Big Sandy River in Oregon dammed in 1912. On the Olympic Peninsula of Washington State the Elwha Dam was built just west of Port Angeles in 1913. The region's rivers were being brought under control, one by one, to benefit economic expansion and rapidly growing communities.[9] Damming rivers was consistent with conservationist strategy, bringing the resource under control and management to provide numerous benefits to business and citizens.

These dams unquestionably brought numerous benefits to their communities. The wealth and abundance of nature in the region subsidized the growth of the economy. Hydroelectricity was used for business and home lighting, for streetcars, and for the creation and expansion of a manufacturing economy. But the cost in lost fish was staggering. In many cases, dams were built illegally. The Oregon Constitution forbade the building of dams without fish passageways, and a Washington State law passed in 1890 banned dams without fish passage. These prohibitions were ignored, and dams blocked salmon and steelhead runs with no state interference. The fish lost access to upstream habitats, and numerous runs were quickly extinguished; other damage was not as immediately evident. For example, dams stopped the flow of rocks, sand, and gravel downstream. Loss of debris diminished the quality of beaches at the mouths of rivers and nearshore habitats where species such as clam, herring, and crab thrived. The downstream rivers' qualities as a salmon nursery deteriorated with the loss of rocks and gravels needed for spawning sites. The loss of spawned out salmon carcasses upstream of dams meant the removal of key nutrients from the riparian ecosystem. Rotting salmon flesh contributes nitrogen, phosphorus, and other elements to the river ecosystem and soils. Additionally, those salmon feed a variety of animals, including bears, raccoons, ravens, bald eagles, among many others. Fish such as trout thrive on the fry and smolts that emerge from salmon redds. There were more subtle, insidious impacts on upstream riparian habitat. Watersheds with healthy salmon runs have more fertile soil with higher rates of plant and tree growth compared to rivers where the salmon have been removed. They also have larger numbers and increased varieties of insects in the river and an increased and more diverse bird population because of the amount of insects available as food. Salmon were and are a keystone species. The river and watershed ecosystems in which they spawn and live are dependent on their existence to function effectively. Losing these fish hurts numerous species across the watershed ecosystem and even into the ocean where species such as orca, sharks, seals, walrus, and others depend on salmon. The rivers helped subsidize economic growth in the region but at a cost to the health of the overall environment, not simply the river ecosystem.

While dam builders ignored the lessons learned in New England, that dams without fish passageways doomed anadromous fisheries, fishery managers quickly adopted the strategy employed in New England—the use of hatcheries. The first one to be built in the Pacific Northwest arose from the decision to employ hatcheries instead of meaningful regulation. Livingston Stone, fresh from failure with hatchery production on the McCloud River in California, arrived in the Pacific Northwest to assist in the construction of the Clackamas Hatchery in 1877. The Oregon legislature's resistance to funding the hatchery forced the U.S. Fish Commission to take control of the facility, which was soon beset by problems such as flooding, sedimentation from logging in the watershed, interference by mills, and salmon runs blocked by downstream dams. These problems reinforced the fish managers' sense of powerlessness in the face of politicians, timber companies, commercial fisherman, and cannery owners unwilling to suffer restrictions on their activities. The fact that fisheries managers still embraced hatcheries as the key strategy for maintaining the productivity of Northwest rivers speaks not simply of their love of science but also of the limits of conservation in that era.

In practice hatcheries resemble farming. Employees harvest male salmon for their milt and females for their eggs, fertilize the eggs by hand, and then raise the fry and smolt until they are large enough to survive in the wild, finally releasing them into rivers and the ocean. This seems an almost perfect blend of careful husbandry and traditional feral livestock practices. The process of raising the salmon from eggs until smolts are big enough to compete in the wild, and then releasing them to grow strong and large off the resources of a wild environment, is remarkably similar to the earlier practices of allowing livestock and pigs to run feral. Unlike those pigs and cows, the homing instincts of salmon would bring them right back to the hatchery where they were born to be harvested and reproduced. In Washington State alone from 1896 to 1915, the total salmon and steelhead fry production increased from 4,500,000 to 1,021,174,416.[10] This practice of creating "salmon without rivers," as fisheries biologist and writer Jim Lichatowich coins it, was a halfway measure that seemed to preserve or even increase salmon runs while allowing for unfettered resource exploitation and development. The view that salmon could be better managed by experts than by nature itself is expressed clearly by Livingston Stone:

> Nature, perhaps more aptly speaking, Providence, in the case of fish . . . produces great quantities of seed that nature does not utilize or need. It looks like a vast store that has been provided for nature, to hold in reserve against the time when the increased population of the earth should need it and the sagacity of man should utilize it. At all events nature has never utilized this reserve, and man finds it already here to meet his wants.[11]

Figure 8.2 Hatcheries to grow fish for release into rivers and lakes came into use in the Northeast as fish runs were devastated by dams, logging, erosion, and overfishing. In the American West during the Progressive Era of the early 20th century, they gained greater currency for conservationists as a way to preserve fish runs while allowing resource exploitation and economic development. As is seen in this photograph of an Oregon salmon hatchery, a number of tanks were used to raise fish at different life stages until they were big enough to be released into rivers. Gerald W. Williams Collection, Special Collections & Archives Research Center, Oregon State University.

Stone not only reminds us of the assumption shared by Americans that abundance was God's gift to their society, but that conservationists have a sacred responsibility to maximize that abundance for Americans. The most enthusiastic advocates even argued that the rivers could be made to produce even more fish as the river watershed ecosystems collapsed. An ideology of abundance so centrally located in the American mythos encouraged conservation while allowing further environmental degradation.

Funding for research was slashed as hatchery budgets exploded. Continued ignorance is at least partially a function of decisions prioritizing production over scientific inquiry and understanding. From a contemporary perspective it is easy to see the failings of these conservationists and criticize their simplistic understanding of a complex salmon ecosystem. Some historians and activists view conservation era fisheries managers as essentially aiding and abetting economic development and environmental destruction. The government sanctioning of the idea that hatcheries effectively replaced rivers removed the burden of moderate land use, stewardship, or even state regulation and protection of watersheds and rivers, permitting the destruction of salmon habitat. However, it must also be noted that the eager belief of conservationists that hatcheries could turn degraded rivers into salmon factories stems from or at least is balanced by their vitriol and anger at practices that destroyed the resource. One fish commissioner fulminated that

> It seems to me to be a crime against mankind—against those who are here and the generations yet to follow—to let the great salmon runs of the State of Washington be destroyed at the selfish behest of a few individuals, who, in order to enrich themselves, would impoverish the state and destroy a food supply of the people.[12]

Rage as they might, their options were limited. The dominance of resource extraction industries in these states along with their power in state government, coupled with a near cultural consensus that development and growth was of the highest priority, prevented conservationists from wielding effective regulatory power in the early 20th century. Their political weakness necessitated a dependence on science and hatcheries to perform their public functions. These conservationists shared the same values as the capitalists—specifically, a strong belief in the validity of strong economic growth and development as integral to American success. They also launched an unprecedented expansion and intercession of state and federal power in environmental and economic issues during this era. The crisis of resource loss necessitated an expansion of government power.

Fighting for Nature

Hetch Hetchy stands as one of the epic environmental battles of American history, one that played across a national backdrop, pitting the archetypal preservationist John Muir against the iconic conservationist Gifford Pinchot. Lying in the balance was the beautiful Hetch Hetchy Valley and the sanctity of Yosemite National Park versus economic and population growth.

The city of San Francisco had purchased land and water rights to build a dam on the Tuolumne River; the resulting reservoir would back water into the Hetch Hetchy Valley inside the park. San Francisco leaders sought a reliable and large source of drinking water, but their request to flood this portion of the park was rejected by Secretary of Interior Ethan Hitchcock in 1903, setting off a conflict that persisted through the following decade. The city could have secured water in other locations and was not limited to this site, but city government, engineers, and supporters insisted on that spot. Pinchot's strongly favoring the flooding of Hetch Hetchy was consistent with his utilitarian view of nature and belief that protected federal lands should serve multiple uses. When John Muir intervened, describing the beauty of the valley and the need for its preservation, Pinchot admitted he had never visited. Muir's impassioned defense came to naught as Pinchot strongly and publicly supported the dam, stating at one point, "the intermittent esthetic enjoyment of less than one per cent is being balanced against the daily comfort and welfare of 99 percent."[13] His backing in place, along with that of a new Secretary of Interior, the city reapplied for a permit in 1908. They did not anticipate the opposition of a new conservationist group.

Created in 1892 by a couple of University of California English professors and John Muir, the Sierra Club was an active mountaineering and preservationist group that allowed membership of both men and women. It became a vocal and assertive organization under the leadership of John Muir. Born in Scotland to religiously strict parents, he grew up in Wisconsin, farming, clearing forest, and observing nature. Muir educated himself in natural science and always felt a passion for geology. As a young man he walked a thousand miles from Indiana to Florida. In 1868 he migrated to Yosemite, California, the place for which he would become synonymous. Working as a sheepherder and at a sawmill, he fell in love with the Yosemite landscape while seeing up close the consequences of grazing and logging on this landscape. Publishing works on geology and biological science, he also wrote articles and books about Yosemite as well as the western and Alaskan landscape. His writing increased tourism and interest while also generating widespread support for parks and a growing preservationist philosophy.

When the Hetch Hetchy fight was joined, the organization was primarily a recreational group with preservation interests. Members journeyed into the Sierras with pack mules hauling tents, fine foods, and drink. The fight to save Hetch Hetchy radicalized the group, at least for a short time, as members who felt the valley could be legitimately flooded to serve the interests of San Francisco without doing damage to Yosemite, left the club. Remaining members lined up in support of Muir's view of Yosemite as sacred space requiring absolute and complete protection. The organization took the fight to a national audience. Pamphlets, editorials, and photographs celebrated the beauty of Yosemite. The material produced by the club was distributed nationally and to members of Congress. In hearings

Figure 8.3 John Muir and Teddy Roosevelt after their three-day wilderness journey in 1903. Muir led Roosevelt on a tour of many of the beautiful sites of Yosemite National Park. In their long, rambling conversations they both developed great respect for each other's knowledge of natural history and love of nature. Muir convinced Roosevelt to expand the boundaries of the park to include state forests under threat of clearcutting. Library of Congress.

over the bill, supporters of the park, including wordsmiths such as the poet Harriet Monroe, editor of *Poetry* magazine, provided eloquent testimony to the beauty and sublime aspects of the valley. The use of promotional material, editorials, publications, and testimony meant that one important element of this fight was that Americans learned to view nature through a new prism, one that was less utilitarian and placed stronger consideration on aesthetics, the idea of sacred landscapes, and national parks as important refuges and recreational sites. These arguments expanded at least some Americans' views of the best uses of nature.

Muir put in 15-hour days to finish his book about the park in which he declared,

> These temple destroyers, devotees of raging commercialism, seem to have a perfect contempt for Nature, and instead of lifting their eyes to the God of the mountains, lift them to the Almighty Dollar. Dam Hetch Hetchy! As well dam for water-tanks the people's cathedrals and churches, for no holier temple has ever been consecrated by the heart of man.[14]

For Muir, influenced by the English Romantics and American Transcendentalists, God resided in nature, so Yosemite was a natural church. Some took offense at Muir's condemnation of those who sought to use the landscape, and as historian Char Miller points out, this was not simply conservationists versus preservationists. People involved on both sides of this fight were generally more nuanced in their views of nature, seeing utilitarian uses as legitimate in cases and supporting preservation in others. Supporting the point that this wasn't simply crass commercialists and utilitarians versus nature-lovers, and possibly reflecting fatigue with Muir's rhetoric, was the comment by conservationist John Burroughs:

> I suppose if there was a lake in that valley and San Francisco wanted to drain it and make meadows in its place, John Muir would howl in the way he does now . . . Grand scenery is going to waste in the Sierras—let's utilize some of it.[15]

Women activists fought hard to protect Hetch Hetchy. The General Federation of Women's Clubs had organized its Forestry Division in 1902. With a national membership of 800,000, the federation was able to bring serious clout to bear in their belief that forests should be preserved rather than simply managed as a resource. They opposed flooding Hetch Hetchy, writing articles, letters, and petitions. In a 1909 Congressional hearing, more than 50 women activists submitted letters of opposition to the proposed dam. While frustrated at their lack of political power because of not having the vote, their impact was strong enough to compel a California congressman to write to Pinchot that the reservoir was threatened by a conspiracy "engineered by misinformed nature lovers and power interests who are working through the women's clubs."[16]

Dam proponents emphasized "hard facts" or what they perceived as a common sense approach to using nature, emphasizing engineering concepts, water capacity, and usage as well as the utilitarian argument that Hetch Hetchy would facilitate the rise of a dynamic, growing city and improve the quality of life for its citizens. Mayor James Phelan articulated the irritation of dam supporters with the nature rhetoric of Muir, Harriett Monroe, and other opponents when he said, "I am sure he [Muir] would sacrifice his own family for the preservation of beauty. He considers human life very cheap and he considers the work of God superior."[17] While witty, the comment is disingenuous because other sources of water were available; human life was not at risk.

The city won, and the valley was flooded. While not preordained by any means, the victory of San Francisco and other dam proponents was the most likely outcome given the attitudes about growth, the appropriate uses of nature, and economic development in that era. John Muir, the Sierra Club, and other park advocates certainly believed in the justice of their cause and fought to win. Viewing the event in its historic context, it is surprising they were able to delay the dam's construction for a full decade. The activist amateurs taking on Gifford Pinchot, engineers, businessmen, and city officials in a society placing great emphasis on economic development achieved more than what one might reasonably consider possible at that time. City engineers, elected officials of San Francisco, and consultants were all paid to conduct studies, lobby, and write reports. Therefore, they had considerably more time and resources to commit to this fight, whereas the amateur opponents had to find time and resources outside their own careers and daily lives to oppose the professionals. While these preservationists brought tenacity and effective strategy to the contest, they still constituted a significant minority of the American population. Increasing sentiment and support for preservation as well as the resources and ability to question and challenge the utilitarian arguments would be necessary to win another such fight.

Parks for People

The decision to create a federal agency to manage the national parks and legislation to give them special status conferring greater protection, arose partly as a result of the defeat in the Hetch Hetchy fight. Activists were determined to prevent such a violation of park boundaries ever happening again. The General Federation of Women's Clubs threw their support behind the bill. According to historian Polly Welts Kaufman,

> the General Federation passed a resolution supporting a National Park Service at its biennial convention held that May [1916] in New York. [Mary Belle King] Sherman presented Stephen T. Mather, soon to become first director of the Park Service . . . Mather praised the clubwomen for helping to crystallize opinion in favor of the service. In addition to lobbying congressmen, members produced a mailing list of 275,000 names for the *National Parks Portfolio,* the Department of the Interior's major publicity effort in support of the bill.[18]

The creation of the National Park Service, accomplished in 1916, was not simply an exercise in preservation. The activists and legislators that pushed this legislation for approximately five years believed that the national parks had to serve utilitarian purposes to a limited degree while also providing iron-clad protection to landscape. Also, the parks were meant for human use and enjoyment. Providing Americans with recreation, exercise, and respite from the stresses of daily life were to be core functions of the national parks.

The first two park superintendents, Stephen Mather and Horace Albright, prioritized landscape and engineering. Park officials pursued the building of good roads, bridges, and facilities in a way that would maximize access and the aesthetic, nationalist experience of viewing monumental nature. As historian Alfred Runte has argued, national parks memorialized America; they served as rock and ice signifiers of America's greatness and future promise. In the 19th century, beginning with President Thomas Jefferson's hope that the Lewis and Clark expedition would discover a wooly mammoth in the West, American leaders and promoters turned to nature to replace lost European history, culture, and traditions. A young Republic, America had no literary tradition to speak of in the early 19th century, no real architecture, no great cathedrals or other monuments to either man or God. But the wealth of nature as manifested by massive passenger pigeon flocks (until exterminated), bison herds stretching for miles (until almost exterminated), the Niagara Falls, the Rocky Mountains, and numerous other examples, served not only as a reservoir of natural resources positioned by God to make America great but also as a replacement to lost culture and history. The national parks protected the transcendent landscapes that in their sublime beauty suggested America's greatness and also helped unite Americans divided by the recent Civil War and Reconstruction, labor-capital conflicts, and other divisive tensions. When Mather and Albright viewed their domain and hired droves of landscape architects and engineers, they did so to frame these natural monuments in a manner that helped Americans celebrate their own identity and nation.

Figure 8.4 In the first decades of the National Park Service, park managers sought to create roads and facilities that provided access to beautiful, monumental nature such as this view of the Grand Teton Mountains. National Archives photo by Ansel Adams.

Ecological issues were largely ignored because of misunderstanding and overconfidence. The Park Service relied on experts in the Biological Survey and state and federal fisheries agencies to manage predator control and fish production. Park leadership in the 1910s and 1920s operated on the assumption that immense tracts of land left relatively untouched existed in their "natural" state and were automatically healthy. The agency did intentionally change the ecology of parks by seeking to provide a more enjoyable nature for tourists—meaning larger elk, deer, and bison herds to observe and more fish to catch in streams and lakes. In Yellowstone National Park, haying operations and winter corrals were employed to resurrect the bison from the handful of survivors saved by the army in the late 19th century to a population of approximately 1,000 in the 1920s. In fact, bison managers were so successful they reached the carrying capacity of the Lamar Valley, where this herd was based, and began culling the herd, selling bison to local meat processors. Elk were fed hay through the winters until the population far exceeded the carrying capacity of the park. This resulted in extensive damage to the willow shrubs along streams, driving down beaver populations and leading to declines in wetland habitat and species dependent on them such as whitetail deer. The desire for a pastoral ideal of recreational abundance compelled park managers to attempt to recreate nature in ways that were ecologically damaging.

Parks against People

National and state parks provided numerous benefits to Americans as well as some protection and preservation of nature. But it is also true that in the creation and management of these parks, marginalized groups such as poor whites, Indians, and Hispanics experienced oppression and unfair exclusion from access to needed resources. Karl Jacoby's *Crimes Against Nature* makes the case that government conservationists at the state and federal level treated local populations poorly when creating state and federal parks, ignoring their land use practices and failing to consider their perspective and needs when creating parks and policy. According to Jacoby, conservationists also unfairly pilloried these groups, overstating their abuses of nature in a way that reflected middle-class and upper-class disparagement of those in the economic underclass. The creation of Adirondack State Park in New York in 1892 shut out the poor whites of that region dependent on forests for firewood, fencing, and habitation, as well as for hunting and fishing for subsistence. The creation of the park and enforcement of laws against hunting and wood harvest effectively criminalized traditional practices. Locals employed a variety of strategies to get around these new laws, but what had been legal and accepted now was criminalized.

The creation of Yellowstone National Park in 1872 undermined the economies of local Blackfeet, Bannock, and Crow Indians as well as whites. Native peoples were in desperate straits by the end of the 19th century. The continual loss of lands combined with the failures of land allotment and reservation agriculture created starvation and frustration on the reservations. The near extermination of the bison and eradication of other game worsened their ongoing hunger and desperation. In contrast, the national park represented a reservoir of natural wealth, filled with elk, deer, fish, and bison and an opportunity to continue traditional cultural practices. The area had been actively used for centuries by Shoshone,

Bannock, Crow, and Blackfeet Indians. The mosaic landscape of the park revealed a long tradition of burning for game, reducing insects, and easing travel through underbrush removal. Native peoples traditionally hunted game such as elk, deer, and bison; harvested fish; engaged in trade with other tribes; and used a variety of ecosystems throughout the greater Yellowstone Valley region. With the creation of the reservation system and the forced relocation of Indians to these impoverished sites in the 1860s and 1870s, it became possible to imagine a tabula rasa landscape wiped clean of the "savage" presence, a national park alienated from Indian land use and inviting tourists to enjoy its "pure" nature. But the poverty of the reservations combined with the presence of diminished but still extant herds of elk and bison in the park meant that these Indians mounted large hunts, leaving the reservations for large stretches of time in the fall to harvest wild game in the park in order to avoid starvation in the coming winter. Not only did park officials and the military strive to keep Indians from entering or using the park or its resources, the official histories and information displays for decades perpetuated the myth of a nature untouched or untrammeled by man, ludicrously asserting that native peoples were frightened by the geysers and mud pots, believing the valley to be populated by evil spirits.

As Native Americans were excluded from the park and confined to reservations hundreds of miles from Yellowstone, local whites living just outside the newly established boundaries proved adept at defying government authority in order to preserve their economic activities and way of life. According to Jacoby,

> The fact that many locals learned how to slip this red tape off their rifles [installed by soldiers trying to prevent hunting within park boundaries] and hunt in the park as before provided an apt illustration of the ability of the region's residents to elude the army's controls. This resistance manifested itself in many ways, most dramatically in the creation of a shadow landscape of surreptitiously erected footbridges and 'unfrequented and little known trails,' used by those who wanted to sneak past the official entrances and gather an illegal load of wood or poach some game.[19]

Many poachers were so active in the park that they built cabins or other structures to use while they hunted within the boundaries. Built in dense forest, these hideouts were scattered throughout the park, "an illicit counterpoint to the officially sanctioned tourist landscape of hotels and campsites that spread across Yellowstone in the late nineteenth century."[20]

The deployment of army troops in the parks curtailed poaching and ended native use of this landscape. Federal troops were used in various parks to control hunting and non-sanctioned uses until the creation of the National Park Service in 1916. National Forests also excluded local populations from land and blocked them from traditional subsistence and economic activities. In the Southwest, Hispanic and Pueblo people lost access to forests and grazing lands, and in the Pacific Northwest, native peoples were no longer able to harvest berries in national forests. Additionally, poor whites across the West, as well as Hispanic and Basque sheepherders, were not able to compete with better financed ranchers for grazing leases. The creation of parks and national forests arose from the need to protect resources and important landscapes in a culture with little commitment to stewardship. To poor whites, Hispanics, Indians, and a few other groups, these new restrictions were part of a pattern of conquest and theft of resources they had experienced for decades or centuries. For better-financed stockmen, mining companies, and logging interests, the new forests and parks reflected government abuse of power and restrictions on economic activities needed by the American people. The expanded government power and determination to more efficiently use or protect nature through the drawing of boundaries was a necessary step against the ruthless economic exploitation of nature in that era.

Women and Healthy Cities

While conservationists strived to protect fish and forests, grassland and elk, in the cities women activists took on the challenge of cleaning and improving the urban environment. Women activists led clean-up and sanitation programs, demanded municipal reforms, built playgrounds, and paved new ground in the study of environment and disease. Because the genesis of American environmentalism is found in this conservation era, recognizing women's activism in the cities acknowledges that urban conservation is as important to understanding shifting attitudes about environment and human relations to ecosystem as is Hetch Hetchy or Gifford Pinchot. Women activists' cleaning of the cities and workplaces also lays a foundation for later environmental justice activism. The rise of cities as centers of exchange and the conversion of resources to manufactured goods and wealth drove the American economy to unprecedented heights. But, as capitalists and farmers failed to employ principles of stewardship in logging, fishing, mining, and farming, so too city managers and businessmen failed to protect the inhabitants of cities as they filled with garbage and disease. The efforts to remedy the environmental impacts in the city are consistent with the work of conservationists in that these reformers sought to ameliorate the hazards of economic growth and did this through research, activism, and an expansion of government power.

City streets were reservoirs of filth and pathogens in the late 19th and early 20th centuries. Residents dumped garbage, including food items, clothing, shoes, furniture, various other items, and ash, directly into streets and alleys. In addition to the approximately 300 pounds of garbage and rubbish produced annually per person by Americans in the cities, the average urban citizen also produced 300 to 1,200 pounds of ashes. One way to deal with waste was to feed it to pigs in the city, but the increased incidence of trichinosis and hog cholera arising from the handling and transport of the waste led to the abandonment of this practice. Municipal governments contracted with private businessmen to remove refuse, but service was spotty and poorer neighborhoods were often neglected. Waste gathered and moved from city streets was commonly dumped in rivers, oceans, and lakes or in massive mounds on open ground in the city. New York City discarded a large portion of its garbage directly into the ocean for years and, according to historian Martin Melosi, in "1886 . . . dumped 1,049,885 of its 1,301,180 cartloads of refuse into the ocean."[21] Much of Chicago's refuse went directly into Lake Michigan.

In addition to personal waste was added the problem of industrial pollution. Manufacturers dumped chemicals, metals, and other waste products directly into alleyways and onto vacant lots as well as into available waterways. Urban governments did not regulate effluent discharge from industry for fear of losing business and jobs. They were compelled

Figure 8.5 In the early 19th century refuse was a persistent problem in American cities, particularly in poor sections of the city suffering from poverty, overcrowded tenements, and limited municipal service. Children had few options for exercise and play. University of Illinois at Chicago.

then to capture clean water from outside the city and pipe it in, which allowed the urban waterways to become quite degraded. The description urban environmental studies scholar Robert Gottlieb paints of one particularly rough part of Chicago works as a general description for the environments of many poor, urban areas:

> [Packingtown] was bounded on the north by the stagnant backwater of the South Fork of the Chicago River, where putrefying refuse filled in the river bed and formed 'long, hideous shoals along the bank.' The decaying organic matter released quantities of carbonic acid gas, which continually broke through the 'thick scum of the water's surface,' causing this section of the river to be named 'Bubbly Creek.'[22]

Gottlieb describes the city garbage dumps in the area, dug originally for the clay to be used in brickyards:

> These vast, open pits were fed daily by horse-drawn wagons carrying trash from throughout the city. This included waste from the packers that was then burned, creating a permanent stretch of fire surrounded by a moat to keep it from spreading.[23]

As if this waste and pollution was not enough, land laying to the east of this neighborhood was used for discarding hair and other waste from the meat packing plants to rot and dry. Typical of cities of this era, there were no trees, grass, or shrubs and limited garbage removal along with no sewer system. "Packingtown had become an urban environmental catastrophe by the turn of the century."[24] While processing corn, wheat, beef, hogs, and lumber had allowed Chicago to take its place as a great industrial city, part of the cost of this wealth was a degraded urban and human environment.

Problems with human waste and industrial pollution in the cities were compounded by the urine, feces, and carcasses of urban beasts of burden. Animal waste from the thousands of working horses in each city filled the streets. According to Melosi,

> In New York City alone there were about 130,000 horses; in Chicago something over 74,000. Engineers estimated that the normal, healthy city horse produced between 15 and 35 pounds of manure and about a quart of urine daily, most of which ended up in the streets. Cumulative totals of manure produced by horses were staggering: 26,000 horses in Brooklyn produced about 200 tons daily; 12,500 horses in Milwaukee produced 133 tons.[25]

This was on top of the waste generated by animals raised by urban dwellers for milk, eggs, and meat. Cities employed street sweepers from the lower class, but they inevitably missed some of the waste. Because many city streets in the late 19th century were not paved, manure was ground into the dirt and mixed with dust to blow into people's faces and settle on their clothes.

The life of an urban workhorse was brutish and short. Because of poor conditions, overwork, and abuse, they were lucky to live more than four years. This meant a lot of horse carcasses to collect and dispose of—15,000 in 1880 in New York alone. Poorer communities experienced the worst of this, struggling the most with piles of garbage, debris, animal waste, and corpses left to rot and fester. Prior to the urban playground and park movement, these sites were often the very places children played, increasing their risk of illness and injury. Regular outbreaks of cholera, typhoid, and yellow fever in American cities contributed to high child-mortality rates and were partially tied to unsanitary conditions in urban environments.

This view of the late 19th century city as an ecological dystopia is challenged by Ted Steinberg in *Down to Earth,* where he argues eloquently for the value of the organic city. While acknowledging problems arising from animal waste and carcasses as well as disease, he also notes that urban residents made use of kitchen gardens and raised a number of domestic animals for food and trade. Moreover, many poor city people earned an income by collecting animal waste for reuse. The sanitation of the city, while making the city healthier in certain ways, also reduced the complexity of the urban ecosystem and removed city dwellers further from their food sources. This rendered them more dependent on long networks of food production and distribution. As a result, over the course of the 20th century, Americans, urban and rural, lost their agricultural knowledge while increasingly harmful and ecologically destructive farming and livestock practices remained essentially concealed from sight. It is interesting to note this argument while also seeing the contemporary, thriving urban farming movement and the removal of ordinances in many cities against poultry raising as forms of city agriculture are reinvented in the 21st century.

But typhoid, cholera, and other diseases were real threats inflicting a disproportionate death toll on the poor and on their children. Women in the cities took up the mantle of urban environmental reform and led the cleanup of urban areas, jumping into action in cities across the nation with the goal of reducing garbage and disease. These women activists applied pressure on municipal governments to provide more funding and urban improvement. One of the most influential of these groups was the Ladies' Health Protective Association of New York City, created in 1884. After organizing to remove a massive manure pile from their community, they then expanded their agenda to include improved sanitation in schools and slaughterhouses, street cleaning, and refuse removal. These women activists framed their efforts as seeking an extension of the cleanliness and order of a well-kept home into the city streets. The way they employed a rhetoric of domesticity to attack this problem was instrumental in building support in the city and passing laws to improve street cleaning and sanitation. But they were much more than urban housecleaners—their reforms would help enlarge government and change the relationship of government to business and nature.

Women activists did not limit themselves to municipal reform. They also sought to educate citizens to dispose of waste responsibly by distributing informational leaflets and strategically situating garbage cans. A woman's club in Florida decorated cans with rhymes such as the following:

> My name is Empty Barrel,
> I'm hungry for a meal.
> Pray fill me full, kind stranger,
> With trash and orange peel.[26]

Education reached more sophisticated levels with the Boston's Women's Municipal League using a traveling exhibit extolling the benefits of sanitation. According to Melosi, it "consisted of models contrasting dirty and clean meat markets, dairies, and tenements and provided information about the prevention and cure of tuberculosis."[27]

The "second city" on the shores of Lake Michigan provided the backdrop and laboratory for major innovations in urban environmental reform and occupational health that proved influential on the local and national level. The source of much of the energy and passion that drove those efforts was Hull House. This settlement house was established west of downtown Chicago in 1889 by Jane Addams and Ellen Gates Starr. It was the model for the settlement house movement that would explode from there, leading to approximately 400 in various cities; women reformers lived in the house and worked in the local community. In Hull House generally young, single, educated women sought opportunities to address injustice in society. They worked with the immigrant community inside Hull House and in the homes and neighborhoods of the immigrants, offering classes in topics such as literature and sewing as well as kindergarten for children. Settlement houses in other cities pursued similar goals, implementing the model of Hull House. Many of these urban reform centers took on improving and cleaning the urban environment, but the

Figure 8.6 Jane Addams and Elizabeth Burke headed as delegates to the Women's Suffrage Legislature in 1911. Addams created Hull House in Chicago and provided leadership and support for a number of municipal reforms in the city that influenced state and national politics. Women activists in settlement houses across the nation not only worked for reform but increasingly demanded the vote to expand their role and influence in American society and because they saw it as their political right. Library of Congress.

Chicago settlement house led the way, creating innovations and demanding reforms that had national implications.

Rotting garbage in the streets was a major problem in immigrant neighborhoods. Investigating the failure to remove refuse from these communities, Addams's frustration mounted until she submitted a bid to collect and remove waste in the 19th ward. Winning the position of garbage inspector for that community, she and other Hull House workers collected and removed waste with "Miss Addams' garbage phaeton" while publicizing the issue to a degree that compelled city government to enact garbage management reforms. Also from the nexus of urban reform led by Hull House, Mary McDowell fought to improve the polluted landscape of Packingtown and the greater Chicago area, earning the sobriquet "Garbage Lady" along the way. Under her leadership several women's groups created waste committees; these groups played an integral role in the creation of the Chicago City Waste Commission in 1913. This commission proved influential in nationwide waste management reforms and marked a transformation of the urban environment and the role of government in cleaning and maintaining city cleanliness.

Florence Kelly cut her teeth on urban reform and launched her career from Hull House, focusing her attention on urban playgrounds and environmental workplace problems. After leading an effort to remove particularly unsafe and unsanitary tenements, she also helped build the first urban playground in the nation. Administered by Hull House employees, this work helped launch a national urban playground movement. Jane Addams added to this effort as well, serving as president of the newly created National Playground Association. Women activists in cities across the country pushed for playgrounds because they believed they were essential to provide an alternative to the polluted sites poor children regularly played in. Regular exercise and separation from garbage would improve their health and the urban environment. Experiencing success with tenement removal and playground construction, Kelly applied herself to the study of workplace environments and their health impacts. In a comprehensive study of the community surrounding Hull House, she was able to show that the practice of "sweating," that is the situating of garment factories in tenements and distributing jobs into tenement homes, extended industrial pollution into living spaces and was "ruinous to the health of the employees." The results of the study were published in *Hull House Maps and Papers* and led to an investigation and subsequent state legislation that provided protections for women and children working at home and created eight-hour days for women and children. Kelly proceeded from this work to become a state and national leader on the issues of workplace reform for women and children.

Figure 8.7 Women reformers at Hull House and across the country not only cleaned the cities but also built the first urban playgrounds in an effort to improve the health of children by giving them places to exercise and play. University of Illinois at Chicago.

Hull House clearly served as an important incubator for ideas and reform, and there is perhaps no better example of this than a pioneer in the fields of occupational health and workplace environment, Alice Hamilton. Seeking a life of service after medical school at the University of Michigan, Alice Hamilton moved into Hull House in 1902, thus beginning a remarkable career fighting for improved public health and occupational safety and health. Working as a professor of pathology at Northwestern University, the settlement house presented a perfect environment for understanding the health threats arising from the urban environment and to tap into and use the reformer energy buzzing in the Chicago reform community. In addition to several other reform activities, including organizing a health clinic for babies, Hamilton investigated a typhoid epidemic and helped discover that sewage

Figure 8.8 An important pioneer in the study of environment and disease, Alice Hamilton also created the field of occupational health in the United States. She did early work as an activist at Chicago's Hull House while also employed as a professor of pathology at Northwestern University. Her study of the impacts of chemicals and metals on workers in particular led to several important reforms and safer workplaces. Library of Congress.

outflow was the key factor in the disease appearing in several communities. Her work on that health crisis also resulted in 11 out of 24 members of the Chicago Board of Health being fired and subsequent improvements in sanitary measures. Hamilton also began her career in occupational health at Hull House, gaining an education about the number of ailments arising from industrial conditions and work environment. Seeing the symptoms of phossy jaw (see Document 8.3) and the unwillingness of American companies to make reforms versus the improvement in conditions already instituted in Europe, she developed a keen interest in investigating occupational environment and health issues.

Appointed in 1908 to the Illinois Commission in Occupational Diseases, Hamilton served as the medical investigator on the impact of industrial pollutants on workers. The study of lead poisoning in particular constituted the focal point of this study. This "shoe-leather epidemiology," as she coined it, meant visits to industrial work sites with close observation of how workers interacted with a number of chemicals and metals, such as lead, zinc, and other health hazards, along with testing for the amount of lead in various materials and manufactured goods. Investigators visited workers' homes, interviewed them, and examined symptoms of poisoning. Likewise they interviewed druggists, pharmacists, and doctors, as well as undertakers, to gain a better understanding of the health impacts of lead and other chemicals and metals on workers. She was able to determine after this investigation, which included inspections of more than 300 work sites, that workers were exposed to lead in approximately 70 different processes. The result of this remarkable study was the passage of groundbreaking laws passed in Illinois in 1911 requiring improvement of the workplace environment and protections for workers. Factory owners were compelled to provide employees working with lead, zinc, brass, and arsenic with monthly medical examinations. In addition, the companies were required to limit exposure to these materials and report illnesses to the Department of Factory Inspection, which could conduct prosecutions.

With the Progressive Era, a period of professionalization of many fields, and an emphasis on science and technology, the amateur reform efforts were supplanted by the emergence of the field of sanitation engineering. Focused on a bacteriological approach, they worked within municipal government to improve access to clean water, to develop sewer systems, and to professionalize garbage clean up and removal. By 1910 more than 70 percent of cities with populations exceeding 30,000 ran their own water supply systems with extensive distribution of water throughout the community via pipes into homes, leading to a dramatic increase in personal water consumption. Major progress was made in the area of sewage removal. According to Melosi,

> In 1860 there were only 136 municipal waterworks in the country; by 1880, there were 598. Increases in sewer lines of all kinds were no less extensive. The miles of sewers increased from 8,199 (in cities of more than ten thousand people) in 1890 to 24,972 (in cities of more than thirty thousand people) in 1909.[28]

Waste removal was professionalized in similar ways with cities using incinerators and reduction to remove and compress garbage and, in the years before World War I, the growing practice of buried garbage, better known today as landfills, to replace above-ground mounding and dumping in oceans, rivers, and lakes.

Women's urban conservation was aimed directly at the consequences and wreckage of urban industrial growth. The untrammeled economic growth and production of manufacturing abundance degraded city environments, and these activists were determined to correct these excesses. The expansion of government's power at the municipal, state, and federal level was in large part due to their demands for reforms. Government increasingly took on the responsibility of protecting its citizens from industrial and environmental threats even as other conservationists expanded government power to protect nature from humans. For the first time in American history a regulatory state was created and some limited restrictions placed on free-market capitalism. Women's activism in urban conservation, as in broader Progressive era reforms, also convinced them they needed the vote and a greater voice in American society and politics. Women viewing the river sewers in their cities, children playing in piles of fetid garbage, and workers crippled by industrial pollutants decided enough was enough and demanded the vote. In this way environmental issues and urban conservation contributed to an expansive political system that included women; America became more democratic as a result.

After three centuries of reckless resource extraction, destructive agriculture, and economic development, Americans finally began dealing with pressing environmental problems on a local, state, and federal level. From waste and disease in the city and workplace to land and resource loss, resolving these problems required investigation, creative thinking, and a willingness to employ the power of government. Regulations on harvest, logging practices, and exposure to chemicals and garbage, as well as expanded protection for state and national parks, all reflect a broad, disparate effort to improve the natural and built environment as well as set limits upon the damage that could be done by industry and individuals. Americans thus entered into a period where they began to reevaluate land and resource use as well as their own ideas about the best uses of nature. In doing so, they inevitably made mistakes and, in some cases, expanded the power of government in a manner that was detrimental to those least able to protect themselves. That said, this conservation era initiated a long century of conservation and environmental activism and reform that would transform the landscape of the nation as well as Americans' views of nature.

Document 8.1 From *My First Summer in the Sierra, John Muir,* 1911

Nicknamed "the Wilderness Prophet" and "Father of the National Parks," John Muir gave voice to a new way of seeing, valuing, and using nature. In his essays and books he described nature's beauty and grandeur and pointed out the sacred in beautiful, transcendent landscapes. His activism, publications, and language helped create a preservationist view of nature and land protection that would gain strength over the course of the 20th century. Muir and the Sierra Club led the fight against the flooding of the Hetch Hetchy Valley in Yosemite National Park. He wrote furiously for long hours to publish *My First Summer in the Sierra* to try and stop the dam project. The following passage is from that classic work.

1. How does faith and knowledge of Christianity inform Muir's appreciation for and understanding of nature? According to Muir, what is the relationship of God and the Yosemite landscape?

2. To what degree does physical labor, knowledge, and direct interaction with the environment inform Muir's appreciation and descriptions?

. . . Wherever we go in the mountains, or indeed in any of God's wild fields, we find more than we seek. Descending four thousand feet in a few hours, we enter a new world—climate, plants, sounds, inhabitants, and scenery all new or changed. Near camp the goldcup oak forms sheets of chaparral, on top of which we may make our beds. Going down the Indian Canon we observe this little bush changing by gradations to a large bush, a small tree, and then larger, until on the rocky taluses near the bottom of the valley we find it developed into a broad, wide-spreading, gnarled, picturesque tree from four to eight feet in diameter, and forty or fifty feet high. Innumerable are the forms of water displayed. Every gliding reach, cascade, and fall has characters of its own. Had a good view of the Vernal and Nevada, two of the main falls of the valley, less than a mile apart, and offering striking differences in voice, form, color, etc. The Vernal, four hundred feet high and about seventy-five or eighty feet wide, drops smoothly over a round-lipped precipice and forms a superb apron of embroidery, green and white, slightly folded and fluted, maintaining this form nearly to the bottom, where it is suddenly veiled in quick-flying billows of spray and mist, in which the afternoon sunbeams play with ravishing beauty of rainbow colors. The Nevada is white from its appearance as it leaps out into the freedom of the air. At the head it presents a twisted appearance, by an overfolding of the current from striking on the side of its channel just before the first free outbounding leap is made. About two thirds of the way down, the hurrying throng of comet-shaped masses glance on an inclined part of the face of the precipice and are beaten into yet whiter foam, greatly expanded, and sent bounding outward, making an indescribably glorious show, especially when the afternoon sunshine is pouring into it. In this fall—one of the most wonderful in the world—the water does not seem to be under the dominion of ordinary laws, but rather as if it were a living creature, full of the strength of the mountains and their huge, wild joy.

From beneath heavy throbbing blasts off spray the broken river is seen emerging in ragged boulder-chafed strips. These are speedily gathered into a roaring torrent, showing that the young river is still gloriously alive. On it goes, shouting, roaring, exulting in its strength, passes through a gorge with sublime display of energy, then suddenly expands on a gently inclined pavement, down which it rushes in thin sheets and folds of lace-work into a quiet pool,—"Emerald Pool," as it is called,—a stopping-place, a period separating two grand sentences. Resting here long enough to part with its foam-bells and gray mixtures of air, it glides quietly to the verge of the Vernal precipice in a broad sheet and makes its new display in the Vernal Fall; then more rapids and rock tossings down the canon, shaded by live oak, Douglas spruce, fir, maple, and dogwood. It receives the Illilouette tributary, and makes a long sweep out into the level, sun-filled valley to join the other streams which, like itself, have danced and sung their way down from snowy heights to form the main Merced—the river of Mercy. But of this there is no end, and life, when one thinks of it, is so short. Never mind, one day in the midst of these divine glories is well worth living and toiling and starving for.

. . . It seems strange to me that visitors to Yosemite should be so little influenced by its novel grandeur, as if their eyes were bandaged and their ears stopped. Most of those I saw yesterday were looking down as if wholly unconscious of anything going on about them, while the sublime rocks were trembling with the tones of the mighty chanting congregation of waters gathered from all the mountains round about, making music that might draw angels out of heaven. Yet respectable-looking, even wise-looking people were fixing bits of worms on bent pieces of wire to catch trout. Sport they called it. Should church-goers try to pass the time fishing in baptismal fonts while dull sermons were being preached, the so-called sport might not be so bad; but to play in the Yosemite temple, seeking pleasure in the pain of fishes struggling for their lives, while God himself is preaching his sublimest water and stone sermons!

Document 8.2 Discussing Efforts to Remove Garbage, from *Twenty Years at Hull House, Jane Addams,* 1910

Jane Addams's Hull House was an integral agent in urban reform and in the rise of Progressive politics in the early 20th century. Established to help Chicago immigrants adjust to life in the city and providing classes in hygiene, literacy, and literature, as well as childcare for families where both parents worked, Hull House was also a locus of women's activism in Chicago and hundreds of other cities as the settlement house movement spread. Disease, malnutrition, and poverty were striking urban problems that Jane Addams and others sought to rectify through municipal reform as well. They also led an effort to create city playgrounds for children who had no avenue for physical exercise.

1. What were the primary health and sanitation problems in the poor sections of Chicago? What groups are criticized in this passage?
2. What steps do the women of Hull House and of the communities take to clean up and improve their urban environment? How do you situate this within the definition of environmentalism in that era and today?

One of the striking features of our neighborhood twenty years ago, and one to which we never became reconciled, was the presence of huge wooden garbage boxes fastened to the street pavement in which the undisturbed refuse accumulated day by day. The system of garbage collecting was inadequate throughout the city but it became the greatest menace in a ward such as ours, where the normal amount of waste was much increased by the decayed fruit and vegetables discarded by the Italian and Greek fruit peddlers, and by the residuum left over from piles of filthy rags which were fished out of city dumps and brought to the homes of the rag pickers for further sorting and washing.

. . . It is easy for even the most conscientious citizen of Chicago to forget the foul smells of the stockyards and the garbage dumps, when he is living so far from them that he is only occasionally made conscious of their existence but the residents of a Settlement are perforce constantly surrounded by them. During our first three years on Halsted Street, we had established a small incinerator at Hull-House and we have many times reported the untoward conditions of the ward to the city hall. We had also arranged many talks for the immigrants, pointing out that although a woman may sweep her own doorway in her native village and allow the refuse to innocently decay in the open air and sunshine, in a crowded city quarter, if the garbage is not properly collected and destroyed, a tenement-house mother may see her children sicken an die, and that the immigrants must therefore, not only keep their own houses clean, but must also help the authorities to keep the city clean?

. . . The Hull-House Woman's Club had been organized the year before by the resident kindergartner who had first inaugurated a mothers' meeting. The members came together, however, in quite a new way that summer when we discussed with them the high death rate so persistent in our ward. After several club meetings devoted to the subject, despite the fact that the death rate rose highest in the congested foreign colonies and not the streets in which most of the Irish American club women lived, twelve of their number undertook in connection with the residents, to carefully investigate the condition of the alleys. During August and September the substantiated reports of violations of the law sent in from Hull-House to the health department were one thousand and thirty-seven. For the club woman who had finished a long day's work of washing or ironing followed by the cooking of a hot supper, it would have been much easier to sit on her doorstep during a summer evening than to go up and down ill-kept alleys and get into trouble with her neighbors over the condition of their garbage boxes. It required both civic enterprise and moral conviction to be willing to do this three evenings a week during the hottest and most uncomfortable months of the year . . .

With the two or three residents who nobly stood by, we set up six of those doleful incinerators which are supposed to burn garbage with the fuel collected in the alley itself. The one factory in town which could utilize old tin cans was a window weight factory, and we deluged that with ten times as many cans as it could use—much less would pay for. We made desperate attempts to have the dead animals removed by the contractor who was paid most liberally by the city for that purpose but who, we slowly discovered, always made the police ambulances do the work, delivering the carcasses upon freight cars for shipment to a soap factory in Indiana where they were sold for a good price although the contractor himself was the largest shareholder in the concern. Perhaps our greatest achievement was the discovery of a pavement eighteen inches under the surface in a narrow street . . . This pavement became the *casus belli* between myself and the street commissioner when I insisted that its restoration belonged to him, after I had removed the first eight inches of garbage. The matter was finally settled by the mayor himself, who permitted me to drive him to the entrance of the street in what the children called my "garbage phaeton" and who took my side of the controversy.

. . . Perhaps no casual visitor could be expected to see that these matters of detail seemed unimportant to a city in the first flush of youth, impatient of correction and convinced that all would be well with its future. The most obvious faults were those connected with the congested housing of the immigrant population, nine tenths of them from the country, who carried on all sorts of traditional practices in the crowded tenements. That a group of Greeks should be permitted to slaughter sheep in a basement, that Italian women should be allowed to sort over rags collected from the city dumps, not only within the city limits but in a court swarming with little children, that immigrant bakers should continue unmolested to bake bread for their neighbors in unspeakably filthy spaces under the pavement, appeared incredible to visitors accustomed to careful city regulations.

Document 8.3 Occupational Illness in Chicago, from *Exploring the Dangerous Trades, Alice Hamilton*, 1943

Alice grew up and was home schooled in Fort Wayne, Indiana, and then earned her doctor of medicine at the University of Michigan. She pursued further studies in bacteriology and pathology in Munich and Leipzig and conducted further postgraduate studies at Johns Hopkins University. She then took a professor of pathology position at Northwestern University and joined Hull House as both resident and worker. Hull House proved integral to her career and pioneering work in occupational epidemiology. Her discovery of the environmental conditions in the workplace leading to a condition called phossy jaw helped create an occupational health regulatory movement during the Progressive era, a movement in which she played a key role at the state and federal level. She also led studies and surveys, of problems such as the impact of lead poisoning, that provided the first physical evidence of negative health impacts and led to workplace reforms. She became the first female faculty hired at Harvard when she joined the newly formed Department of Industrial Medicine in 1919.

1. What is phossy jaw and what are its causes? How does Hamilton characterize American responses to this ailment?
2. What role does living in Hull House and interacting with the workers and their families play in Hamilton's interest and study of these occupational health issues?

It was also my experience at Hull-House that aroused my interest in industrial diseases. Living in a working-class quarter, coming in contact with laborers and their wives, I could not fail to hear tales of the dangers that workingmen faced, of cases of carbon-monoxide gassing in the great steel mills, of painters disabled by lead palsy, of pneumonia and rheumatism among the men in the stockyards. Illinois then had no legislation providing compensation for accident or disease caused by occupation. (There is something strange in speaking of "accident and sickness compensation." What could "compensate" anyone for an amputated leg or a paralyzed arm, or even an attack of lead colic, to say nothing of the loss of a husband or son?)

Phossy jaw is a very distressing form of industrial disease. It comes from breathing the fumes of white or yellow phosphorus, which gives off fumes at room temperature, or from putting into the mouth food or gum or fingers smeared with phosphorus. Even drinking from a glass which has stood on the workbench is dangerous. The phosphorus penetrates into a defective tooth and down through the roots to the jawbone, killing the tissue cells which then become the prey of suppurative germs from the mouth, and abscesses form. The jaw swells and the pain is intense, for the suppuration is held in by the tight covering of the bone and cannot escape, except through a surgical operation or through a fistula boring to the surface. Sometimes the abscess forms in the upper jaw and works up into the orbit, causing the loss of an eye. In severe cases one lower jawbone may have to be removed, or an upper jawbone—perhaps both. There are cases on record of men and women who had to live all the rest of their days on liquid food. The scars and contractures left after recovery were terribly disfiguring, and led some women to commit suicide. Here was an industrial disease which could be clearly demonstrated to the most skeptical.

. . . All this I had learned, but I had been assured by medical men, who claimed to know, that there was no phossy jaw in the United States because American match factories were so scrupulously clean. Then in 1908 John Andrews came to Hull House and showed me the report of his investigation of American match factories and his discovery of more than 150 cases of phossy jaw. It seems that in the course of a study of wages of women and children made by the Bureau of Labor, under Carroll Wright, investigators came across cases of phossy jaw in women match workers in the South. This impelled Wright to institute an investigation in other match centers. Andrews was asked to carry it out and did so, with a result most disconcerting to American optimism. Some of the cases he discovered were quite as severe as the worst reported in European literature—the loss of jawbones, of an eye, sometimes death from blood poisoning.

This episode in the history of industrial disease is very characteristic of our American way of dealing with such matters. We learned about phossy jaw almost as soon as Europe did. The first recognized case was described by Lorinser of Vienna in 1845; the first American case was treated in the Massachusetts General Hospital only six years later, in 1851. But while all over continental Europe and England there was eager discussion of this new disease, many cases were reported and all sorts of preventive measures proposed, practically nothing was published in American medical journals from 1851 to 1909, both laymen and public health authorities contenting themselves with the assurance that all was well in our match industry. When, however, the facts were at last made public in 1909, action was prompt. A safe substitute for white phosphorus had been discovered by a French chemist, the sesquisulphide, the American patent rights for which had been bought by the Diamond Match Company. This company with rare generosity, waived its patent rights and allowed the free use of sesquisulphide to the whole industry, and this made it possible for Congress to pass the Esch law, which imposed a tax on white-phosphorus matches high enough to cover the difference in cost between them and sesquisulphide matches. So phossy jaw disappeared from American match factories.

Notes

1 Quoted in Paul Schullery, *Old Yellowstone Days* (Albuquerque: University of New Mexico Press, 2011), 143, 144.

2 Ibid., 143, 144.

3 Ibid.
4 Quoted in Douglas Brinkley, *The Wilderness Warrior: Theodore Roosevelt and the Crusade for America* (New York: Harper Perennial, 2009), 195.
5 Quoted in Stephen Fox, *The American Conservation Movement: John Muir and His Legacy* (Madison: The University of Wisconsin Press, 1981), 124.
6 Quoted in Brinkley, *The Wilderness Warrior,* 413.
7 Ibid., 410.
8 Nancy Langston, *Forest Dreams, Forest Nightmares: The Paradox of Old Growth in the Inland West* (Seattle: University of Washington Press, 1995), 90.
9 Duncan Hay, *Hydroelectric Development in the United States, 1880–1940* (Washington, DC: Edison Electric Institute, 1991), 100–103; Report of Committee of the National Electric Light Association, "Western Hydroelectric Transmission Developments," *Journal of Electricity, Power and Gas* 34:23 (June 5, 1915): 444–447.
10 John N. Cobb, *Pacific Salmon Fisheries: Appendix III to the Report of U.S. Commissioner of Fisheries for 1916* (Washington, DC: Washington Government Printing Office, 1917), 244.
11 Quoted in Jim Lichatowich, *Salmon Without Rivers: A History of the Pacific Salmon Crisis* (Washington, DC: Island Press, 2001), 129.
12 Leslie Darwin, *Thirtieth and Thirty-First Annual Reports of the State Fish Commissioner to the Governor of the State of Washington, April 1, 1919, to March 31, 1921* (Olympia, WA: Frank M. Lamborn, 1921), 15.
13 Quoted in Char Miller, "A Sylvan Prospect: John Muir, Gifford Pinchot and Early Twentieth-Century Conservationism," in Michael Lewis, editor, *American Wilderness: A New History* (New York: Oxford University Press, 2007), 144.
14 Quoted in Fox, *The American Conservation Movement,* 144.
15 Ibid.
16 Quoted in Polly Welts Kaufman, *National Parks and the Woman's Voice: A History* (Albuquerque: University of New Mexico Press, 1996), 32.
17 Quoted in Fox, *The American Conservation Movement,* 142.
18 Kaufman, *National Parks and the Woman's Voice,* 35.
19 Karl Jacoby, *Crimes Against Nature: Squatters, Poachers, Thieves, and the Hidden History of American Conservation* (Berkeley: University of California Press, 2001), 108.
20 Ibid.
21 Martin Melosi, *The Sanitary City: Urban Infrastructure in American from Colonial Times to the Present* (Baltimore, MD: The Johns Hopkins University Press, 2000), 180.
22 Robert Gottlieb, *Forcing the Spring: The Transformation of the American Environmental Movement* (Washington, DC: Island Press, 2005), 103.
23 Ibid.
24 Ibid.
25 Martin Melosi, *Garbage in the Cities: Refuse, Reform, and the Environment* (Pittsburgh, PA: University of Pittsburgh Press, 2005), 20.
26 Quoted in Ibid., 100.
27 Ibid., 101.
28 Ibid., 76.

Further Reading

Brinkley, Douglas. *The Wilderness Warrior: Theodore Roosevelt and the Crusade for America.* New York: Harper Perennial, 2009.

Chase, Alston. *Playing God in Yellowstone: The Destruction of America's First National Park.* New York: Harcourt, Brace, Jovanovich Publishers, 1987.

Crane, Jeff. *Finding the River: An Environmental History of the Elwha.* Corvallis: Oregon State University Press, 2011.
Fox, Stephen. *The American Conservation Movement: John Muir and His Legacy.* Madison: The University of Wisconsin Press, 1981.
Gottlieb, Robert. *Forcing the Spring: The Transformation of the American Environmental Movement.* Washington, DC: Island Press, 2005.
Hays, Samuel. *Conservation and the Gospel of Efficiency: The Progressive Conservation Movement, 1890–1920.* Pittsburgh, PA: University of Pittsburgh Press, 1959.
Jacoby, Karl. *Crimes Against Nature: Squatters, Poachers, Thieves, and the Hidden History of American Conservation.* Berkeley: University of California Press, 2001.
Judd, Richard. *Common Lands, Common People: The Origins of Conservation in Northern New England.* Cambridge, MA: Harvard University Press, 2000.
Kaufman, Polly Welts. *National Parks and the Woman's Voice: A History.* Albuquerque: University of New Mexico Press, 1996.
Langston, Nancy. *Forest Dreams, Forest Nightmares: The Paradox of Old Growth in the Inland West.* Seattle: University of Washington Press, 1995.
Lewis, Michael, ed. *American Wilderness: A New History.* New York: Oxford University Press, 2007.
McEvoy, Arthur F. *The Fisherman's Problem: Ecology and Law in the California Fisheries, 1850–1980.* New York: Cambridge University Press, 1986.
Melosi, Martin. *Garbage in the Cities: Refuse, Reform, and the Environment.* Pittsburgh, PA: University of Pittsburgh Press, 2005.
Spence, Mark. *Dispossessing the Wilderness: Indian Removal and the Making of the National Parks.* New York: Oxford University Press, 1999.
Taylor, Joseph. *Making Salmon: An Environmental History of the Northwest Fisheries Crisis.* Seattle: University of Washington Press, 1999.
Worster, Donald. *Rivers of Empire: Water, Aridity, and the Growth of the American West.* New York: Oxford University Press, 1985.

Timeline

Creation of the First Wilderness Area	1924
Creation of the L-20 Regulations	1929
Publication of Aldo Leopold's *Game Management*	1930
Creation of the Agricultural Adjustment Administration	1933
Creation of the Civilian Conservation Corps	1933
Beginning of the Dust Bowl (the storms began)	1933
President Franklin Delano Roosevelt takes office	1933
Creation of the Tennessee Valley Authority	1933
Passage of the Duck Stamp Act	1934
Beginning of the Planting of the Great Plains Shelterbelt	1934
Creation of the Wilderness Society	1935
Completion of Hoover Dam	1935
Creation of the Soil Conservation Service	1935
Completion of Bonneville Dam	1937
Expansion and Designation of Olympic National Park	1938
Creation of the U Regulations	1939
Completion of Grand Coulee Dam	1941

9

Restoring and Transforming the Land in the 1920s and 1930s

A day in the life of a mother living on the Great Plains during the Dust Bowl was one of unstinting labor. Her day would start with a bucket of water from the well and cleaning the dirt off of dishes that had been cleaned the night before. After a storm the dirt lays thick and red, black, or brown, depending on the storm's origin, on every surface. The floors, the table, the counters, every surface covered in inches of dirt, and inside the cabinets and closets the dust has slipped in, covering dishes and clothes. After sweeping out the dust and shaking dirt out of all the clothes, linens, towels, and rugs, the dishes in the cabinet need to be cleaned again. She worked carefully trying to avoid painful shocks from the heavy static charge in the air.

Food was scarce, so she picked wild berries, dandelion leaves, sheep sorrel, and lambs quarter. Where she could grow a garden she did, watering carefully from the well, tenderly nursing each plant along, praying a storm didn't appear from the West and dump a blanket of dirt or kill the plants with static electricity. If the government bought dying cattle to keep a rancher going and killed them outside of town, she joined the other people in carving meat from the carcasses on the canyon floor. The woman canned everything: peaches, tomatoes, and pickles. The family needed this addition to the meager amounts of federal relief, and canned goods could be traded for clothing, tools, or medicine or sold. What a few short years earlier had been a supplement to the family income, which came primarily from the sale of wheat, now became the key source of food and an important source of money. She harvested eggs from her chickens and churned butter from her dairy cows, feeding her family and bartering for other food, goods, and even medical care. In a barren landscape of dust and death during the middle of the nation's greatest economic crisis, eggs and butter held great value.

Cleaning, cooking, raising, and storing food for her family and for barter were more than enough to fill her day, but she was not done. Sheets had to be wetted and hung over doors and windows to block sifting dust. Children needed to be given their dose of castor oil or some other home remedy to fend off the pneumonia that raged on the Plains, and the dying child tended to as the fever raged and consumed their young body. All of this and reminding the children to mind their manners and do their homework, while mending their threadbare clothes to make it through another week or day of school.

Finally, counseling or sitting quietly with the husband after another day without work or a day plowing land drier than old bone, without a cloud in the sky, together despairing and not knowing, never truly knowing, whether it was right to stick or better to pack the truck and try their luck in California.

At day's end the collapse into bed and anything but the quiet of night. Listening to a child's wracking coughing, wondering if it had gotten worse; hearing and trying to ignore the constant howl of the wind, the roof seeming to try and lift right off the house's frame; thinking about an eternity of dirt; and yearning for sleep in order to face another day.

The Great Depression devastated many Americans' lives, and in some communities the impact of ecological deterioration and economic collapse was so severe that people were thrown into what looked like a pre-capitalist economy. Forced to adapt, they turned to the government and neighbors and relied on credit at local stores. Americans also reverted to old ways of living and eating. Hunting, fishing, gardening, and the gathering of wild foods were integral to success in many communities, and women's work in the family and in the community often meant the difference between life and death. After decades of abundance and the belief that Americans were blessed in a way that other nations were not, the crises of the 1930s made many more citizens open to ideas of conservation, restoration of nature, and an increased federal presence in their daily lives.

World War I seemingly brought an abrupt end to the Progressive movement, and in the years following the war America roiled with labor tensions, race riots, a rural economic recession, and increasing xenophobia expressed in immigration quota laws and the rise of the Ku Klux Klan of the 1920s. American historians interpret this era as one in which the federal government reverted to a more conservative, business-friendly model and a period dominated by urban economic growth, prohibition, and the rise of organized crime. While the assumption that conservatism predominated and progressivism was over may be correct, in the areas of ecology, wildlife management, and park construction and expansion, it is clear that conservationists continued to challenge assumptions about proper nature management. They also critiqued earlier models of park development and national Forest Service policies while introducing revolutionary reforms through the 1920s and into the 1930s.

What is remarkable is much of this work was done within the context of a suffering American economy and populace, in a time when abundance could no longer be taken for granted. The plummeting American economy was caused by several factors. Disparity between pay and corporate profits, speculation in the stock market by corporations and banks, increasing personal debt, and an agricultural economy struggling since the end of World War I were but a few of the reasons for the depression that began in 1929 and reached its nadir in 1932 and 1933. Ironically, it was abundance itself that played an integral role in the collapse. The production of manufactured goods few Americans could afford to buy and the bountiful harvests of various commodities drove prices down and forced companies and farms out of business, costing hundreds of thousands of Americans their jobs. Certainly, it must be remembered that the reluctance of businessmen to increase wages in 1920s was a factor in the economic collapse as well.

When President Franklin Delano Roosevelt took office in March 1933, the nation's unemployment rate was at least 25 percent and more likely approached 35 percent. In addition to the high number of people completely unemployed, many of the fortunate job holders worked fewer hours for lower wages. Most states had shut down their banks, and factory production had ground to a halt. With his bank holiday and the rapid introduction and passage of the Emergency Banking Act, FDR was able to build strong support in the Republican and Democratic parties for what became known as the New Deal. While the primary goal of the New Deal was to get the economy moving again by fixing structural problems, providing relief, and stimulating the economy, many of the programs were designed to also modernize the economy through dam building and hydroelectric production, while other programs sought to restore the environment via erosion control and tree, grass, and shrub planting, along with other measures. At the same time that the federal government took on nationwide land and habitat restoration, conservationists worked on their own and through government to protect remaining habitats and restore degraded land. A preservationist impulse gained momentum in the 1930s, distinct from earlier conservationist thought, in which an increased emphasis in the importance of habitat, wilderness, and wildlife protection drove new policies and activism. The New Deal also transformed the landscape in ways both useful and damaging through the construction of dams across the nation. The Tennessee Valley Authority (TVA) managed the construction of several dams in the Southeast, providing hydroelectricity for homes and industrial manufacturing, and the American West was the beneficiary of federal government largesse with the construction of several major dams, including the Hoover Dam on the Colorado River and the Bonneville and Grand Coulee dams on the Columbia River. These dams did great damage to salmon fisheries in the West and undermined tribal culture and economies. At the same time the dam projects provided both irrigation for agriculture and urban development, while also generating millions of watts of electricity for manufacturing and homes. They would prove critical to the transformation of the western landscape and economy.

In the politically and economically conservative 1920s, some Americans began to question land-use and wildlife management policies and institute revolutionary changes. Through the dramatic expansion of federal government power that occurred during the 1930s under FDR and the New Deal, American society began to renegotiate its relationship with nature and lay the groundwork for much of the environmental activism of the post-war era.

Creating a More Perfect Nature

Conservationists originally sought an ordered landscape of desired and useful animal and bird species and, as in forestry and fisheries, embarked on ill-considered extermination programs to "improve" nature. It is in the eradication of predator species that the arrogance, false assumptions, cultural values, and shaky science of the conservationists are most clearly revealed. There was no room in the managed western environment for non-desired, "destructive species" such as wolves, cougars, bobcats, coyotes, golden eagles, prairie dogs, bald eagles, and many others. Ranchers and farmers had removed animals threatening their economic success for decades by the late 19th and early 20th century. Many communities had funded and organized massive extermination efforts. The Laramie County Protective Association paid over $1,150 in 1890 and 1891 for the hides of 231 wolves killed in that county. Earlier, the first form of government in the Willamette Valley of Oregon had risen around the effort to create a wolf eradication program. The blame for wholesale extermination should not be born by conservationists alone; this lays too much blame at the feet of government. Private individuals, ranches, farming consortiums, and local governments also worked diligently to remove undesired and feared species that competed with their economic activities. But federal agencies provided a level of funding and government support that pushed these animals to the brink of extinction.

Figure 9.1 With little knowledge of ecology, and no understanding of the role wolves played in a healthy ecosystem by culling elk herds and keeping herbivores from overgrazing the landscape, park managers sought to eliminate predators such as the wolf, mountain lion, bobcats, and others in order to increase the population of desired animals like elk. In this photo, soldiers and Yellowstone National Park employees proudly display a wolf skin. Courtesy of the Yellowstone National Park Archives.

When it came to predator extermination, however, the federal government played its part well. Under military control only 14 wolves had been killed in Yellowstone over 32 years. But the National Park's obsession with predator removal meant that at least 120 were killed within 7 years following the 1915 decision to move aggressively against them. As many as 1,300 coyotes were killed as were numerous mountain lions. This cleansing of nature of its undesirables (similar to the removal of Indians and poor whites from the park) was conducted from behind a veil of lies. The acting Superintendent of the National Park Service had declared publicly that the killing of predators was forbidden and only done when absolutely necessary; in the same period he also issued an order for their complete extermination. Not only did the park cooperate with the Bureau of Biological Survey in the killings, but it also encouraged employees, including plumbers and mechanics, to shoot, trap, and kill predators, supplementing their wages by selling their hides and pelts. It is believed that by 1926 the park's wolf population had been exterminated and most likely all the mountain lions as well.

With the annihilation of the larger predator, coyotes were able to expand their range dramatically. As a result the Biological Survey shifted its focus to these predators, both because of the believed threat to cattle and sheep but also to ensure a continued flow of federal appropriations. By 1923 more than half the agency's budget was designated for predator removal. The use of strychnine poison resulted in some gruesome deaths and widespread annihilation. "In a period of five weeks two Utah hunters put out a poison line approximately 300 miles long in a great loop and around their first two stations on their return found about 40 dead coyotes."[1] In 1923 more than 1,700,000 baits were used by the survey with that number easily doubling to more than 3,500,000 in 1924. Because magpies ate poisoned baits intended for coyotes, the logic of extermination required removal of the ubiquitous black and white western bird as well to improve the efficacy of coyote removal. Hunters also sought the eradication of magpies because of the threat they presented to the Hungarian partridge, Mongolian pheasant, and California quail. These invasive species were introduced for hunters into the Midwest and West far from their origins in Europe, Asia, and California. Their ground nested eggs were easy pickings for the native magpie. Conservationists reduced magpie populations to protect introduced non-native species to make nature better serve human interests and needs.

Smaller predators also made the kill list. River otters were exterminated on Fish Lake in Yellowstone National Park because they ate trout planted there for tourists. Park managers sought to recreate nature in order to benefit man and fit a preconceived architecture of abundance serving recreation and leisure. Similarly, pelicans carried a parasite detrimental to Yellowstone Lake trout, so eggs in pelicans' ground nests were smashed by park employees. Fisher, bobcats, martens, and numerous other competitors with man were also eagerly eliminated on public lands across the American West.

Predator removal created a massive hole in the food chain with unexpected and disastrous consequences among prey populations. Rodents in some ecosystems constitute a major part of the coyote's diet. With the coyote removed there was little left to check rodent population growth, and their numbers exploded in some areas. In southern California, millions of mice consumed crops and dangerously slickened roads due to the layer of bodies smashed by cars. In the San Juan Mountains of Colorado, porcupines no longer eaten by coyotes, wolves, and mountain lions also experienced an exponential increase in numbers, debarking and killing multitudes of trees in the process.

Predators were not the only species targeted for eradication in order to create a better nature for people and their preferred animals. Prairie dogs of the Great Plains and ground squirrels throughout the West were almost wiped out through extermination efforts. One advocate of prairie dog and ground squirrel elimination stated that "A county completely cleared of a rodent pest is a beautiful and convincing sight."[2] The same individual spoke in very precise terms of extermination efforts.

> The black tail prairie dog . . . has been completely exterminated from the state [Arizona]. This rodent infested 650,000 acres in Cochise and Graham counties. It took three years' time, 83,826 quarts of poisoned grain, 1920 quarts of CS2 gas and $75,381.00 or $0.11½ per acre to finish the job. The last dog was killed on June 25, 1922, and is on display at this conference.[3]

Prairie dog and ground squirrel towns, colonies, and dens were incompatible with horse herds, livestock ranching, and sheep herding as these animals broke their legs in holes or through collapsed tunnels and died. The prairie dog functions as a keystone species in the Great Plains environment, and the rapid and drastic diminishment of this species rippled out in numerous ways. They provide food for approximately 150 species, and their colonies function as independent habitats rich and diverse in numerous species. Some of the many animals hurt by prairie dog and ground squirrel eradication were golden eagles, owls, rattlesnakes, coyotes, the black-footed ferret, salamanders, and many more.

Resource management in the Kaibab Forest on the northern rim of the Grand Canyon provides an example of the type of disaster that can emerge from the hubris of conservationists and wildlife managers seeking to recreate nature to better serve humans without considering or comprehending the complexity of the environment. Set aside as a game preserve in 1906, the area hosted a deer population of approximately 4,000. Extermination of predators such as wolves, coyotes, and mountain lions in the preserve set loose an explosion of deer numbers. Historian Donald Worster explains the annihilation:

> From 1906–1923 government hunters ranged the area, killing all the predators they could find, and they worked, as usual, with deadly thoroughness. During the period from 1916 to 1931, they trapped or shot 781 mountain lions, 30 wolves, 4,889 coyotes, and 554 bobcats.[4]

With the deer population climaxing at almost 100,000 by 1924, starving, desperate animals stripped bushes, trees, and grass, denuding the landscape. With the exhaustion of food sources, the population then collapsed as deer starved. Sixty percent of the population died in the winters from 1924 through 1926, stabilizing at around 10,000 by 1939. This event and many others like it allowed wildlife biologists and public lands managers to begin to perceive the critical role predators play in a variety of ecosystems.

The Fight for Predators and Wildlife

Even as predators were driven to the very cusp of extinction in the continental United States, a dissenting force arose in the 1920s and 1930s. Private citizens, forest service and national park employees, and others began to fight for the right of predators to exist while providing increased scientific evidence for the significance of predators in the food chain and in the functioning of a healthy landscape. Mammalogists began challenging the war on predators at their annual conference in 1923, publicly criticizing the Biological Survey's extermination policies. A continuing debate over predator control emerged from this protest with papers exploring the issue at following conferences as well as numerous articles published in *The Journal of Mammalogy* on both sides of the issue. Through the 1920s and 1930s, biologists working for research institutions, parks, the Forest Service, and even the Biological Survey began to articulate a more balanced ecological view, increasingly questioning predator control and seeking to redress earlier mistakes in game management to establish a more effective natural balance. One clear example of this was the reduction of the Yellowstone elk population. The combination of predator destruction and earlier feeding operations had facilitated explosive growth from the once small herd, and by the 1930s the animals had overgrazed their habitat to such a degree that the elk and many other species faced certain starvation. Not without controversy, the killing of thousands of elk over several years helped to improve natural habitat and stabilize the population.

While biologists in government agencies challenged the dominant thinking on predators, private citizens began organizing effectively in defense of wildlife, prey and predator both. One of the great agents of change in wildlife protection was activist Rosalie Edge. Her decades of labor, particularly via the Emergency Conservation Committee (ECC), resulted in numerous benefits for both wildlife and Americans. She brought serious political acumen and organizing skills from her years of work in the suffrage movement. Edge's conservationist career began with a campaign against the Audubon Society. Audubon had played an integral role in ending the harvest of certain bird species' plumage for the millinery trade in the late 19th century with the passage of the Lacey Act in 1900. From the perspective of wildlife supporters, however, by the 1930s Audubon had become too chummy with hunting organizations and arms and ammunition manufacturers, preventing the organization from actively promoting bag and harvest limits desperately needed in the 1920s and 1930s as bird numbers plummeted. Willard Van Name, a biologist working for the American Museum of Natural History, wrote a pamphlet titled "A Crisis in Conservation," pointing out the society's cozy relationship with commercial duck hunters. Van Name was forbidden by museum directors, who also held leadership positions in the Audubon Society, from publishing the piece under his own name. In defiance he sent it to Edge to distribute, and an important wildlife advocacy group was born. While Edge was unable to immediately move Audubon to change its policies, the journalist, writer, and noted James Madison biographer Irving Brant joined them, and they created the ECC. The three of them, with Edge as the dynamo that kept them in motion, used regular publication of thousands of pamphlets over its lifetime on numerous environmental issues, a deep mailing list of powerful and influential conservationists and politicians, and political lobbying in Congress and at the scene of environmental problems to fight for wildlife and nature in multiple locations across the nation.

Finally forcing reforms within the Audubon Society, the ECC took on several other campaigns in the 1930s. Informed by a park employee of Yellowstone National Park's policy of smashing pelican eggs to protect trout, the group published a pamphlet on the practice and brought it to a quick end. Upon learning of a Forest Service plan to log off a grove of sugar pines on land close to Yosemite National Park, the ECC published two pamphlets and gained Senate support via introduction of a bill to provide protection for the trees. With support flagging, Edge took matter into her own hands and traveled to California to build a local constituency for the bill, which eventually passed.

The ECC expended considerable energy into defending parks, particularly from logging by the U.S. Forest Service and excessive road and facility construction by the U.S. National Park Service (see document 9.3), and fought to expand park boundaries to protect wildlife and forest ecosystems. One hallmark victory was the securing of park status and boundary expansion for the Olympic National Monument in 1938. This designation as a national park and redrawing of its boundaries to include several hundred thousand more acres protected key low-elevation river valleys with old growth forest that also functions as critical habitat for the resident elk population. It was the slaughter of almost 250 elks in a four-day hunt that triggered this effort. Over the five years of lobbying and battling with regional lumber companies, politicians, and the Forest Service, park advocates organized a carefully stage-managed visit by FDR to the Olympic Peninsula of Washington State. This included arranging to have a group of school children standing by the road in Port Angeles with signs asking that he save the park. This and other efforts by the ECC and local conservationists convinced the president to convert the monument into a park and expand its size dramatically. Preserving this amount of ecosystem and creating Olympic National Park served both preservationist and ecological goals. Not only would it preserve elk and landscape but important salmon rivers like the Hoh, Bogachiel, and Sol Duc benefited from habitat preservation. This has allowed salmon, steelhead, bald eagles, and numerous other species to persist or thrive in those watersheds.

These activists did not merely insist on increased government protection. Edge put her own money where her mouth was to stop the mindless slaughter of predatory birds at Hawk Mountain on the Appalachian flyway in eastern Pennsylvania. As hawks, falcons, and other raptors flew directly over the site by the hundreds of thousands, hunters used the high elevation site to slaughter the raptors in the hundreds or thousands by the day.

Figure 9.2 The expansion and protection of national parks during the New Deal, such as FDR's expansion, Olympic National Park, provided numerous ecological benefits not understood at the time. The key issue driving park expansion and protection on the Olympic Peninsula was preserving the Roosevelt Elk (named after the president's cousin, Teddy Roosevelt). But in protecting that animal's habitat they also helped preserve numerous other species, such as the salmon runs on the hundreds of park streams. What biologists have only come to understand completely in the last 30 years are the numerous beneficial impacts of spawning salmon to riparian ecosystems. Not only do spawning fish provide food for species such as bears, ravens, raccoons, eagles, among others, but their rotting flesh also contributes to the health of the overall ecosystem. The transfer of energy and nutrients from the ocean to these river and lake ecosystems by spawning salmon increases plant growth, contributes to bug growth and population, feeds small fish and other aquatic creatures, and even benefits birds feeding on the abundance of bugs created in an ecosystem enriched with spawning salmon. National Archives.

Sometimes the hunters collected bounties for birds; other times they simply left the carcasses scattered across the slopes of the mountain to rot. When she heard of this in 1934, Edge leased 1,400 prime acres of land at the site and hired wardens to keep gunners off the land; the slaughter ceased. She then purchased the land and deeded it to the Hawk Mountain Sanctuary Association, created in 1938. The site has been protected ever since; it is an important location for bird research and is quite popular with birders.

The federal government expanded its conservation role greatly with numerous measures to protect waterfowl. Loss of wetlands to farming and development and the brutal nationwide drought of the 1930s inflicted great damage on duck populations with numbers dropping from 100 million in 1930 to as low as 20 million by 1934. Under the leadership of Ding Darling at the helm of the Biological Survey, further restrictions were passed on hunting while 31 refuges totaling 840,000 acres were created protecting breeding habitat. The passage of the Duck Stamp Act in 1934 created a $1 federal hunting license, the source of ongoing funds for duck restoration and preservation, and crucial to the success and restoration of duck populations. During FDR's New Deal administration remarkable success was achieved in helping ducks recover. Their numbers rose from that low of 20 million to 40 million in 1937 all the way to 70 million in 1941. A pivotal reason for this success was the ongoing acquisition and protection of habitat. By 1940 FDR's administration had created 159 new refuges containing over 7.5 million acres. Through these remarkable steps he continued the work of his cousin Teddy, and the federal government's increased commitment to protecting habitat created a lasting benefit for wildlife and those who enjoy wildlife.

Producing Less for Growth

It is appropriate, perhaps, that one of the more successful projects of the New Deal was the Agricultural Adjustment Administration (AAA). Time and again in American history, soil exhaustion and bad agricultural practices led to economic and cultural difficulties. In this particular case, American farmers produced too much food. The AAA provided subsidies to farmers to leave lands fallow for crops such as cotton, tobacco, wheat, and corn, or for livestock, such as pigs. The goal of this program was to reduce gluts in these crops and raise prices. Over time, farmers participating in this program were required to practice conservation measures in order to collect subsidy payments. Unfortunately, these cash distributions from the federal government tended to benefit large landowners and absentee landowners over those working smaller pieces of land. Worster provides a telling comparison from western Kansas: "The president of the Santa Fe Bank in Sublette, for example, raked in $4270 on his 1933 contract, while a Mennonite man living on a 160-acre farm got $23."[5] The program caused hundreds of thousands of sharecroppers and tenant farmers to be thrown off the land as the soil they had worked was selected to lay fallow. Sixty percent of the cotton farms in Oklahoma, Texas, and Arkansas were run by tenant farmers and sharecroppers. Soils were highly eroded and provided limited returns. Owners of the land were relieved to have the federal government provide them with the safety net of subsidies as they kicked tenants off the land, used the money to mechanize their agriculture, and looked to improve their own status. One consequence of this agricultural subsidy program was to create a massive dislocation of populations of poorer Americans, forced to move west or into the cities in pursuit of jobs for survival. According to one landowner,

> In '34 I had I reckon four renters and I didn't make anything. I bought tractors on the money the government give me and get shet o' my renters. You'll find it everywhere all over the country thataway. I did everything the government said—except keep my renters. The renters have been having it this way ever since the government come in. They've got their choice—California or WPA.[6] (See Document 9.2.)

Given no subsidy themselves, dislocated sharecroppers and tenant farmers organized themselves into the Southern Tenant Farmers Union and demanded relief or land. Due to violent force and suppression of the organization by powerful farmers and a weak response from the federal government, their only option was to migrate out of the region; many of them migrated to California in pursuit of jobs, an event made famous by John Steinbeck's *The Grapes of Wrath*.

Fighting for Soil in the Dust Bowl and Palouse

Following the collapse of the cattle industry and the end of open-range ranching in the late 1880s, a land rush onto the southern Plains commenced. Public land obtained through the Homestead Act was cheap, as was acreage sold by landowners, such as the XIT Ranch of Texas, which offered up parcels at $13 an acre in an effort to recover from the cattle crisis. Besides the Jeffersonian ideal of the yeoman farmer, manifest destiny, the desire for social mobility, and general land hunger, teleological theories of farming and rain drew people onto a landscape only recently referred to as the Great American Desert. A common and purportedly scientifically proven belief was the idea that the rain followed the plow. Not only because breaking the soil released water into the atmosphere, creating clouds, but also because productive use of the land was part of God's plan. These beliefs must have assuaged the concerns of new arrivals from the wetter, greener eastern lands, the South, and the Midwest when they first viewed the mostly treeless and seemingly barren landscape they were going to call home.

The implementation of dry land farming techniques allowed them to sleep better at night, find success in an arid environment, and prepare the ground for one of America's great ecological disasters. This consisted of "deep plowing in the fall, packing the subsoil, frequently stirring up a dust mulch, and summer fallowing—leaving part of the ground each year to restore moisture."[7] Wheat replaced corn as the popular grain of the Plains, and mechanization increased the scale of production, which required greater amounts of land, to the degree that almost all of the grass cover of the southern Plains would be gone by 1930. Wheat prices were high prior to World War I, but the needs of soldiers and the inability of Europe to produce crops because of agricultural fields turned to battlefields drove the price of wheat to unseen realms, with the U.S. government promising farmers a minimum of more than $2.00 a bushel in an effort to spur production. In 1917, 45 million acres of land grew wheat; in 1919 that number was 74 million. For the period of 1917–1919 the nation's farmers produced 952 million bushels of the "golden stream," 38 percent more than for the four-year period of 1910–1913. This increase in production could not simply be accounted for by farmers switching to wheat alone. In the states of Kansas, Colorado, Nebraska, Oklahoma, and Texas, 11 million acres of grassland were destroyed and replaced by wheat.

This stunning increase was facilitated by extensive agriculture (more land) and intensification of farming through mechanization. The introduction of gas and diesel tractors and combines increased productivity as well as costs. The value of farm equipment on a typical Kansas farm more than tripled in the decade of 1910–1920 when mechanization took hold. Whereas in 1830 it required 58 hours to harvest an acre of wheat, mechanization pushed that time down to approximately 3 hours on the southern Plains in 1930. No longer needing horses to pull plows and combines, superfluous pasturage could be used to grow wheat, too. Likewise, with no requirement to grow hay for winter feeding, hayfields could also be made into tributaries of the river of wheat flowing out of the Plains. While this process seemed to promise greater wealth for the individual farmer, mechanization meant the loss of jobs for the crews of men normally hired to work fields and harvest wheat. Increased yields were one result of mechanization; so were layoffs and a greatly simplified, homogenized landscape. With the loss of pastures and hayfields went also the loss of habitat for numerous species, including skunks, quail, western meadowlark, mice, hawks, coyotes, and so on.

The scale of some of these industrial farms was astounding for that time but fairly typical today. One was more than 54 square miles or 34,500 acres and required 25 combines to harvest the wheat. Worster describes the process on a local and broader level:

> Near Perryton, Texas, H.B. Urban, an altogether typical wheat farmer of the day, arrived in 1929 and cranked up his two internationals; each day he and his hired hand broke out 20 acres, until virtually his whole section of land was stripped of its grama and buffalo grass. In thirteen southwestern Kansas counties, where there had been 2 million crop acres in 1925, there were 3 million in 1930. During the same period farmers tore up the native vegetation on 5,260,000 acres in the southern plains—an area nearly seven times as large as the state of Rhode Island. Most of the freshly plowed ground went into wheat, so that over the twenties decade the production of that cereal jumped 300 per cent, creating a severe glut by 1931. That was how men prepared for the days to come. When black blizzards began to roll across the plains in 1935, one-third of the Dust Bowl region—33 million acres—lay naked, ungrassed, and vulnerable to the winds. The new-style sodbusters now had their turn at facing disaster.[8]

The Dust Bowl was not simply a natural disaster in the sense that most Americans think of it. It was created by trying to apply the model of industrialized capitalism developed in a eastern, well-watered agriculture to a drier landscape that experiences cyclical droughts. Stripping land of all its cover left it vulnerable when the skies dried up. Those dust storms were something to behold and created terror and death for residents of the region. Varying in color based on the origins of the storm, brown, red, black, and yellow walls of dust would bear down on communities with little warning, sometimes catching farmers still on their tractors or tending livestock. Families were trapped on roads and livestock caught outside; many were suffocated by the dust. The suspended dirt in the air was so heavy at times that

Figure 9.3 This image of a dust storm approaching and enveloping a town was a regular occurrence for people of the Great Plains during the Dust Bowl years. The storms dumped dust, killing livestock and crops and creating a mess for people to clean up. Sometimes people became disoriented and got trapped and died in the storms as well. Courtesy of the Western History Collection, University of Oklahoma Libraries, Snowden Dwight Flora Collection, Flora 77.

people couldn't see their hands in front of their faces. When the storm passed crops would be covered in drifting dust, several feet deep in places, killing the plants that promised a future. The top soil was stripped away in many places, leaving only hardpan underneath, soil that was difficult to work and low in fertility. Deaths did not stop once the storms passed. Many children died of pneumonia worsened by the dust in their lungs, and animals expired from respiratory problems arising from dirt-filled lungs and from starvation. Storms rolled through year after year, and it was only a serendipitous moment that caught the official attention of the federal government in Washington, D.C. The introduction of a bill for relief from the Dust storms occurred as one of the major storms arrived on the East Coast, dropping Great Plains soil on congressmen who had ignored the crisis to this point. Once fully aware of the environmental and economic problems in the southern Great Plains, the federal government applied its money, labor, and ingenuity to the fight to save the soil and the communities dependent on that soil by means of 1935 drought relief legislation.

While dust bowl sufferers turned to their own limited resources to persist until the rains returned, the federal government provided assistance in numerous ways, trying to reform ranchers' and farmers' destructive practices. Contrary to rural and western mythology of independence from government assistance, southern Plains farmers demanded help from the government. According to Worster,

> It may seem ironic that wheat farmers in Kansas, who had become so fully a part of the economic system, would want protection from the penalties for failure. But the experiences of overproduction and dust storms were sufficiently traumatic to

Figure 9.4 The level of the dirt on this man's farm can be seen against the fence. He is trying to dig it up and build it again on top of the new layers of dirt. While federal government aid was essential to surviving the Dust Bowl, individual and community perseverance through times of great stress and doubt played a key role as well. Library of Congress.

> produce a revised maxim for business farming in the decade: do not interfere with us when we are making money but rescue us when we are going bankrupt.[9]

The purchase of cattle on the edge of starvation (and therefore rendered unmarketable) provided small amounts of cash to ranchers for beef that were going to die anyway. In some communities, desperate residents went to the arroyos and draws where the beef were slaughtered and carved what thin pieces of meat they could from the carcasses. In the latter part of the New Deal, in 1936 and afterwards, the federal government paid ranchers to defer the use of damaged grasslands, for managing water use more efficiently, and to reduce herds.

The government provided subsidies to farmers of the Plains to take land out of use while training and paying them to incorporate soil conservation measures into their work. Listing was used to break up chunks of soil to hold dirt, and farmers were also employed to practice contour farming and strip planting. The federal government also bought 6 million acres of devastated land, planted it in grass, and then leased it to ranchers for forage only. Farmers also gained employment planting shelterbelts of trees across

Figure 9.5 Public efforts to support land conservation efforts. Library of Congress.

the Plains. Farmers and ranchers adopted stewardship practices at the behest of government education and funding.

While the Palouse region of southeastern Washington State and a small portion of the panhandle of Idaho did not suffer the same ecological and manmade crisis of the Dustbowl region, the same processes that built both wealth and lost soil on the southern Plains drove the agricultural economy of the Palouse. The soils of the Palouse are loess, a fine, glacial-tilled soil that blew into the region over thousands of years, creating hills and dunes of the rich, fertile, and easily eroded soil over a foundation of thick basalt. When agriculture began in earnest in this area in the 1880s, farmers raised a variety of crops for local and distant markets. Lentils, peas, wheat, and barley were commonly produced in this area with cattle grazing on native grasses and haying to support cattle constituting a significant portion of land use. Just as on the Plains, this all changed during World War I and the 1920s.

World War I generated increased crop production in the Palouse as it did on the Plains with acreage devoted to wheat increasing by 28 percent from 1909–1919. Pressure for increased production came not only from the government and the opportunity for high profits but also from publications such as *Washington Farmer.* According to historian Andrew Duffin, this magazine and other media urged farmers to maximize their land use. Encouraging them to produce more for the war effort, farmers were told to carefully survey their acreage for unused soil.

> It lies in the fence corners, along the ditches or around old buildings or barnyards. Very often it is the best land on the place, at least after having lain idle for years. It ought to be put to work. There is profit in it; there is loss if it is not utilized.[10]

Figure 9.6 Erosion was not limited to the Great Plains and the Palouse. As evidenced by this photo from Georgia in the 1930s, farming practices across the country led to severe soil loss. Library of Congress.

It is the use of every square inch of farmland that Aldo Leopold would criticize so adamantly in his later essay collection, *Sand County Almanac.* Every inch of land put into production meant increased erosion as the soils blew and washed away year after year.

As was the case on the Plains and across the nation, Palouse farmers aggressively mechanized agricultural labor. Tractor purchases in Whitman County doubled from 1925 to 1930, and approximately equal increases in petroleum-powered combine and grain truck purchases meant that farmers accrued more debt not only for the new equipment but also for the purchase of more land, which was the logical outcome of the increased capacity of mechanized farming; this added to their debt burden. Even if they had wanted to leave the strips to grass and habitat, they would not have been able to. The bank had to be paid.

Farmers put nearly every inch of the land into tillage, plowing, and planting right up to and through the numerous streams and rivers as well as working the ridges and hills where the soil lay thinner on top of the basalt foundation, and was more prone to erosion. The lack of a strong conservation ethic or philosophy of land stewardship contributed to this aggressive land use. The effect was to create one of the most highly eroded landscapes in America.

As in other rural areas across the nation, Palouse farmers eagerly took advantage of AAA subsidies. Out of 2,837 farmers in Whitman County (the heart of the Palouse wheat economy), 2,667 agreed to reduce land use by a total of just more than 400,000 acres for an increased payment of 29 cents per bushel for the wheat they grew. Nationwide, subsidies for crop reduction did resolve one fundamental problem in American agriculture and the economy: the removal of huge crop surpluses stifling commodity prices. Erosion was worsened by subsidies as the unused land was largely left fallow in summer, unplanted and vulnerable to wind and rain. An incident of short, heavy rainfall in the summer of 1931 illustrates how quickly precious soil could be lost. According to Duffin,

> station personnel estimated that on fallowed land throughout the region, erosion removed an average of two inches of topsoil (about 275 tons per acre), with some areas losing up to six inches of soil. For the roughly twenty thousand Palouse acres in summer fallow in 1931, about 5.5 million tons of soil was removed in one torrential downpour. The crew also noted that on fields covered by wheat or grasses, almost no erosion occurred.[11]

Regardless of clear evidence, farmers in this region resisted simple stewardship measures until the government created incentives for them to do so.

Washington State University extension agents and employees of the federal Soil Conservation Service (SCS) traveled the region trying to convince farmers to not leave summer land fallow and to plant more legumes for the nitrogen benefits, and to hold soil. Moreover, researchers and the government taught and encouraged efforts to build soil through tree planting, mulching of stubble, and contour plowing. The creation and implementation of the one-way disc plow versus the traditional moldboard plow reduced soil loss and was used to till old wheat stalks into the soil for increased fertility and to slow erosion. Spreading straw and seeding by punching holes in the soil instead of breaking the soil surface were other strategies encouraged and sometimes employed to prevent soil loss to erosion. Despite incentives and agricultural benefits, the voluntary nature of these programs meant that by 1936 most Palouse farmers were working the land in the same way they had been five years earlier, and erosion continued to plague the region. The increased use of diesel-powered caterpillar tractors, popular in the hilly country of the Palouse because of their weight and low-center of gravity, worsened erosion as water rushed swiftly off hard-packed tread trails.

New Deal programs did not successfully transform farmers' land-use practices. The government focused too much on persuasion and subsidies, trying to nudge farmers along to proper soil conservation through enticements and education. Worster points out that farmers' behavior was not changed in any substantive way, and they quickly returned to destructive practices when the rains returned and prices for crops rose. Duffin argues a more resigned position, stating that the soil agencies should have "permanently retired" the highly erodible soils of ridges and hills. He also states that "Farmers had no ideological impetus to conserve soil in part because stewardship had been lacking all along. To expect them to suddenly conjure a new land ethic during the Depression was quixotic at best."[12] As long as farmers could expect the government to step in and help them in times of crisis, when high prices returned it was too hard to resist returning to destructive habits.

The crises in the Palouse and southern Great Plains brought on by farmers' destructive habits necessitated an expansion of federal power to help them survive as well as to conserve soil and restore its fertility. At the demands of farmers and ranchers themselves, the federal

Figure 9.7 Erosion remains a problem throughout the Palouse as farmers continue practices such as cultivating steep slopes that the federal government attempted to end during the New Deal. Sheeting erosion is clear in this contemporary photo of a Palouse wheat field. U.S. Department of Agriculture.

government applied pressure, funds, and training to convince and assist farmers to incorporate better stewardship into their agricultural practices. This was accomplished more successfully on the Great Plains than on the Palouse, likely because of the more devastating nature of the Dust Bowl. But financial incentives or free labor, such as that provided by the Civilian Conservation Corps (CCC) discussed later in this chapter, were the key instruments of limited change. With the erosion slowed and rising commodity prices with the outbreak of World War II, it was too easy to return to irresponsible land-use practices. The programs were not without impact as some conservation practices continued, tree belts grew and held soil, and the SCS maintained its efforts to control soil loss. But without ongoing federal money and with debts to pay, many farmers returned to plowing and planting most of their land, albeit with some of the measures such as listing and contour plowing better integrated. The expanded federal government that so many farmers and ranchers criticize so adamantly today is at least partially a consequence of their own ecological and economic impacts.

Young Men and Nature

The CCC has endured in the American memory as one of the most popular and successful of the New Deal programs because it provided jobs and helped restore the environment. The idea for the CCC sprang from President Roosevelt's experiences as a young man planting forests in New York with the Boy Scouts as well as his background planting forests on his Hyde Park estate. The majority of logging companies in the nation at that time stripped the land and moved on to new forests. By the 1930s almost all private land in America had been logged as had a great deal of public land in the West. For example, out of approximately 18 million acres of original pine and oak forest in Texas, less than a million acres remained standing in 1930. Very little deforested land had been replanted.

In other regions land exhaustion and overuse combined with overgrazing led to widespread erosion of soils across the nation. The Corps was conceived and organized to bring restoration to these deforested and eroded landscapes while relieving unemployment. They are best known for their work in restoring forests, but the dominant image of men stooped over small trees elides the range, scale, and variety of restoration work conducted by the CCC.

Created in 1933 during FDR's "One Hundred Days," the period during which he pushed through several bills designed to turn around the stalled economy and fix many structural problems in the American economy, the CCC began with a planned enrollment of 250,000 men, growing to 2.5 million by the termination of the program in 1943. In the beginning the CCC only hired young white men, ages 18–25. They were paid $30 a month, of which $25 was sent to their families. Organized into units run by the Army, the men were sent to camps far from home. NAACP protests against the "white only" policy led to acceptance of African-Americans into the program. Camps were integrated early, but race tension and conflict resulted in segregated camps. Native American protests for opportunities in the program led to the establishment of camps and projects on reservations. In those cases, Indian enrollees remained near to work on land and soil conservation.

One misconception is that CCC enrollees only worked on public lands. In reality they provided extensive service on private farmlands as well as in national forests and national parks, employing a variety of strategies to stop the loss of soils to erosion. One approach to protect exposed soils was to plant species that grow quickly and hold well. The use of invasives such as kudzu and tamarisk, as well as native species such as Virginia creeper, worked effectively to hold soil and in many places have transformed the local ecosystem, as anyone from Mississippi who has seen an old sharecropper shack swallowed by kudzu can attest to. Gullying was a particular problem across the country, particularly in the Georgia hill country, the Palouse of southeastern Washington, and the arroyos and plains of the Southwest. CCC workers built more than 6 million check dams designed to "check" or slow the flow of water. The dams were rudimentary, constructed of rock, timber, wire, and other available materials, but they worked in providing time for water to be absorbed into the soil, reducing its scouring ability and also allowing sediments to be deposited behind the dams rather than lost downstream. As soils built up behind these dams, they were seeded with vegetation to hold the soil and refill the ravine. In addition to those constructed on public lands, more than 300,000 of them were built on private lands. An essential part of the CCC mission was education and training, and the hope was that as enrollees learned to use tools such as check dams and plantings to control erosion, they would take these strategies back to their own communities and farms and implement them there.

One enrollee wrote home describing his labors, explaining that one learns how to

> keep rains from washing away the topsoil, how to heal scalded spots on the hillsides, how to build terraces and outlet channels and terrace outlets. You'd learn how to plan

> strip-cropping and run the contour lines for the farmer to follow . . . plus gullies and sod gully banks, dig diversion ditches; run contour furrows and build water spreaders; move old fences and build new ones; learn about crop rotation, and planting orchards on the contour; . . . learn about meadow strips and vegetated waterways where the excess water can run from the fields without stealing the soil.[13]

Workers drug brush up onto hills and dunes to protect growing grass. Terraces were dug by CCC men, and these effectively slowed the loss of soil as water and sediments had time to pool and deposit on each bench as it worked its way downhill. In the Palouse, they actually carried dirt from ravine and valley bottoms to the tops of hills and ridges, attempting to reestablish soil on top of exposed basalt. These efforts failed. Strip cropping was also implemented on private land. This was simply the practice of alternating sections of land growing crops with strips cultivated with grains, rye, and kudzu, species that grow quickly and densely, firmly holding dirt where it belongs. These strips contained the flow of water and dirt from tilled land. Another response to the agricultural land crisis was the use of contour farming by the USDA, SCS, and the CCC. This method simply had farmers and government employees plow parallel to the slopes of hills and across them rather than tilling soil straight up and down slopes. This method was taught and adopted throughout the nation in areas struggling with erosion. This helped conserve soil by slowing water flow downhill. CCC workers helped create more than 150,000 miles of contour furrows.

The work undertaken by these young men paid off dramatically in the short term. The yield of cotton crops on Texas terraced land increased by an average of 68 pounds per acre. Similarly, grain sorghum yield increased by 128 pounds per acre where CCC terracing work was performed in Texas. According to historian Neil Maher,

> when the Corps helped farmers to combine both terracing and contour furrowing, field output was even more impressive, increasing wheat yields on the Southern Great Plains by 3.5 bushels per acre, bean production in New Mexico by 78 pounds per acre, and grain sorghum growth in Texas by an astounding 262 pounds per acre.[14]

The introduction of basic soil conservation practices and stewardship principles led to increased abundance.

While some enrollees worked to conserve soil, others fought to save forests. Because of earlier predator extermination, porcupine populations exploded, and they presented a threat to forest health in some areas. In camps in northern Arizona and Colorado, CCC enrollees included killing porcupines among their duties. In the Coconino National Forest they killed these animals through shooting and also by soaking pieces of wood in a formula of strychnine and salt and nailing the bait to branches about 10 feet above the ground. The bigger threat was from forest fires. Drought had hit the whole nation, not simply the southern Great Plains. The result of the droughts and decades of fire suppression by the Forest Service were large fires on public and private lands in the West; more than 4 million work days were logged by enrollees fighting fires. They were pressed into service to fight fires and also to build an infrastructure of fire fighting and suppression. While this was seen as an entirely positive development at the time, in fact these efforts furthered ecological damage arising from fire control policies. Regardless, in service of that goal they

Figure 9.8 Because of drought and Forest Service fire suppression polices that made forests highly vulnerable to fires, CCC enrollees fought fires across the West. From the Gerald L. Williams Collection, Special Collections & Archives Research Center, Oregon State University.

built 3,479 fire towers and houses and connected towers with more than 65,000 miles of telephone line.

They planted trees, also. More than 3 billon trees and possibly as many as 4.3 billion were planted by these young men nationwide. In 1936 alone more than half a billion seedlings were transplanted from nurseries. From the red clay hills of northern Mississippi and the mountains of the Great Smokey Range to the national forests of the West, by 1938 around 1.3 billion trees had been planted. The Regional Forester report of 1942 for the Southern Region summarized the multiple benefits of the CCC:

> At the beginning of the CCC program in 1933 there were about forty-seven million acres in the South receiving fire protection, and by January 1, 1942, this had increased to seventy-five million acres. The CCC has opened new portals to millions of outdoor

Figure 9.9 CCC enrollees filled and planted this coal strip mine in Kansas. Restoration over mined, logged, and farmed lands that were ecologically degraded and abandoned constituted a major portion of CCC work. Library of Congress.

> recreation seekers at a time when Uncle Sam wants every worker to get the rest and relaxation necessary to send him back to his job doubly efficient. Approximately eight million trees for America's future needs were planted by the CCC in this Region on some eight hundred thousand acres of land left bare by the timber-hungry pioneers who built the West. That planting means protection for watersheds, refuge for wildlife, lumber for peace and war, jobs for local people.[15]

Forest planting brought numerous short-term and long-term benefits, as did building roads, parks, and outdoor recreation areas. This generally positive work did cause some environmental problems, however, leading Aldo Leopold, Robert Marshall, and other key conservationists to oppose the CCC and build a stronger commitment to wilderness and habitat preservation.

Conservation Hinting at Environmentalism

Aldo Leopold was trained in the forestry program at Yale and began his career in the Forest Service in 1909. While an ardent conservationist in his early years, a committed Pinchot acolyte dedicated to hunting, predator control, and management of nature for human use, he early on demonstrated a sympathy for preservationist values that hinted at the radical development of his thinking in later years. As a young man he commented on Yosemite tourists:

> The tourists all gape at Yosemite but what none of them see is the fifty miles of foothills on the way in. They are almost a relief after the highly frosted wedding-cake (and the wedding guests) on the other end.[16]

This comment also functioned as a criticism of the management of the national parks under Stephen Mather, when commercialization of the parks was used to draw visitors. In the 1920s, Leopold began arguing for setting aside primitive areas to function as wildernesses within Forest Service holdings. The construction of hotels and permanent roads would be blocked on this land while cattle could still be run. This denoted a startling advance in thinking about land use within the Forest Service for that period. Leopold notes one key moment when he began to consider the rights of nature. Having just shot a wolf, he saw the dying light in her eyes and wrote of this moment: "I realized then and have known ever since, that there was something new to me in those eyes."[17]

Several years later he conducted a national wildlife survey, wrote and published the wildlife biology textbook *Game Management* in 1930, and took a professor position at the University of Wisconsin in Madison in the 1930s. In conducting his survey and writing his textbook, Leopold helped lay the foundation for a new discipline, wildlife biology. Leopold supported the CCC in its beginning years but grew frustrated with errors and environmental impacts. Noting the lack of coordination between specialized crews, Leopold used the example of one road crew damaging a trout stream while another worked to improve stream habitat. He also commented on the removal of an eagle tree as part of an effort to improve a forest in northern Wisconsin in an article for *American Forests* magazine: "I cite in evidence the CCC crew which chopped down one of the few remaining eagle's nests in northern Wisconsin, in the name of 'timber stand improvement.'"[18] Similarly, efforts to drain tidal swamps on the East Coast to control mosquitoes also troubled Leopold and others because of the loss of habitat for fish, waterfowl, and other species. Rosalie Edge also criticized drainage for mosquito eradication in no uncertain terms. Looking at demands for mosquito control on Long Island and other parts of New York and the draining of saltwater and freshwater marshes, she writes,

> as soon as it becomes known that Long Island townships have much money to spend on mosquito control, an outcry arises from upland communities, insisting that their mosquitoes also be destroyed. The situation in the uplands is entirely different from that of the seaside; in one the marshes are salt and in the other they are fresh. The various species of mosquitoes are not the same. While the mosquitoes of the salt marsh easily fly twenty-five miles, or more, the fresh marsh mosquito does not travel more than a mile away from its breeding-place. What do county officials or project makers care for such elementary facts? Fresh water marshes, miles distant from any town, are drained without regard to their importance as breeding places of valuable birds and furbearers.[19]

Leopold, Edge, and others were pleased at the progress accomplished under this conservationist president but expressed unease at the poor planning, the lack of habitat and landscape study, and an increasing reliance on scientific technology. In developing his critique

Leopold employed the new concept of ecology and asserted the need for a reorientation toward a land-use philosophy that supported multiple uses and the health of ecosystems. Scientists also lent their critical voices to opposing CCC practices. Conservation biologist Leonard Wing wrote in a 1936 *American Forests* article, "Lumber-farmers armed the CCC with saws, axes, shovels and matches and sent them on the trail of 'weed trees.'"[20] He noted that this destroyed important habitat, removing food for deer and grouse, leaving "no tender bark for rabbits, no holes for birds, no drumming logs for partridges."[21] The desire to restore natural abundance and improve access to recreational areas, all laudable goals, was having a negative impact on existing habitat.

The American problem of loving our nature to death really started with the popularity of automobile touring, which became all the rage following WWI as Americans loaded flivvers to the bumpers to travel to parks and forests. National Parks were popular destinations with visits growing from 240,000 in 1914 to more than 3.5 million by 1931. As historian Paul Sutter writes, "After World War I, nature tourism and outdoor recreation became cultural imperatives, crucibles of national character that filled the vacuum created by the vanishing frontier."[22] The increasing popularity of parks and national forests also caused ecosystem decline as campers dumped garbage and refuse, trampled meadows and pastures, and muddied creeks and streams. Accidental fires were started by sloppy weekend campers. Maybe more importantly, the increasing web of roads threatened the health of large chunks of intact ecosystem, de facto wilderness in some places. These numerous impacts triggered increasing calls for enhancing the protection of nature through the creation of wilderness areas in both national forests and national parks. Leopold had led the initial opposition to roads and the creation of wilderness after his assignment to work in and manage sections of national forest in the Southwest. While working in the Carson National Forest in northern New Mexico, his concern grew over the rapid pace of recreational use and road construction to serve it. Large roadless areas in that forest were being fragmented by roads into smaller and smaller sections. He wrote, "part of the lost areas were justifiable sacrifices to timber values; part, I think, were the victims of poor brakes on the good roads movement."[23] In 1921 he published "The Wilderness and Its Place in Forest Recreation Policy" in the *Journal of Forestry* where he offered the revolutionary argument that some areas of the forest should be set aside for those who would hike and camp, hunt and fish, but where roads, buildings, development, and even economic activities such as logging would be forbidden. Leopold insisted this should not be seen as non-use, as might be argued from a traditional utilitarian perspective, but was consistent with the principle of highest use. This innovative wilderness philosophy and continued lobbying by Leopold led to the creation of the Gila Wilderness Area inside Gila National Forest in 1924. Leopold offered a middle position between preservation and conservation and launched a movement in the Forest Service that gained strong support. He published further articles making the case for wilderness. Sutter says that these publications

> lamented the demise of uncharted places and the spiritual loss that occurred as remote corners of the earth were mapped and charted . . . [they] shared a concern about the impacts of roads, cars, and a new modern culture of outdoor recreation.[24]

National attention and support led to the setting aside of more wilderness areas and the creation of the L-20 regulations setting aside and protecting "primitive areas."

Another progressive Forest Service employee took up the mantle of wilderness preservation and also published articles extolling wilderness as a refuge from civilization, as recreational space deserving of protection. Robert Marshall was a powerful advocate for wilderness. In response to charges that creating wilderness areas was elitist and kept working-class people out of nature reserved for use only by upper class hikers, campers, and fishermen, Marshall employed a minority rights argument, asserting that setting these areas aside would not hurt the rights of the majority who preferred road and automobile-based recreation but would protect the rights of a minority to use nature in a more intimate manner. During the interwar period wilderness advocates used this argument repeatedly in defense against charges of elitism. The result of such efforts as Marshall and Leopold's and growing support for wilderness was the creation of the U Regulations in 1939, expanding the power of the Forest Service to designate wilderness areas. Ongoing threats to the health of nature led to the expansion of federal power to set aside and protect areas of ecological value. This growth of federal power constituted a necessary next step in the evolution of the relationship between government and nature.

Both the principles of ecology and criticism of the CCC spread across the nation and influenced national policy as the National Park Service hired wildlife biologists to oversee the CCC on its lands and try to protect habitat for wildlife. The national media picked up on these criticisms of the Corps and in the process educated Americans on the ideas of ecology, wildlife biology, and habitat management. As a result, the CCC inadvertently helped advance ecological consciousness in ways that Maher asserts contributed to the flourishing environmentalism of the post-WWII era. The efforts to conserve and restore nature and extend access to it through roads, trails, campgrounds, new and expanded parks, and other amenities helped popularize outdoor recreation and increased the appreciation of nature in the late 1930s through the early 1940s. As Maher points out,

> Whereas less than 3.5 million people visited national parks in 1933, by 1938, just three years after the Corps expanded its work into the nation's parks, that number had skyrocketed to 16 million and rose again to 21 million by 1941, an overall increased during the 1930s of approximately 600 percent.[25]

State parks also saw increased crowds of visitors. While the CCC built trails, planted trees, cleared habitat, conserved soil, and endured mounting criticism, environmental activists organized to stop the most destructive practices. The Audubon Society launched a campaign in 1935 with members penning their complaints to FDR and the Corps commander. These critiques focused on predator control programs, road construction through habitat, and the draining of wetlands. The focus of the criticisms was on preserving ecosystems. This rapid evolution in conservation thinking must have shocked leaders such as Leopold, Murie, Ding Darling, and so forth, who had fought so long and hard to gain support. There were numerous campaigns by local groups nationwide. In Minnesota the Izaak Walton League led an effort to block road projects threatening duck habitat. Ongoing frustration led activists to create the National Wildlife Federation. In such efforts an emergent environmental consciousness are made evident, and these would emerge strongly in post-war conservationist and environmental activism.

The various conservation strategies, ranging from holding and rebuilding soil in erosion plagued areas, including forest replanting, construction of state and national park features, and demonstration farms, not only inculcated varying degrees of stewardship but also helped build support for the federal government as well as expand and strengthen FDR's New Deal coalition. As it turned out, these agencies and programs conserving and transforming a degraded landscape increased the power and reach of the Democratic Party. The flow of money into communities, from cash spent on tools, equipment, food, and lumber, as well as the creation of jobs in local communities, subsidies to farmers, and increased yields from crops, increased tourist dollars, and various other economic benefits, generated increased support for the party and the New Deal. Moreover, the physical evidence of tree belts, restored soil and productive farms, erosion control measures such as check dams, and state and national parks served as a reminder of the benefits of good government. The federal government taught many Americans to be better stewards of the land while also triggering a rising preservationist and ecological consciousness helping to lay the foundation for the powerful environmentalism to emerge after World War II.

New Deal Hydraulic Landscapes

The southeastern states were one of the most impoverished regions of the nation in the 1930s. Most of the population lived without electrical power, and almost a third suffered from malaria. Overfarming led to widespread soil exhaustion while erosion and deforestation worsened soil loss and removed habitat. This undercut a key component of the diet of this poor, rural culture: game food. The creation of the TVA in 1933 under the leadership of Senator George Norris of Nebraska and President Roosevelt represented a dramatic increase in the federal government's role in economic development and rural life in this region. Organized as a federally owned corporation, the intent of the TVA was to modernize the region through the construction of dams and generation of hydroelectricity, the manufacture of fertilizer, and the training of regional residents in soil conservation and tree replanting as well as federal restoration projects conducted in those areas.

The success of the TVA helped generate further support for major projects and strengthen the ambition of the FDR administration to create extensive western hydraulic landscapes. These massive dams may be the most persistent symbols of New Deal labor, federal power and management, and environmental control remaining from that era. These gleaming monuments to either progress, hubris, or expansive federal power, depending on your perspective, still mark the western landscape. Hoover Dam on the Colorado River south of Las Vegas, providing the energy for neon lights, miniature Europes in the city, slot machines, and an ongoing western American bacchanal was the first such project.

The Hoover Dam began as Boulder Dam during President Hoover's administration. Ironically, the president for whom it was named had resisted large appropriations for the project, but that changed when FDR took office. Built by the Bureau of Reclamation, it stored and controlled Colorado River water to provide reclamation for Imperial Valley, California, control flooding, and generate hydroelectricity. That electricity and municipal water made the rise of Las Vegas possible. Completed in 1935 with Public Works Administration funds, the success of the project transformed the Bureau of Reclamation into a powerful agency

Figure 9.10 The Hoover Dam as photographed by Ansel Adams. Damming the Colorado River in 1935, it provided irrigation for more than 1 million acres in California's Imperial Valley as well as the water and hydroelectricity that enabled the rise of Las Vegas in a desert landscape. The success of the project opened the federal floodgates for dam projects and dollars all across the West. National Archives.

of relentless dam construction as the spigot of federal dollars opened up. This initiated a roughly two-decade span of unlimited dam construction for this agency and the U.S. Army Corps of Engineers, who also quickly initiated major projects and tapped the flow of federal dollars. Funneling water into the Imperial Valley, Hoover Dam irrigated more than 1 million acres of former desert into an agricultural cornucopia of fruit, wheat, almonds, and other highly valued foods. While the Imperial Valley Project was unprecedented in its scale and use of federal dollars, the next major project for the Bureau in California far surpassed that. The Central Valley Project (CVP) was much larger and complex in nature and reflects the momentum of the Bureau and its aspirations to create a massive agricultural society in the arid, interior West. As Worster writes, "the CVP also marked the virtual abandonment by the Bureau of its original self-justification, the myth that it was set up mainly to be a builder of homes for the homeless urban masses."[26] The key feature of this project was the Shasta Dam, begun in 1938 and completed in 1944. At 602 feet high, it is the largest hydroelectric dam in California. One of its primary functions was to use hydroelectricity to lift and move the Sacramento River. The Contra Costa Channel ran water to San Francisco Bay, providing irrigation for farmers threatened by the invasion of salt water into a water table that been heavily pumped for agriculture. The Delta-Mendota Canal ran the other direction, transporting water from the Sacramento River to the San Joaquin River and dumping water into that river to provide for irrigation needs and to mitigate against other Bureau projects for the San Joaquin. Not done muscling the landscape into a more useful configuration, the Bureau then built the Friant Dam on the San Joaquin to move water south down a 150-mile canal to the area around Delano and Bakersfield to expand irrigated agriculture there. While the Bureau argued these projects were necessary to provide water to farmers who were experiencing dry wells and over-tapped rivers, in truth the CVP provided water for more than 3 million new acres of farmland.

Further north, in the arid region of central Washington State, the Grand Coulee Dam was born because of the powerful ambition on the part of businessmen and boosters as well as FDR's interest in a monolithic project symbolizing the success of his administration and providing not only jobs but also national hope and pride through the control of nature. This region was not desperate for more hydroelectricity, and certainly America generated more than enough agricultural goods. Many Midwestern and Southern congressmen and senators were befuddled by western irrigation agriculture projects, when only a few years earlier the federal government was paying farmers to grow less. Approximately 70 percent of the population lived without power far from transmission lines, and there was little manufacturing in the region. The primary intent behind the dam was to provide water for irrigation, a job hindered by the fact that the river sits 500 feet below the Columbia Plateau itself; much too deep for the pumping technology of the day. Even raising the reservoir more than 300 feet would still require the dam to generate a great amount of power just to fuel pumps to move water to the fertile plains above. Five hundred feet wide at its base, spanning more than 5,200 feet across the coulee and requiring more than 12 million cubic yards of cement, the Grand Coulee Dam was the largest concrete structure in the world upon its completion in 1941, remaining so until the construction of the Aswan Dam on the Nile River in 1970. This monumental project necessitated great labor and ingenuity. The river bed was filled with deep layers of sand, gravel, and rock. An approximate amount equivalent to 80 million wheelbarrow loads was dug and hauled out of the dam site. Preparation of the site was meticulous, even requiring the brushing of sediments from base rock in order to ensure a good seal between it and the poured concrete. Cold water was piped through the dam as cement was poured in blocks to hasten hardening, and slowly the dam rose to stop and back up the Columbia River with an upstream reservoir 150 miles in length.

No effort was made to save the salmon and steelhead that traveled far up the Columbia to spawn in it and its tributary streams. 1,100 miles of spawning streams were lost, amounting to millions of lost fish. The dam was simply too high for existing technology to lift salmon upstream at a cost that would not have made dam construction too expensive. Moreover, upstream river bottom habitat or riparian zones, with wild game population such as quail, deer, and other species was lost. For tribal peoples such as the Coeur D'Alene, Spokane, Colville, and many others, this meant an instant loss of a valuable resource and a violation of treaty agreements. Lost salmon protein was replaced to some degree by the agricultural system made possible by irrigation from the reservoir. Around 640,000 acres of land were watered by this project, and the desert was transformed into apple, cherry, and apricot orchards, wheat fields, and vineyards that over time have become predominantly owned and controlled by large, industrial farming operations.

The creation of a vast irrigated agriculture in the American expense by the New Deal government transformed the western environment and the economy over time. The rise of agricultural communities in western deserts meant the loss of dry land habitat and increased food production for local consumption and export. This artificial abundance resulted in multiple social impacts as well. It meant the further entrenchment and strengthening of large landowners in rural communities. The government's largesse in providing taxpayer financed water did not lead to a yeoman's paradise of small irrigated farms as originally intended, but rather greater power for already existing landowners, and marginalization and "peonage" for lower-class agricultural workers. One direct consequence was the creation of a permanent underclass of Hispanic and Mexican farm laborers. The expansion of hydroelectric benefits to citizens of the Southeast and the West and an expanding manufacturing economy because of this electricity also were clear benefits stemming from the damming of rivers. It made the rise of Southern cities and the creation of the sunbelt possible. Finally, the construction of this western landscape of dams, power plants, and transmission lines would prove critically important in the coming global conflict.

Solutions to ecological crises such as the Dust Bowl, deforestation, and erosion across the country, as well as managing agricultural production, involved creating and expanding federal agencies. Similarly, increased government commitment to wildlife protection and larger, better protected national parks not only reflected a greater commitment to ecological health but also increased government power. Finally, the construction of major dam and irrigation projects and the extension of electric transmission lines across the Southeast and the West increased the power of the Bureau of Reclamation and the U.S. Army Corps of Engineers and the federal government overall. The stunning popularity of these programs addressing ecological issues or transforming nature to serve societal needs meant that a vast majority of Americans were willing to accept increased government power. In the need to save farmers and ranchers and protect and restore nature, and by harnessing southeastern and western rivers, FDR led and oversaw a dramatic expansion of federal power that most Americans gladly supported.

By the end of the New Deal, effectively in 1938 due to FDR's court-packing scheme and an effort to reduce deficit spending, the American landscape had been transformed in numerous ways. The soil was holding in the southern Plains. In fact, by the beginning of the war, with the return of the rains, agriculture was again productive in that region. Billions of trees had been planted, and millions of acres of land had been set aside as wildlife refuges, reforested, or included within expanded national park boundaries. The CCC not only had helped conserve soil, increase harvest yields, plant trees, and generate opposition that led to an emerging environmental movement, but they also laid the infrastructure of parks, cabins, picnic shelters, and trails that would increasingly bring Americans into the great outdoors, particularly after the end of World War II. This would contribute strongly to the post–World War II environmental movement. Finally, the transformation of nature through soil conservation, forest plantings, and the construction of dams laid down an unanticipated but deeply needed foundation of agricultural production and hydroelectrical generation so necessary to Americans and their allies as the world slipped into war over the next few years.

Document 9.1 The Soil Soldiers: The Civilian Conservation Corps in the Great Depression, *Leslie Alexander Lacy*, 1976

The following are poems, excerpts from letters, and published articles by CCC enrollees and camp newspapers. It was common in the camps for men to publish newspapers with jokes, stories, and advertising local events.

1. Historian Neil Maher argues in *Nature's New Deal* that the CCC conserved men's bodies through labor, food, and the building of morale. What evidence for this argument do you see in the following passages?
2. Examining the language of the following poems, how does physical experience of the landscape, through work, leisure, and discomfort, inform the writers' knowledge of place? Also, what evidence and examples of pride for the work they have done do you see in these primary sources?

Stumps

I hope that I shall never see,
A stump outside the CCC;
A stump whose wiry roots are found
Deep in the earth's tenacious ground,
A stump at which I slave away,
All during a torrid summer day,
Stumps are dug by guys like me
And others in the CCC

The Father of the Forests

When the great experiment has ended,
And the echoes of axes have died,
When the last of the trees have been planted,
And accomplishment fills us with pride;
We'll as say goodbye—and regret it,
Stand by for a decade or twain,
To rejoice in the fact that as builders
We have not bullied in vain.
We are hopeful that those who're to follow,
Through years that shall pass just as these,
Will derive a bountiful blessing,
From our temples—builded of trees.
For trees are a boon to a nation,
Regardless how great or how small,
Rich gems of nature's adornment,

Extending a solace to all.
We shall not be flaunting a banner,
Our wonderful deeds to proclaim,
We shall not be shouting from hilltops
Words that will merit us fame;
For when the trees have foliaged and flourished,
We shall smile—and yet we shall know,
That although we labored and planted,
It was God who made them to grow.

Last Will and Testament

We, the departing members of Co. 129, do herewith and hereby make the following bequest:

We leave you—
The campsite; the buildings and all
The mountains as background; the river at the door.
The sweet-smelling pastures; the tinkling brook,
The road up the mountain wrought by our own sweat.
We leave you—
The climate; the bitter cold;
The wondrous spring days, the clear, cool fall;
The hot dry summer, the rain on the roof,
The mud underfoot, the snow, waist deep, the ice
On the slopes;
It's part of a pattern we'll never forget.
We leave you—
The winds that whistle at night,
That bite, bite and bite on a cold winter's night,
And stab through your clothes on the hill at work,
The grateful breeze after a hot summer's day;
The warm breeze that heralds spring and summer's sway;
We knew them all.
We leave you—
The sounds of a busy camp;
The roar of the blast, the drone of the trucks,
The ring of the ax, the screech of the saw,
The fall of a tree whose day is past,
The blast of the whistle, the bugler's note,
The trampings of feet, the barracks' din,
And all of the radios mingling in.
The snores, grunts, and mutterings in the barracks at night.
They're all there, with others, too numerous to mention.
We find it quiet outside.
. . . But the best that we can leave to you,

Is the many friends, both good and true.
Some are gone; some still here;
We'll remember when we're out of here.
We had our fun now; now we've got to go.
No regrets; still it's hard to leave.

Document 9.2 "Migratory Farm Workers in California," from *On the Dirty Plate Trail: Remembering the Dust Bowl Refugee Camps, Sanora Babb*, 1938

As probably best known from John Steinbeck's *The Grapes of Wrath*, hundreds of thousands of Americans, "Okies" as described by the author and also Southerners and Texans, flooded into California. Most of them sharecroppers and tenant farmers as well as small farmers, the combination of the the Dust Bowl, economic crisis, foreclosure, and the Agricultural Adjustment Act had forced them from the lands they had worked before to move west. The agriculturalists of California's growing farm economies in the Central and Imperial Valleys and other locations solicited these itinerant laborers to work their fields. The massive dislocations of population threatened to overwhelm local resources and threatened the control of large landowners over labor and access to property. Much as the federal government provided taxpayer subsidized irrigation to California farmers, it stepped in to build refugee camps and provided limited relief.

1. How does the creation of these camps reflect failures in land use, government policy, and the expansion of hydraulic landscapes in the West?
2. What was the condition of farm laborers described in this piece? How were they treated? What were the planters' attitudes about the laborers and President Roosevelt?

There are three types of camps in the state in which these workers live.

(1) Government camps, of which there are only seven completed, providing modern sanitary facilities and wooden tent floors. These are "demonstrational" camps, and most of them are open only a few months a year.

(2) The private grower camps, operated on the property of [the] grower, with either free tent space, rented ground space or rented cabins, or tents. The grower furnishes water (sometimes five miles away) and poorest kinds of toilets. The workers are usually forced to move as soon as crops are harvested. In some of the private camps, a certain number of workers are housed the year round at high rentals. In all these camps the grower maintains one or more camp spies.

(3) Squatter camps, in which the majority of the migratory workers live. There are now about 2,000 of these in the state. These camps are on private property without consent of the owner, and absolutely barren of any sanitary facilities.

"Without consent of the owner" means that the land is owned by an absentee, not that they are welcome.

While the small farmer is almost extinct in California, most of the ones existing are somewhat sympathetic to the farm workers because they have begun to realize that they are little more than farm laborers themselves. Most of them borrow from the finance companies to plant their crops, and the returns go first to the company. Sometimes the small farmer owes the company a balance. I heard one of them say that he had shipped oranges for three years and was in the red every year after hard work and sacrifice. He said, "I tell you, there's something wrong, and I'm ready to get my gun out if I just knew who to shoot." I also overheard an argument in a café, which almost ended in a fight, between a big grower and a small farmer. The big grower was angry because the small farmer was voluntarily paying 35 cents an hour to orange pickers. He said "Eighteen cents an hour's enough for them; it's too good, they're not used to anything better. You're just stirrin' up trouble for all of us, you goddam [sic] fool!" The small farmer said: "A man with a family can't live on the wages big farmers pay; it's all he can do to live on 35 cents. I'll pay 45 cents, even 50 cents if I have to. It'd be better for all of us." The big grower said: "If you got no better sense than that, there's a way to show you some one of these days. Wait till someone kills that goddam [sic] Roosevelt, and we'll show you what we'll pay these tramps."

In August and September, in the fruit, the migratory workers make the most money of the year. During January, February, and March, the lot of them are at their lowest ebb. There is almost no work for any great number of them during these months, and since they do not make enough in other months to save for the three workless ones, they must either get some kind of relief or starve. Families who have been in the state a year or more may (with a great deal of difficulty) draw relief from the SRA. Those who have been in the state less than a year have had no way of getting through unemployed times until the recent federal aid through the Department of Agriculture, and this is a temporary grant, ending in May.

When I went to Tulare County, therefore, I was able to find most of the men and women in their tents, although they went out some part of every day hunting for odd jobs weeding spinach, picking scrub cotton and oranges (mostly done there by high school students of the surrounding towns). I was able to go into the country with a rural rehabilitation director at large, and on two days a week to help him interview farm workers for their relief grants. Since many of them did not yet know there was any help forthcoming, we went into the country searching for families in tents, old barns, sheds, any shelter they could find. The rehabilitation director went out to find them because he was sincerely concerned. In this way, I saw many families who had had no work, and were too weak and hungry to look any more. Some of them had no beds and were sleeping in old barns on hay or on the bare ground. One pregnant woman with nine children was sleeping in a shed with several calves in order to keep warm. They had no covers or furniture of any kind and had lost their car. Many families were lying in bed unable to get up. Many of them were on the last of their food with no way of getting more. Some of them had already received their first checks, and bought a supply of flour, lard and potatoes—their unchanging menu. A number of the people had to be forced to sign for relief; they thought it was charity, and kept repeating they had never had to ask for help in their lives. This was always the first thing the men said when they came to sign for relief in the office.

Document 9.3 "Roads and More Roads in the National Parks and National Forests," from ECC Pamplet #54, *Rosalie Edge*, 1936

Rosalie Edge was a critical figure in wildlife protection in the 1930s and later. She, along with Willard Van Name and Irving Brant, formed the Emergency Conservation Committee (ECC). This group provided critical information and lobbying on a vast array of conservation and wildlife issues.

1. What does Edge argue about the negative impacts of road building and National Park and National Forest projects overall?
2. Describe the author's attitude about car-based nature tourism? What problems arise from this way of interacting with nature?

. . . Turning to government-owned lands, we find that work relief has entered our National Parks and Forests in force. Each one of these has its C.C.C. camps; and road-building is again the chief employment of the hundreds of men thus introduced into the wilderness. Can anyone suppose that a wilderness and a C.C.C. camp can exist side by side? And can a wilderness contain a highway?

It is conceded that the National Parks must have roads. The Parks are recreational and educational centres for all the people; and admirably do they fulfill these functions. On the other hand, no one who knows the National Parks is so naive as to believe them to be wilderness areas. They have within their borders great hotels and acres of well-equipped camps. The crowds that visit them are splendidly handled; but the management of thousands of visitors makes it necessary to have offices and living quarters for a large personnel, besides stores, parking houses, docks, corrals, and garages; all of which encroach upon the wilderness. Virgin timber has been felled to build hotels, and valuable trees are cut each year for firewood. In the past, grazing has injured both the forests and meadows; and logging operations have been extensive within the Park boundaries. Some primitive areas, however, still exist in almost all the Parks. These should be guarded as the nation's greatest treasure; and no roads should be permitted to deface their beauty.

The Park Service is eager to prevent repetition of the vandalism that has ruined Park areas in the past; but great pressure is brought to bear by commercial interests that press to have new areas opened in order to obtain new concessions. In addition, there is thrust upon the Park Superintendents the necessity to employ C.C.C. men, whether or not their services are needed; and the wilderness goes down before these conquerors. The support of the public at large must be added to the efforts of the Park Service in order to save the most beautiful of the wild places.

. . . In the Parks we find hotels and other buildings in a style according, as much as possible, with the surroundings—how shocked we should be to find a skyscraper in a National Park! We need to develop roads that shall be suited to Park purposes and not to bring into their solitudes the great boulevards that are appropriate only where the population is densely crowded. Engineers are not trained in esthetic values; and when producing a triumph of their profession they give small heed to the beauty of the flora, or the interest of the other features of the landscape on which they lay their heavy hands. In the Yellowstone Park a road was last summer, quite needlessly, carried over a thermal spring. What is one less hot spring to a road-engineer? The Yellowstone Park has many hot springs—but now it has one less. A road, suitable for the transport of great loads, is not needed in the Parks; but around the camp fires any evening one may hear the boast: "We drove all the way up without changing gear," or "We never dropped below forty." Our Parks should not be desecrated for the whims of such drivers; obstacles might well be put in the way of fast driving in order to induce the tourists to contemplate the wonders of the forests and mountains spread out before them. Why cut away the crest of each rise, leaving ugly cuts with sides so steep that they cannot support plant life? A continuous easy grade is not essential for driving which is almost entirely recreational; and much primitive beauty is lost through exalting every valley and bringing every mountain low. Even the wilderness not traversed by roads is not safe from the despoiler. High up on Ptarmigan Pass in Glacier Park we met a tractor widening a so-called trail to the width of a wagon road, and watched the C.C.C. men stoop and pick out small stones with their hands. They were making a Rotten Row of a trail across what is still happily a great wilderness of virgin forest.

Last summer we stood at the top of Logan Pass and watched the cars come sweeping to the summit. They might pause for five minutes in the great parking place, decorated with landscaped beds of shrubs bordered with stone copings, which belittle what was once one of the most glorious points of the Rocky Mountains. Many people did not leave their cars, others stepped down for a few minutes to look, and to wonder that such height could be reached without a heated engine. A ranger invited and even pleaded with the sightseers to go with him on a short walk to see the secluded wonder of Hidden Lake. "You can have no idea standing here," he said, "what a wonderful thing it is to go there . . . a very little way. . . . " While he spoke, his voice was drowned in the whirr of the self-starters. The little group of nature-lovers who followed him discovered the loveliness of the lake and saw, besides Rosy Finches and White-tailed Ptarmigan. They did not miss the company of the motorists who were by that time far in the valley below, rushing on in their enjoyment of perpetual motion.

Notes

1 Michael Robinson, *Predator Bureacracy: The Extermination of Wolves and the Transformation of the West* (Boulder: The University Press of Colorado, 2005), 170.
2 Quoted in Ibid., 199.
3 Ibid., 199, 200.
4 Donald Worster, *Nature's Economy: A History of Ecological Ideas* (New York: Cambridge University Press, 1994), 271.
5 Donald Worster, *Dust Bowl: The Southern Plains in the 1930s* (New York: Oxford University Press, 1979), 158.
6 Quoted in Richard White, *It's Your Misfortune and None of My Own: A New History of the American West* (Norman: University of Oklahoma Press, 1991), 481.
7 Worster, *Dust Bowl,* 87.
8 Ibid., 94.
9 Ibid., 154.

10 Quoted in Andrew Duffin, *Plowed Under: Agriculture and Environment in the Palouse* (Seattle: University of Washington Press, 2007), 58.
11 Ibid., 82, 83.
12 Ibid., 100.
13 Leslie Alexander Lacy, *The Soil Soldiers: The Civilian Conservation Corps in the Great Depression* (Radnor, PA: Chilton Book Company, 1976), 81.
14 Neil M. Maher, *Nature's New Deal: The Civilian Conservation Corps and the Roots of the American Environmental Movement* (New York: Oxford University Press, 2008), 66.
15 Lacy, *The Soil Soldiers,* 151.
16 Quoted in Stephen Fox, *The American Conservation Movement: John Muir and His Legacy* (Madison: The University of Wisconsin Press, 1981), 245.
17 Ibid.
18 Quoted in Maher, *Nature's New Deal,* 165.
19 Rosalie Edge, "ECC Pamphlet #54," written and distributed in 1936.
20 Quoted in Maher, *Nature's New Deal,* 167.
21 Ibid.
22 Paul Sutter, "Putting Wilderness in Context: The Interwar Origins of the Modern Wilderness Idea," in Michael Lewis, editor, *American Wilderness: A New History* (New York: Oxford University Press, 2007), 170.
23 Quoted in Aldo Leopold, *American Canopy: Trees, Forests, and the Making of a Nation* (New York: Scribner, 2012), 299.
24 Sutter, "Putting Wilderness in Context," 175.
25 Maher, *Nature's New Deal,* 73.
26 Worster, *Rivers of Empire: Water, Aridity, and the Growth of the American West* (New York: Oxford University Press, 1985), 240.

Further Reading

Chase, Alston. *Playing God in Yellowstone: The Destruction of America's First National Park.* New York: Harcourt, Brace, Jovanovich Publishers, 1987.

Deitrich, William. *Northwest Passage: The Great Columbia River.* New York: Simon & Schuster, 1995.

Duffin, Andrew. *Plowed Under: Agriculture and Environment in the Palouse.* Seattle: University of Washington Press, 2007.

Fox, Stephen. *The American Conservation Movement: John Muir and His Legacy.* Madison: The University of Wisconsin Press, 1981.

Maher, Neil M. *Nature's New Deal: The Civilian Conservation Corps and the Roots of the American Environmental Movement.* New York: Oxford University Press, 2008.

Reisner, Marc. *Cadillac Desert: The American West and Its Disappearing Water.* New York: Penguin Books, 1993.

Robinson, Michael. *Predator Bureaucracy: The Extermination of Wolves and the Transformation of the West.* Boulder: The University Press of Colorado, 2005.

Sellars, Richard West. *Preserving Nature in the National Parks: A History.* New Haven, CT: Yale University Press, 2009.

Worster, Donald. *Dust Bowl: The Southern Plains in the 1930s.* New York: Oxford University Press, 1979.

———. *Rivers of Empire: Water, Aridity, and the Growth of the American West.* New York: Oxford University Press, 1985.

World War II Timeline

Domestic and International	
Japanese Invasion of China	1937
German Invasion of Poland	1939
German Invasion of the Soviet Union	June 1941
Lend-Lease Act	October 1941
Beginning of Hydroelectric production at Grand Coulee Dam	October 1941
Pearl Harbor	December 7, 1941
Declaration of War against Germany, Italy, and Japan	December 1941
Preliminary Work on Nuclear Weapons Research and Development	December 1941
Creation of War Production Board	January 1942
Invention of Synthetic Napalm	April 1942
Beginning of Manhattan Project	June 1942
Hanford Site Created	January 1943
Fire Bombing of Hamburg	July 1943
Ending of Naples Typhus Outbreak	December 1943
Firebombing of Japan	November 1944–August 1945
Fire Bombing of Dresden	February 1945
Battle of Iwo Jima	February–March 1945
Battle of Okinawa	April–June 1945
Trinity Atomic Bomb Test	July 1945
Dropping of Atomic Bombs	August 6th and 9th, 1945

Abundance and Terror 10

Americans in World War II

Events were unfolding quickly on the morning of July 16, 1945. As the components of the atomic bomb nicknamed "Little Boy" were being loaded quietly in the pre-dawn hours aboard the USS Indianapolis, observers at the Trinity test site on the White Sands Proving Ground in southern New Mexico were preparing to view the first detonation of an atomic weapon.

The physicists gathered that morning had devoted their lives in the war years to designing and building a weapon that would ensure Allied victory and bring the bloody war to a swift end. Remarkable amounts of resources and some of the world's greatest scientific thinkers had been committed to the Manhattan Project and creation of an atomic weapon. Now they nervously waited to see what their labor had wrought. There was some anxiety due to the unknown—what would happen upon detonation? Physicist Enrico Fermi irritated some of his fellow physicists with his offer of a wager over whether the explosion would ignite the atmosphere and incinerate the planet. He was mostly joking, but for soldiers who overheard him it must have been terrifying. Advised to lie down in the sand with their heads turned away from the blast, observers refused. They wanted to truly see. Some did put on suntan lotion in the dark, while others donned dark glasses and heavy gloves.

The bomb lit up the Jornada del Muerto in the early morning hours, and the physicists carefully recorded and collected data on the shape and dimensions of the explosion, the speed with which it grew and collapsed, the temperatures, and the range to which materials were consumed, burnt, or charred by the heat. But their emotional responses better reflect the terrifying new era of war which they had now unleashed. Physicist Isidor Rabi wrote, "It blasted; it pounced; it bored its way right through you. It was a vision which was seen with more than the eye. It was seen to last forever. You would wish it would stop."[1] J. Robert Oppenheimer, a leader of the Manhattan Project and physicist, wrote of the moment,

> we waited until the blast had passed, walked out of the shelter and then it was extremely solemn. We knew the world would not be the same. A few people laughed. A few people cried. Most people were silent. I remembered the line from the Hindu scripture, the *Bhagavad-Gita:* Vishnu is trying to persuade the Prince that he should do his duty and to impress him he takes on his multi-armed form and says, 'Now I am become Death, the destroyer of Worlds.' I suppose we all thought that, one way or another.[2]

Even as the physicists quietly contemplated the dimensions of their work and this moment, the use of the new weapon was already underway. As Richard Rhodes writes so eloquently, "four hours after the light flung from the Jornada del Muerto blanched the face of the moon, the Indianapolis sailed with its cargo under the Golden Gate and out to sea."[3]

With the dawn-shattering attack on Pearl Harbor on December 7, 1941, America was thrust completely into the fighting of World War II. The nation was already gearing up for war, using New Deal infrastructure to produce war materiel for the Allied Powers and accumulating American military supplies. Although much remained unknown about the coming years, President Franklin Delano Roosevelt understood that America's fundamental contribution would be an actualization of both Germany's and Japan's greatest fears, a fully mobilized American wartime industrial economy. Natural resources and prosperous industrial capitalism enabled the United States to produce far beyond early expectations and dramatically more than allies and foe alike. This "arsenal of democracy," as President Roosevelt termed it, constituted the primary contribution of Americans to the Allied effort to stop the Axis forces. The wealth of nature and the industrial capitalist economy built with that wealth would allow the United States to not only win the war, but in so doing become the dominant economic and political power in the world.

Resources and production proved critical, but the war could not be won simply by providing fighters, bombers, ships, and tanks to allies. American soldiers, marines, sailors, and pilots had to fight, die, and win to achieve final victory over the Germans, Italians, and Japanese. In so doing they encountered landscapes and environments foreign to them and dangerous in numerous ways. From the jungles and rocky atolls of the South Pacific to the rugged mountains and river valleys of central Italy, landscape and environment played a crucial role in worsening the conflict, in some cases driving the fighting to levels more brutal than in other American combat theaters. The coarsening of violence in conjunction with innovations in long-range aviation warfare, incendiary bombs, and atomic weapons led to a type of war entirely unanticipated in 1941 and impossible without the natural wealth and industrial production of the American war machine.

It was fortunate that President Roosevelt's administration devoted so much time, manpower, and money toward infrastructure construction during the New Deal. Dams of the Tennessee Valley Authority and in the American West proved absolutely crucial to wartime production. The results of New Deal programs, including more than 800 new airports, thousands of highway miles, reclamation agriculture, restored farm soils, and so forth, helped the nation mobilize more quickly for war. FDR avoided some of the problems of World War I, rejecting government control or management of the economy. Instead, the federal government, through agencies such as the War Production Board and the Reconstruction Finance Corporation, directed resources to needed areas, financed plant construction, and provided powerful economic incentives for wartime industry. FDR encouraged the patriotism of industrialists and opened up the American economy with measures such as cost-plus contracts, shortened amortization schedules on capital equipment for wartime production, and low-interest loans in addition to direct government financing of plant construction. Early in the war President Roosevelt shocked the nation by announcing that he wanted to see the manufacture of 50,000 planes. When asked by a reporter by what date, FDR—unprepared for the question—replied that he wanted that number every year. In 1939 U.S. manufacturing had produced fewer than 6,000 planes. The president and the rest of America must have been surprised that by war's end the nation far surpassed even the outrageous goal of 50,000 a year, building almost 300,000 fighters, bombers, and cargo planes overall by the summer of 1945. In fact, by 1944 production was so high that the Roosevelt Administration began intentionally scaling back war production to avoid wasting resources.

The region that most benefited and was most dramatically transformed by this wartime economy was the American West. It was valued for a number of reasons. Its proximity to the Pacific made for logistically simpler production and shipment of goods to the war against Japan. The New Deal-era grid of hydroelectric dams and transmission lines offered an abundance of cheap energy for the production of planes, bombs, and ships, and for plutonium processing. With the provision of $2 billion for improvements in power production and distribution, a sixfold increase in hydroelectricital production in the Northwest was achieved during the war years. The wide open spaces and isolation provided some protection from prying eyes, particularly in regards to the Hanford site in eastern Washington and to Los Alamos, New Mexico, where critical nuclear weapons processing and research were

Figure 10.1 Like the B-17 Flying Fortress, the M4 Sherman tank epitomizes American industrial might as well as the U.S. military approach to the war. Whereas the Germans produced a variety of tanks of varying size and strength, the U.S. primarily produced 50,000 Shermans with slight variations for differing combat conditions and uses. The 75 mm guns of the Sherman were unable to penetrate the front armor of two types of German tanks they faced in Europe, and they were forced to work in teams and with tank destroyers to destroy the German tanks from flank shots and from behind. The great advantage of using predominantly one tank model was the ease and speed of production with several American corporations assembling the tanks. When needing to make field repairs, it was simple to strip a tread of parts from another damaged or destroyed tank and swap the parts in to get fighting again. Also, the number of Shermans produced created a great advantage on European battlefields. Library of Congress.

conducted. This was true also for aviation factories, shipyards, munitions and bomb production, and other wartime manufacturing. The availability of land and natural resources, as well as limited industrial development, also increased the West's appeal. The relatively small business community meant little resistance from boosters and businessmen to government programs as well as a strong desire for the prosperity that federal dollars would bring. The federal government flooded the region with money, approximately $70 billion spent on aluminum plants, ammunition orders, shipyards, transmission lines, and so forth. California alone received 10 percent of all federal money spent on wartime production, lifting it from being the ninth largest state manufacturing economy at the start of the war to second by 1945 and launching it to its post-war economic dominance. This economic development caused a massive wave of internal migration, thereby accelerating population growth in the West while greatly expanding the African-American population on the West Coast.

A major reason for the West's transformation was the expanded federal government role in driving industrial manufacturing in the region. According to historian Gerald Nash, "the Defense Plant Corporation supplied the capital for 96 percent of new rubber plants, 58 percent of new aluminum plants, 90 percent of new magnesium plants, and 71 percent of the aircraft factories."[4] Almost all these facilities were built in the West. Several aluminum plants were built along the Columbia River using electricity from the Grand Coulee Dam completed in March 1941. That aluminum was instrumental in the aviation manufacturing of B-17 and B-29 bombers at the Boeing plants of Seattle and Renton, Washington. Those facilities employing aluminum made with Columbia River electricity also used power from the same river for their manufacturing, at one point turning out 16 B-17s every 24 hours. A river whose key importance a short 10 years earlier was its salmon production now played an integral role in the greatest war in human history. The sacrifice of these runs for agricultural abundance now shifted to producing prodigious amounts of electricity to generate a wealth of wartime manufactured goods.

The rise of Boeing Aviation from a company of 4,000 employees to one employing more than 50,000 by wartime's end is one of the celebrated stories of wartime productivity. Boeing produced more than 5,000 B-17s in its Seattle-area facilities and more than 6,000 overall. These bombers proved absolutely central to Allied strategy, constituting a third front of massive area bombing of both Europe and Japan. Marc Reisner suggests in *Cadillac Desert* that America and its allies might not have won the war without the hydroelectricity produced by New Deal dams. He overstates his case, but Reisner is correct in strongly emphasizing their role in military success. The completion of the dam in 1941 was propitious because of the desperate need for hydroelectricity. Historian Donald Worster argues in *Rivers of Empire* that controlling of nature facilitated the control of man. What better example of this is there than the hydroelectricity produced by rivers, controlled by federal power, and used for the production of bombers; the ships in Henry Kaiser's shipyards in Portland and Vancouver; and maybe most important of all, the processing of uranium and production of plutonium at the secret Hanford site in the desert of Eastern Washington. The harnessing of the Columbia River contributed greatly to the defeat of the Axis forces, putting the United States and the Allies in a position to dictate terms to those prostrate nations.

Someone had to bring workers, technical know-how, and federal capital together to generate the material of conquest and dominance. Henry Kaiser is one of the remarkable businessmen of the World War II effort. Having secured federal monies to build the

Hoover, Shasta, Bonneville, and Grand Coulee Dams, he was positioned to take the energy produced by those dams and federal money to construct the machines of war. He was nicknamed "Sir Launchalot" for the almost 1,490 ships produced by his facilities in Richmond, California; Vancouver, Washington; and Portland, Oregon. This number amounted to almost a third of the ships produced overall by the United States during the war. Appropriately enough, the vast majority of these were Liberty Ships, cargo vessels for the hauling of troops, supplies, and war equipment.

Those shipyards were financed with federal money, and the U.S. government even built a new steel mill in southern California, the first west of the Mississippi River. By the end of the war Kaiser employed some 250,000 workers, many of them relocated from the rural West and the South. Lest one assume that ship production was always a coastal industry, Denver offers an exception. Mile-High City workers built 31 prefabricated destroyers and 301

Figure 10.2 Liberty cargo ships lined up and ready to enter service. Shipyards created an assembly line process for building these ships in sections then assembling them into the final product. Liberty ships were used for hauling troops and supplies in the war effort. National Archives.

landing craft in sections. Submarine chasing vessels were built on the edge of the Great Plains as well. They were loaded and shipped by rail over the Rocky Mountains, assembled at the Mare Island facility in California, and launched into service. In Wisconsin the Manitowoc Shipbuilding Company built 10 Gato class submarines, which were pivotal in operations against the Japanese Navy. Built in entirety in the upper midwestern state, they then were transported in floating drydocks to New Orleans and put into service.

This monumental level of production meant that the United States and the other Allied powers gained strength as the Axis forces fell inexorably behind. A moment in the war helps to clarify the exponential difference in wartime industrial capacity between the United States and its enemies. After American victory in the naval Battle of Midway in June 1942, in which the United States secured naval superiority in the Pacific, the Japanese were able to build only 6 more fleet carriers for the remainder of the war while the United States manufactured 17 fleet carriers, 10 medium carriers, and—to add to that profusion—86 escort carriers. Similarly, as Germany and Japan found their fighting abilities limited by access to petroleum, the United States was able to extend its reach further and with greater force because of an abundance of that energy source.

The industrial extraction of western resources by Americans was a process approximately a century old when the war began, and despite all the exploitation of forests, minerals,

Figure 10.3 This woman working on a plane's electrical system exemplifies the role of women in wartime industry. With the flood of men into military service, women were hired to work in wartime manufacturing. African-Americans were hired as well. Both groups did encounter resistance from unions and factory owners, but federal government pressure and wartime needs allowed them to obtain better paying jobs than before the war. Library of Congress.

and farmlands, vast reservoirs of resources remained relatively untouched and, in some cases, undiscovered when wartime production began. Federal agencies launched surveys of the West for mineral resources essential to the war effort. Gerald Nash explains that

> when Japanese occupation of the Dutch East Indies cut off a major source of tin, for example, the [Geological] Survey was able to locate new supplies in western states. When in 1942 German submarines in the Atlantic and Spanish Dictator Franco's light diplomatic alignment with Germany made the shipment of tungsten from Spain highly erratic, the Survey's scientists found vast new deposits in Idaho, and lesser ones in California, Washington, Nevada, Colorado, Arizona, and Utah.[5]

Copper and chrome were discovered in sufficient amounts to serve wartime needs, and prospecting extended onto Indian reservations with new discoveries of lead, coal, and zinc. These proved critical to wartime industries and helped fuel and supply the post-war consumer economy.

The West also was transformed by the number of military installations created in the region. The expansive public lands, isolation, and proximity to the Pacific made it logical to expand the military presence there. The creation of Fort Irwin in the Mojave Desert of California immediately set aside 1,000 square miles (or 640,000 acres) of desert landscape. That post is the same size today. The creation of the White Sands Missile Range protected more than 2 million acres of desert and mountain habitat, and the Marine Corps training post Camp Pendleton, outside San Diego, was established in 1942. Creation of that military site set aside 125,000 acres, including prime beachfront property, riparian zones, a wildlife corridor along a river, and a mosaic of grasses, brush, and forest. In fact, that post currently protects threatened and endangered species such as the California least tern, the least Bell's vireo, and the Pacific pocket mouse, among others. The building of hundreds of posts and bases across the West generated an influx of money into largely rural communities, stimulating the local agricultural and service economies. Moreover, the continuation of many of these military posts after the end of the war greatly expanded the federal presence in the West and contributed to population growth and corresponding environmental damage from suburban sprawl. But an unanticipated benefit arose from the posts' creation. Military training requires large stretches of undeveloped beach, forests, mountains, and desert. As a result, these military sites became large, unofficial refuges for wildlife, particularly bases such as Fort Irwin, California; Camp Pendleton, California; or Naval Air Station Whidbey in Washington. Irwin preserves fragile desert environment, and Whidbey includes forests surrounded by suburbs and farms and productive beaches that would otherwise be developed. Where rapid development consumed the local landscape, these posts have protected habitat. Certainly, one does not contemplate Sherman or M-1 tanks rolling over tortoises on Fort Irwin a model of wilderness, and many of these bases struggle with intractable toxic waste problems. However, they have created vast sections of fairly intact ecosystem that protects various endangered and threatened species. Consider the stretch of the Columbia River next to Hanford. Due to the need for secrecy, no dam construction or development of any other sort has been allowed on that stretch of river. As a result, it is one of the few places on the Columbia where the river bottom is still conducive to spawning salmon; this section of the river still hosts a strong run of wild Columbia chinook salmon, one of the last on the river.

Hanford was one of the locations that ensured American victory, and the production of the atomic bombs that destroyed Hiroshima and Nagasaki was made possible by American resource and capital wealth. Historian David Kennedy sums it up quite precisely:

> The Manhattan Project thus stands as the single best illustration of the American way of war—not so much for the technological novelty of the bombs, or the moral issues they inevitably raised, but because only the Americans had the margins of money, material, and manpower, as well as the undisturbed space and time, to bring an enterprise on the scale of the Manhattan Project to successful completion.[6]

The processing plutonium was a laborious process requiring extensive labor, was highly expensive, and necessitated extensive facilities, isolation, clean water, and electricity in order to eke out a mere dime-size pellet of bomb-ready materiel from two tons of uranium. Richard Rhodes, author of *The Making of the Atomic Bomb,* describes the creation of Hanford:

> The site cost of about $5.1 million, was contained within the eastward excursion of the Columbia: some 500,000 acres, about 780 square miles . . . Roads were sparse on the roughly circular thirty-mile tract. A Union Pacific railroad line crossed one corner; a double electric power line of 230 kilovolts traversed the northwest sector on its way from Grand Coulee Dam to Bonneville Dam.[7]

The Hanford site encapsulates why the West was essential to wartime manufacturing and science. The isolation of the desert provided protection from prying eyes or possible attacks. The natural resource of the Columbia River's flow ensured enough water to cool the fusion piles. And the recently built Grand Coulee Dam provided the energy needed for processing and separating fissionable material. In fact, two of the eight generators at Grand Coulee Dam were set aside solely for a "mystery load" (Hanford) consuming 55,000 kilowatts of energy.

The Oak Ridge site in Tennessee was made possible by the New Deal's creation of regional hydroelectricity from the Tennessee Valley Authority. That facility also reflected America's natural resource and financial wealth with workers developing the 59,000-acre facility with 50 miles of railroad and 300 miles of roads while constructing several buildings to be filled with the equipment and laborers for isolating the U-235. Hanford and Oak Ridge were the largest of several facilities located across the country, with a total of 130,000–150,000 employees struggling to bring this new weapon technology to completion. In contrast, Germany and Japan both abandoned their nuclear weapons programs at least partially because of the lack of resources.

Roosevelt's vision of a great arsenal of democracy was accomplished by exploiting America's rich natural wealth, increasing industrial production, and bringing to bear the energy produced by the TVA and western dams. In 1944 alone, the United States produced more than 96,000 aircraft, more than the combined production of Germany, Japan, and Great Britain in that year. Via lend-lease, a program that provided war materiel to allies with no charge, Americans provided vital support to the Soviet Union in its death struggle with Germany, supplying the Soviets with almost 2,000 locomotives, 7,699 miles of railroad track, just under 80,000 Jeeps, and 350,000 trucks. This, plus steel, minerals for war manufacturing, and 7,000 aircraft helped the Soviets both wage war and produce their own machines of war. Military historian John Keegan writes of the importance of American industrial production and lend-lease in securing victory:

> At the end of the war, the Soviet forces held 665,000 motor vehicles, of which 427,000 were Western, most of them American and a high proportion the magnificent 2½-ton Dodge trucks, which effectively carried everything the Red Army needed in the field. American industry also supplied 13 million Soviet soldiers with their winter boots, American agriculture 5 million tons of food, sufficient to provide each Soviet soldier with half a pound of concentrated rations every day of the war.[8]

The United States also provided the Soviets 540,000 tons of rails and 11,000 freight cars, high-grade petroleum, and three-quarters of Soviet copper. The Soviet Union was not the only ally to benefit from American natural resource wealth, agricultural abundance, and industrial surfeit. The percentage of U.S.-provided military equipment used by Great Britain was 11.5 percent of their total in 1941 and increased consistently so that by 1944 almost one-third of British military equipment was provided by Americans. Throughout the war, nearly one-third of the food consumed by the British was given to them by the United

Figure 10.4 B-17 fuselages lined up for assembly in a Seattle Boeing plant. Library of Congress.

States. By war's end the American economy produced a total of 299,293 aircraft; 40 billion bullets; almost 90,000 tanks; and much.[9] The resource wealth and industrial production of America was mighty indeed.

That production of wartime goods greatly contributed to America's military success in the conflict is beyond question. The idea that this constitutes America's most important contribution to the war effort might be contested but not reasonably. At the outset of the war American leaders assumed that the United States would need to train, organize, equip, and deploy 215 divisions of men for victory in the war effort. This number was arrived at partially on the assumption of Soviet defeat. But American industrial productivity and the value of those weapons on the battlefield, as well as stiff Soviet resistance and defeat of the Germans on the eastern front, meant that government and military planners

Figure 10.5 Popular actress Rita Hayworth doing her part for the war effort. While natural abundance was critical to wartime success, Americans did make sacrifices to support the war. Rationing for items such as coffee, meat, rubber tires, gasoline, and other goods created difficulties for Americans at home. Civilians also participated in drives to collect material such as steel and copper. National Archives.

could reduce the American human commitment to 90 divisions. Suffice it to say that America's natural resources and stunning levels of manufacturing based on that natural capital allowed it to reduce the number of American men sent into harm's way and kept its allies in the war until America was able to make significant military contributions. The reduced loss of American lives in the war compared to the horrifying losses of so many other countries made the transition to a peace-time economy smoother, better enabling America to take its place on the world stage as an international power.

Terrain and Terror in the South Pacific

War is fought in nature and environmental factors contribute to failure and success as well as to the way that fighting develops and worsens over time. John Dower argues in *War Without Mercy* that racism explains the brutality of the combat waged in the South Pacific island campaign and the firebombing of and use of atomic bombs on Japan. He also recognizes the role of landscape, ranging from rocky, barren terrain to dense, tropical jungles as well as a desire for vengeance for Pearl Harbor and the general momentum of war. Eugene Sledge served in the Marine Corps on Peleliu and Okinawa and wrote *With the Old Breed,* one of the great memoirs of the island campaign. Among his descriptions of terrain and combat were several horrifying passages such as the following:

> Added to the awful stench of the dead of both sides was the repulsive odor of human excrement everywhere. It was all but impossible to practice simple, elemental field sanitation on most areas of Peleliu because of the rocky surface. Field sanitation during maneuvers and combat was the responsibility of each man. In short, under normal conditions, he covered his own waste with a scoop of soil. At night when he didn't dare venture out of his foxhole, he simply used an empty grenade canister or ration can, threw it out of his hole, and scooped dirt over it next day if he wasn't under heavy enemy fire.[10] (See Document 10.2.)

To understand the horrors of war it is essential to also see how landscape and environmental conditions contributed to such a state of anger and despair that troops on both sides committed great atrocities.

In the Pacific the Allies needed to break through the tough perimeter of islands established by the Japanese surrounding the home islands. Doing so would enable the launch of bombing raids and preparation and conduct of an invasion into the heart of the nation. For its part, Japan intended to fight to the bitter end, killing as many Allied soldiers and marines as possible in the hope that this would wear down American will so that the invasion of the homeland would be abandoned. Japanese troops were indoctrinated with hatred and fear of American troops as well as with a bastardized version of the Samurai bushido teaching them that the highest form of service was to their emperor and that surrender brought shame onto themselves and their families. Even if they considered surrender, Japanese soldiers were taught that American marines killed their own families in a brutal initiation rite and systematically executed and cannibalized prisoners of war. In addition to these factors, the broken, twisting landscape folded with ridges, ravine, and hills meant that every turn, every step over a ridge line or entry to new terrain made ambush easy for the

Figure 10.6 Marines catching their breath during the fighting on Peleliu. Note the damage to the trees and other plants and land surrounding them. National Archives.

defending Japanese. The fierce resistance—and the environment that made it possible—contributed to the brutality of this fighting.

Fearing machine gun, howitzer, rifle, and mortar fire with each step and absorbing heavy casualties, American marines and soldiers stewed in their terror. Enraged at the conditions and combat, they matched the Japanese ferocity. Moreover, Americans' racist views of Japanese as inferior animals or insects, lingering anger from Pearl Harbor, the reports of the atrocities committed on the Bataan Death March, and the executions of captured American pilots by the Japanese created a level of hatred for the enemy not matched elsewhere else in American combat theaters in World War II. Japanese soldiers launched their attacks from tunnel systems, hidden caves and foxholes, and steel-reinforced caves, employing the terrain to maximum advantage. The tunnels and steel-doored caves could not be effectively destroyed by aerial bombing or the heavy barrages of artillery and rockets from the naval fleets. Even after days and weeks of aerial and naval bombing, much of the Japanese defenses remained unscathed on fiercely contested islands such as Iwo Jima and Okinawa. American combatants responded to the fierce Japanese defense and horrific violence by using napalm in the tunnel systems; backpack- and tank-mounted flamethrowers on caves, tunnels, and other positions; and, in some cases, executing prisoners and even committing atrocities on dying Japanese soldiers. Accounts of the removal of gold teeth from the mouths of still-living Japanese soldiers, killing of prisoners, and the collection of ears and bones testify to the hatred and anger of American soldiers fighting the Japanese. The rugged landscape, enabling a fierce resistance, rendered the fighting in the South Pacific Islands horrifying in terms of conditions, casualties, and atrocities committed.

If ambushes and suicidal Japanese troops were not enough to contend with, malaria and other diseases took a heavy toll on American troops in the South Pacific Islands early in the war. The terrain of combat increased the number of mosquitoes carrying malaria: foxholes, rutted dirt roads pooled with water, holes blasted in the ground from mortar and artillery shells—all created prime breeding habitat for mosquitoes, and they flourished. Before measures were implemented, troops suffered disease losses 8 to 10 times as high as from battle. Some units experienced such extreme rates of malarial infection that they were evacuated before even seeing combat. In response, the U.S. military implemented a remarkably effective anti-malarial campaign. The military required the use of anti-malarial medications such as quinine and atabrine, implemented regular repellent use, and trained troops to avoid malarial sites when possible. A publicity campaign educated and indoctrinated soldiers in a hatred of mosquitoes and a belief in total eradication. Cartoons such as "Private Snafu vs. Malaria Mike" along with posters, matchbooks, and manuals drummed home the

Figure 10.7 Fierce fighting in the Island Campaign and Japanese refusal to surrender from cave and tunnel systems led to extensive use of flamethrowers in combat. National Archives.

message of malaria and insect eradication. Anti-mosquito posters often portrayed the insect with "Japanese features" while arguing for extermination.

Of Bugs and Men

The increased power and popularity of the chemical industry and the concerns over the impact of typhoid (carried by lice) and malaria (carried by mosquitoes) on soldiers meant that great resources were committed to improving pesticides and dispersal methods. The war created a revolution in insect extermination with short- and long-term impacts for disease management, agriculture, and environmental damage. The ability to eradicate led to calls for the extermination of entire species, a refrain that would continue until the early 1960s and the publication of *Silent Spring.*

At the beginning of the war the traditional mosquito removal method involved spraying oil in larval sites, draining areas where water collected, and dispersing poison where mosquitoes reproduced. While this effectively eliminated mosquitoes at the larval stage, the mobile military of World War II needed to destroy adult mosquitoes as soldiers and marines moved into new, infested areas. One major innovation, a revolutionary step forward in pesticide dispersal, was the creation of an aerosol system using Freon for dispersal from a metal cylinder—the can of bug spray was invented. The size of the can enabled soldiers to carry them in their rucksacks, and the capacity of each container to cover 150,000 cubic feet rendered them quite effective. First used in 1942, by 1943 600,000 of these cylinders were being manufactured and distributed. Hence, the bug spray system still in use today (minus the Freon) was developed due to the exigencies of war. Lice were targeted similarly. With Rockefeller Foundation support and the testing of the new pesticide on conscientious objectors, a louse powder system was created that allowed soldiers to carry the pesticide with them and apply it to clothing and bodies. Pyrethrum, made from overseas Chrysanthemum flowers, originally was the active agent in both pesticides discussed here, but naval warfare reduced the amount imported into the states. A replacement needed to be quickly found.

DDT filled the breach. Natural resources and industrial production were crucial to U.S.-Allied victory in World War II, but American science and industrial ingenuity also created the capacity to replace natural goods with synthetics. Japanese occupation of rubber-producing regions and German and Japanese naval interference with maritime trade drastically reduced the amount of rubber imported into the states. The United States implemented a rapid rubber-tree planting program, but more important, immediate, and successful was the invention of synthetic rubber. This also contributed to the rise of petrochemical economies in places like Houston, Texas. This economy rooted in wartime would generate great abundance, wealth, and environmental damage in the post-war years. Similarly, chemists created a seemingly magical solution to bugs and disease: dichlorodiphenyltrichloroethane (DDT). While originally invented in 1874, the poison's efficacy at killing insects was not discovered until the war. Those studying DDT were shocked by its effectiveness. According to historian Edmund Russell,

> tests in Orlando made DDT look "magical": it killed lice and mosquitoes at low doses for a long time. According to one story, ducks flying from treated to untreated ponds

> carried enough DDT on their bodies to kill larvae in the untreated ponds. DDT powder killed lice for four times longer than pyrethrum powders did. Sprayed on walls of buildings, DDT killed adult mosquitoes for months.[11]

This poison killed mosquito larvae at 1 part of DDT to 100 million parts of water and was "100 times as toxic to mosquito larvae as phenothiazine, the most effective synthetic organic larvicide previously known."[12] The imperatives of war made the field testing, deployment, and production of DDT occur much more quickly than might have been the case in peacetime. The desperate American need to be rid of typhoid and malaria opened the valves of federal dollars and gained the critical support of federal researchers, laboratories, and industry.

In addition to these portable cylinders, the military conducted aerial spraying for malaria throughout the South Pacific. A typhus outbreak in Naples in December 1943 enhanced DDT's reputation. Although older pesticides were deployed with great success during this crisis, the use of the "miracle" poison created positive publicity for DDT and marked the first time a typhus outbreak had been stopped in winter. The army's chief of preventive medicine stated, "The wartime development of effective repellents and insecticides will probably constitute the biggest contribution of military medicine to the civilian population after the war."[13] While some may argue that improvements in antibiotics might represent the "biggest contribution," certainly the ability to destroy insects and reduce disease was good news to many. In fact, from the end of the war through the mid-1950s, a concerted effort to wipe out malaria through extensive DDT use achieved at least temporary success in many tropical regions. However, one weakness of the poison became clear as insects began developing immunity to the pesticide, and the mosquitoes that carry malaria recovered.

A piece in a 1944 issue of *The New Yorker* celebrated DDT in rhyme:

> Little insect, roach, or flea,
> Have you met with DDT?
> In the foxhole, up the line,
> DDT gets eight in nine.
> In the tank, beside the gun,
> DDT means battles won.
> . . . Do you fly or do you crawl?
> DDT will fix you all.
> Beetle, borer, bedbug—these
> Horrors worse than DDT's—
> On the leaf, in corn, in bed
> DDT has knocked 'em dead.[14]

In fact, the efficacy of these pesticides in annihilating whole populations of dangerous insects influenced attitudes about extermination of enemies. Increasingly, chemical industry advertisements and American propaganda depicted the enemy, particularly the Japanese, as insects subject to extermination. Dower writes that toward the end of the war,

> 'exterminationist' figures of speech did indeed become a stock way of referring to the killing of Japanese, not only in battle but also in the cities of Japan's home islands. In the steaming combat zones, the Japanese came to be regarded as almost another form of jungle pest. 'Well, which would you druther do—exterminate bug-insecks or Japs!?' asked a sergeant in a comic strip in *American Legion Magazine.* A squad mate spraying bugs replied there wasn't much difference, 'but slappin' Japs is more satisfyin'!'[15]

Note the language in the caption accompanying the "Louseous Japanicus" image and the language of extermination used against the Japanese. Many other cartoons, comics, and chemical industry advertisements trafficked in similar images of the Japanese as ants, spiders, and lice. The solution was extermination.

The obsession with poisoning the enemy became so marked that even key members of the Manhattan Project and of the military considered a scheme to use radioactive waste to kill off Germans. Certainly, problems with developing a fast-fission atomic bomb and fears of German nuclear weapons development also factored into the discussion. Physicists Enrico Fermi, Edward Teller, and Robert Oppenheimer all discussed—and to some degree advocated—potential use of the waste of radioactive fission to kill a large number of Germans by mixing the poisonous waste with dust or liquid and dropping it on food supplies. Teller and Oppenheimer decided that Strontium 90 would be most effective because the human body takes it up in place of calcium deposits it in bone, causing cancer and other major illnesses. Oppenheimer wrote,

> I think that we should not attempt a plan unless we can poison food sufficient to kill half a million men, since there is no doubt that the actual number affected will, because of non-uniform distribution, be much smaller than this.[16]

Rhodes writes that,

> there is no better evidence anywhere in the record of the increasing bloody-mindedness of the Second World War than that Robert Oppenheimer, a man who professed at various times in his life to be dedicated to *Ahimsa* ('the Sanscrit word that means doing no harm or hurt,' he explained) could write with enthusiasm of preparations for the mass poisoning of as many as five hundred thousand human beings.[17]

A careful reading of the whole quote provided by Rhodes suggests that Oppenheimer was counseling patience with the plan until enough poison was developed for a massive operation, not necessarily because of "bloody-mindedness." It may have been a delaying tactic within a government bureaucracy increasingly paranoid, fraught with fear of Germany, and determined to inflict heavy casualties. Regardless, the considered use of the nuclear poison by leading scientific minds correlates with Russell's argument that nuclear weapons are merely a continuance of chemical warfare and suggests that Dower might be partially wrong in his assertion that Americans would never have dropped a nuclear weapon on Germany.

Rain of Fire

On the nights of February 14 and 15, 1945, almost 1,250 British and American heavy bombers dropped 3,900 tons of incendiary and high-explosive bombs on the German city of Dresden, an event made infamous by Kurt Vonnegut's *Slaughterhouse 5*. Some of the high-explosive bombs weighed 2,000 pounds and were designed to destroy entire city blocks while creating a rush of air to fan the flames started by the 200,000 incendiary bombs also dropped on the city. British Lancasters and American B-17s, many no doubt built with aluminum and electricity from the Columbia River, pounded the city, turning it quite literally into a fiery hell where approximately 25,000 people, mostly civilians, died. The impetus for this raid began months earlier, in the summer of 1943, when a colonel in the American Chemical Warfare Service convinced Colonel Curtis E. LeMay of the American Army Air Force to experiment with incendiary bombs on German industrial targets in France with an escalation of their use in the week-long bombing of Hamburg, Germany, in July 1943. With the destruction of major swaths of this industrial city and the deaths of approximately 42,000 people, the rush was on to increase incendiary bombing. The

Figure 10.8 Heavy bombing of the Germans and Japanese constituted a "third front" in the war. This bombing raid on a German munitions plant cost the lives of 800 bomber crewmen. National Archives.

amplified use of incendiary bombs correlates approximately with American acceptance of and increased use of area bombing instead of sticking with the earlier stated American goal of only conducting precision bombing of industrial, military, and transportation infrastructure targets. According to Russell, "In July 1943, the Eighth Air Force dropped 250 tons of incendiaries. In 1944, it dropped more than 5,000 tons per month. By December 1944, incendiaries made up 40 percent of American bomb loads."[18] This capacity to extend the nation's military reach so far beyond its own borders and with such devastating effect originated in natural resource wealth and industrial production.

While Hamburg and Dresden experienced the horrors of fire-bombing to devastating effect, the United States dramatically increased its use of incendiary bombs in the massive area bombing of Japanese cities. Russell compares the scale and speed of use of incendiary bombs and destruction between Germany and Japan: "In Europe, it took three years for Allied bombers to kill 300,000 people, injure about 780,000 more, and render 7.5 million homeless. In Japan, American bombers wounded and killed roughly the same number of people in months."[19] Civilians represented a much higher proportion of the casualties in Japan than in Germany. Having ascended to the rank of general, Curtis LeMay arrived in the Marianas Islands in February 1945 to direct the extensive mass-bombing campaign on Japan now primarily using incendiary bombs on towns and cities largely constructed of wood. Night raids (which Americans had criticized the British for earlier in the war) meant widespread area bombing with no pretense to precision targeting. Whereas Dower and Paul Fussell argue that this is proof of the increased brutality of the war in regards to Japanese and ideas of extermination, Keegan suggests that the Japanese dispersal of war manufacturing into a number of cities and towns, versus the centralized model practiced by the Germans and Americans, contributed to the decision to conduct this widespread area bombing campaign. Regardless of the reasoning, it meant terror and destruction for the Japanese. The bombing of Tokyo on March 9, 1945, employed 3,345 aircraft, primarily long-range B-29 Superfortresses, dropping incendiary bombs to create a firestorm that left almost 90,000 Japanese dead, hundreds of thousands more wounded, and 267,000 buildings destroyed. This was but one of many attacks on the city. The deaths were harrowing: some were burnt to death, others died from inhaling scorched air and boiling in ponds and canals where they sought refuge from the flames. As the campaign continued into the summer, approximately 300,000 Japanese were killed while more than 10 million were left homeless and 2 million buildings were destroyed. Amidst this carnage, one city, Hiroshima, had been intentionally left off the targeting list. Nagasaki had only been lightly bombed.

This use of fire on Germany and Japan was essentially a form of chemical warfare, according to Russell. The logic of annihilation of whole communities is consistent with the philosophy of chemical weapons use, and chemists played a key role in developing and producing incendiary bombs. He writes, "Physics may have been the pinup science of World War II, but chemistry was the workhorse science of urban destruction."[20] The use of nuclear weapons was unprecedented and to date has not been repeated. But the adoption of mass area bombing and fire bombing—particularly under the leadership of General LeMay, who would move on to command the Strategic Air Command during the Vietnam War, and of Robert McNamara, who analyzed, planned, and organized the Dresden raid and B-29 bombing missions over Japan under LeMay, constituted a new type of American warfare to be continued into the post-war era. McNamara served as Secretary of Defense to both President John F. Kennedy and President Lyndon B. Johnson. Under the leadership of McNamara and LeMay, a type of war based on the wealth of munitions arising from American resource wealth and the productivity of industrial capitalism joined with the concept of annihilation of the enemy would continue into the Vietnam War.

Of course the ultimate weapons of annihilation were the atomic bombs developed in the Manhattan Project. The manufacture of weapons of unprecedented power and impact from material acquired at the atomic level represents not only an escalation of warfare but also a process of both environmental manipulation and destruction far beyond anything known before in the annals of war.

In the early morning of August 6, 1945, the Enola Gay dropped "Little Boy," an atomic bomb with the force of 15,000 tons of TNT, on Hiroshima. Original estimates, although flawed, put the immediate death toll at 70,000. But the numbers mounted as Japanese continued to die from severe burns, radiation sickness, and cancer. The number of dead reached 140,000 by the end of the year and 200,000 by 1950. Almost every building in a five-square-mile area was demolished and hundreds of thousands sick and injured. With little response from the Japanese government but given little time to respond or surrender, a second plane took off early on August 9 with the 10,000 pound "Fat Man" in the bomb bay. Running into rough weather and anti-aircraft fire at the primary target of Kokura, the pilot

Figure 10.9 The atomic bombing of Nagasaki. National Archives.

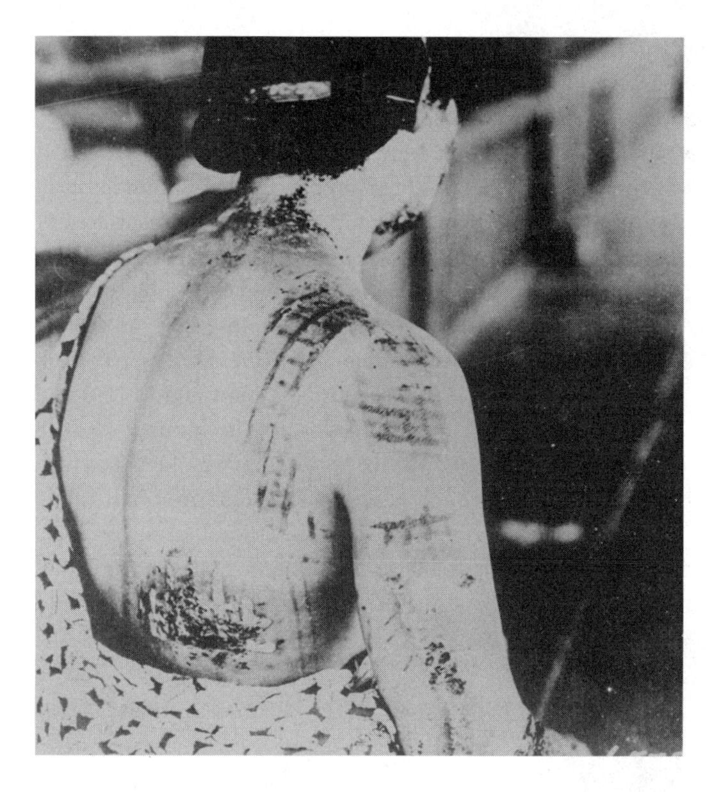

Figure 10.10 The burn pattern from her kimono is clearly evident on this atomic bomb survivor's body. National Archives.

decided to bomb the secondary target of Nagasaki instead. This bomb, equivalent to 21,000 tons of TNT, immediately killed 40,000 Japanese. In both of these explosions, the temperature at the center of the blast reached several million degrees centigrade; people were completely vaporized within a half-mile zone in Hiroshima. Their shadows are imprinted in cement and stone, still visible today. Long-term health impacts were numerous and included hundreds of thousands of Japanese suffering from leukemia, multiple myelomas, lymphomas, and a type of eye damage referred to as "atomic bomb cataracts," caused when people looked directly at the blast. Children in the womb when the bombs detonated later showed physical and mental development problems.

Japan had much to regret as its citizens surveyed a landscape of rubble and charred corpses. A new type of war, far more devastating than any before and predicated on American abundance and wealth, left the once powerful nation a shambles.

Conclusion

World War II was a global conflict in which environment played a crucial role. The rugged terrain and jungles of the South Pacific Islands led to the fiercest fighting of the war for American troops. Of course, terrain and landscape feature prominently in all conflicts, as do disease and agriculture. In this instance, however, the difficulties of the South Pacific Islands slowed the progress of war and coarsened the conflict. The consequence of this is at least three-fold. First, it hardened American attitudes, those of soldiers and civilians. This meant not only less mercy on the battlefield but also increasing support for an unprecedented level of total war, such as area bombing and fire-bombing. Additionally, while attitudes hardened, American scientists, politicians, and military leaders gained time to develop and improve a category of war originating both from the natural resource abundance of America and from the ingenuity and speed of government-sponsored, -funded, and -directed research and industrial production. Finally, time also benefited America, and therefore the Allied powers, by allowing American industrialists to implement innovations while producing ships, bombers, bombs, and other goods more efficiently, drastically increasing the war materiel available as the Allies advanced on Germany and on Japan through the islands of the South Pacific.

The nature of warfare that emerged over the course of the conflict, not only where Americans were involved but also on the eastern front and other locations, led to extensive environmental damage. The destruction of cities, roads, and railroads; the flooding of

Figure 10.11 This scene of devastation in Nuremberg, Germany, the result of Allied aerial bombing, was typical of the European landscape, great parts of the Soviet Union, Japan, China, and many other locations around the world. National Archives.

agricultural lands; the demolition of dams; and the overharvest of natural resources to support the war effort are the most obvious consequences. But the war also meant the deaths of millions of livestock—sheep, horses, and cattle. Moreover, the destruction of infrastructure and the impoverishment of Europe, for example, led to devastating disease outbreaks as well as widespread starvation.

While the rest of the world suffered at the end of the war, America entered a period of unprecedented prosperity. Winston Churchill stated simply, "The United States stands at this moment at the summit of the world." This assertion is evidenced by U.S. control of more than half the world's manufacturing capacity; more than half the world's electricity production; two-thirds of the planet's gold; and twice as much petroleum as every other nation of the world in total, as well as the lead in electronics production and weapons production; a massive shipping fleet; and vast amounts of surplus at the close of fighting. Moreover, because of full employment during the war in America, high wages, relatively low inflation, and excess income because of extensive overtime and a paucity of civilian goods available for consumption during the conflict, the American population entered the post-war era flush and poised to consume. The gross national product more than doubled from $90.5 billion to $211.9 billion, and personal savings increased from just under $7 billion in 1939 to $36 billion in 1945. The director of the Office of War Mobilization and Reconversion summed it up: "The American people are in the pleasant predicament of having to learn to live 50 percent better than they have ever lived before."[21] In comparison, the British, fellow victors in this conflict, did not completely end food rationing until 1954. This consumer generation, giving birth quickly to its baby-boom generation, launched an era of consumption and high living standards unprecedented in world history.

Document 10.1 From *Annual Report of the Secretary of the Interior, Harold Ickes*, 1942

Harold Ickes was one of the key figures in the New Deal and wartime administrations of President Franklin Delano Roosevelt, serving 13 years as the Secretary of the Interior from 1933 to 1946. He also ran the Public Works Administration (PWA) while in charge of the Department of the Interior. As director of the PWA, he was a strong advocate for public power versus private power and played a key role in the development of dam projects in the West.

1. How does Ickes characterize American exploitation and use of natural resources compared to the efforts of countries with less natural wealth? How is this critical to the successful waging of war?
2. What is Ickes's opinion on the role of the West and New Deal programs in wartime success?

It is worthwhile, it seems to me, at a time of crisis, to review our effort of the last year and to appraise our performance under fire, so that we may evaluate our progress and reset our sights. We entered the fiscal year under the stress of a defense program and ended it under the greater pressure of war. The task of this Department was to convert its custodianship of the Nation's natural resources from a peacetime administration to that of war use.

This was an eye-opening proceeding. For years, this Nation had been deluded with the idea that it was practically self-sufficient, that its industrial processes were the world's best, and that its supplies were practically inexhaustible. A painful hangover resulted from that spree. We woke up to find out that we did not have enough steel to do the job; we did not have enough aluminum; we were short of power; we lacked magnesium; our supplies of manganese were too far away to do us much good; our supply of timber and lumber did not hold out; our fisheries and other food resources could not be operated on the old basis nor supply enough to meet demand; our coal supply became endangered and the chaos of war tied our petroleum service up into knots.

In short, we discovered, that so far as our natural resources were concerned, we had been doing everything in the easy way. We had, up to the last year or two, been skimming the cream—and the cream ran out. While the enemies we despised had been making the most of their meager resources—and making them do the job—we had constructed our whole economy on our fat. The former "had-not" nations turned to their secondary ores and low-grade minerals, and by sweat and effort learned how to use them. We neglected our secondary sources of supply—the low-grade deposits of minerals, for example—and went merrily on our way, using only the best, and therefore the most profitable.

. . . A few facts show the extent to which we have provided additional resources or led the way in wielding those at hand against the enemy with greater effect. Generating capacity on reclamation projects increased by 43 percent. From Columbia River projects alone, nearly 2,000,000,000 kilowatt-hours of power poured into war plants. By a process developed in our laboratories, it appears that enough manganese can be extracted from low-grade domestic ores to make 87,000,000 tons of steel annually. More than a million tons of bauxite, the common source of aluminum, and other valuable ores have been found in our search for strategic metals . . .

Outstanding was the Bureau's [of Reclamation] contribution of hydroelectric power—the potent energy that turns the machines that turn out the guns. In both the Pacific Northwest and the Southwest, Reclamation's giant power plants were the bulwark behind industries already working at top speed or mushrooming into existence for the production of war material. Energy poured out of these reclamation plants and others in 11 States for manufacturing and mining; for copper, steel, aluminum, magnesium; for bombs, planes, and ships.

. . . Aside from the generation of power, and the release of water itself for domestic and industrial uses, the regional production of food, forage, and fiber on the reclamation projects in the West supplied urgent needs. This production reduced the burden on transcontinental railroads and highways for the movement of men and equipment. It meant speedy delivery of supplies, and the saving of freight cars, of steel for rails and equipment, of fuel for engines, and of gas, oil, and rubber for trucks.

Stored water irrigated more than 3,000,000 acres of productive land. Irrigation district officials collaborated with the Bureau in an intense effort to get the most from high-production reclamation farms. The gross value of the 1941 crops on land served with reclamation water was $159,885,998, a 35 percent increase over the $117,788,677 of 1940.

Document 10.2 From *With the Old Breed: At Peleliu and Okinawa*,[1] *E. B. Sledge*, 2007 (original date, 1981)

The Battle of Peleliu occurred from September to November 1944, following the Marianas Islands campaign. Predicted by American commanders to be a four-day battle, this conflict became one of the bloodiest of the war for Americans. The 11,000-man Japanese garrison, learning from earlier island battles, lightly contested the American beach landings in amphibious landing craft but were dug into the rocky terrain of valleys, mountains, and ridges further inland. The First Marine Division suffered one-third casualties for the overall battle but in some specific fights experienced casualty rates of 60 percent. The entire Japanese garrison died except for approximately 20 who were captured. The Japanese commander committed ritual suicide rather than surrender. Unfortunately, this victory was of little strategic value (a point that had been argued by Admiral Nimitz against General McArthur prior to the battle) as the airfield was not used for bombing runs.

1. What were the elements of the landscape, both natural and manmade, that made life difficult for U.S. marines and soldiers at Peleliu?
2. How did the challenges of this environment (be sure to also include those elements created by the troops themselves) affect the attitudes of those serving in this battle?

The word passed along the line to us told that when the men of 2/1 moved up toward the Japanese positions following preassault artillery fire, the enemy fired on them from mutually supporting positions, pinning them down and inflicting heavy losses. If they managed to get onto the slopes, the Japanese opened point blank from caves as soon as our artillery lifted. The enemy then moved back into their caves. If Marines got close enough to an enemy position to attack it with flamethrowers and demolition charges, Japanese in mutually supporting positions raked them with cross fire. Each slight gain by the 1st Marines on the ridges came at almost prohibitive cost in casualties. From what little we could see of the terrain and from the great deal we heard firsthand of the desperate struggle on our left, some of us suspected that Bloody Nose was going to drag on and on in a long battle with many casualties.

. . . Occasional rains that fell on the hot coral merely evaporated like steam off hot pavement. The air hung heavy and muggy. Everywhere we went on the ridges the hot humid air reeked with the stench of death. A strong wind was no relief; it simply brought the horrid odor from an adjacent area. Japanese corpses lay where they fell among the rocks and on the slopes. It was impossible to cover them. Usually there was no soil that could be spaded over them, just the hard jagged coral. The enemy dead simply rotted where they had fallen.

1 Excerpt from *With the Old Breed at Peleliu and Okinawa* by E. B. Sledge, copyright © 1981 by E. B. Sledge. Used by permission of Presidio Press, an imprint of Random House, a division of Random House LLC.

. . . It is difficult to convey to anyone who has not experienced it the ghastly horror of having your sense of smell saturated constantly with the putrid odor of rotting human flesh day after day, night after night. This was something the men of an infantry battalion got a horrifying dose of during a long, protracted battle such as Peleliu. In the tropics the dead became bloated and gave off a terrific stench within a few hours after death.

. . . Added to the awful stench of the dead of both sides was the repulsive odor of human excrement everywhere. It was all but impossible to practice simple, elemental field sanitation on most areas of Peleliu because of the rocky surface. Field sanitation during maneuvers and combat was the responsibility of each man. In short, under normal conditions, he covered his own waste with a scoop of soil. At night when he didn't dare venture out of his foxhole, he simply used an empty grenade canister or ration can, threw it out of his hole, and scooped dirt over it next day if he wasn't under heavy enemy fire.

But on Peleliu, except along the beach areas and in the swamps, digging into the coral rock was nearly impossible. Consequently, thousands of men—most of them around the Umurbrogol Pocket in the ridges, many suffering with severe diarrhea, fighting for weeks on an island two miles by six miles—couldn't practice basic field sanitation. This fundamental neglect caused an already putrid tropical atmosphere to become inconceivably vile.

Added to this was the odor of thousands of rotting, discarded Japanese and American rations. At every breath one inhaled hot, humid air heavy with countless repulsive odors. I felt as though my lungs would never be cleansed of all those foul vapors. It may not have been that way down on the airfield and in other areas where the service troops were encamped, but around the infantry in the Umurbrogol Pocket, the stench varied only from foul to unbearable.

In this garbage-filled environment the flies, always numerous in the tropics anyway, underwent a population explosion. This species was not the unimposing common housefly (the presence of one of which in a restaurant is enough to cause most Americans today to declare the place unfit to serve food to the public). Peleliu's most common fly was the huge blowfly or bluebottle fly. This creature has a plump, metallic, greenish-blue body, and its wings often make a humming sound during flight.

The then new insecticide DDT was sprayed over combat areas on Peleliu for the first time anywhere. It supposedly reduced the adult fly population while Marines were still fighting on the ridges, but I never noticed that the flies became fewer in number.

With human corpses, human excrement, and rotting rations scattered across Peleliu's ridges, those nasty insects were so large, so glutted, and so lazy that some could scarcely fly. They could not be waved away or frightened off a can of rations or a chocolate bar. Frequently they tumbled off the side of my canteen cup into my coffee. We actually had to shake the food to dislodge the flies, and even then they sometimes refused to move. I usually had to balance my can of stew on my knee, spooning it up with my right hand while I picked the sluggish creatures off the stew with my left. They refused to move or be intimidated. It was revolting, to say the least, to watch big fat blowfies leave a corpse and swarm into our C rations.

Document 10.3 From *The Atomic Bombings of Hiroshima and Nagasaki, Manhattan Engineer District*, June 29, 1946

Following the bombings of Hiroshima and Nagasaki, the surrender of Japan, and the occupation of that nation by American troops, studies were conducted to determine the impact of the atomic blasts. The in-depth study examined structural damage, the range of fires, death and injury rates, as well as the lingering impacts of radiation poisoning.

1. How destructive were the detonations at the center versus at the periphery of the blast zone? What is the point of this analysis conducted by the army?
2. What are the health impacts of the atomic bombs on survivors? Analyze the language used to describe the various conditions.

In considering the devastation in the two cities, it should be remembered that the cities' differences in shape and topography resulted in great differences in the damages. Hiroshima was all on low, flat ground, and was roughly circular in shape; Nagasaki was much cut up by hills and mountain spurs, with no regularity to its shape.

In Hiroshima almost everything up to about one mile from X was completely destroyed, except for a small number (about 50) of heavily reinforced concrete buildings, most of which were specially designed to withstand earthquake shock, which were not collapsed by the blast; most of these buildings had their interiors completely gutted, and all windows, doors, sashes, and frames ripped out. In Nagasaki, nearly everything within ½ mile of the explosion was destroyed, including heavy structures. All Japanese homes were destroyed within 1½ miles from X.

Underground air raid shelters with earth cover roofs immediately below the explosion had their roofs caved in; but beyond ½ mile from X they suffered no damage.

In Nagasaki, 1500 feet from X high quality steel frame buildings were not entirely collapsed, but the entire buildings suffered mass distortion and all panels and roofs were blown in.

In Nagasaki, 2,000 feet from X, reinforced concrete buildings with 10" walls and 6" floors were collapsed; reinforced concrete buildings with 4" walls and roofs were standing but badly damaged. At 2,000 feet some 9" concrete walls were completely destroyed.

. . . Heavy fire damage was sustained in a circular area in Hiroshima with a mean radius of about 6,000 feet and a maximum radius of about 11,000 feet; similar heavy damage occurred in Nagasaki south of X up to 10,000 feet, where it was stopped on a river course.

In Hiroshima over 60,000 of 90,000 buildings were destroyed or heavily damaged by the atomic bomb; this figure represents over 67% of the city's structures.

In Nagasaki 14,000 or 27% of 52,000 residences were completely destroyed and 5,400 or 10% were half destroyed. Only 12% remained undamaged . . .

Percent Mortality at Various Distances

Distance from X, Percent Mortality in Feet	
0–1000	93.0%
1000–2000	92.0
2000–3000	86.0
3000–4000	69.0
4000–5000	49.0
5000–6000	31.5
6000–7000	12.5
7000–8000	1.3
8000–9000	0.5
9000–10,000	0.0

Radiation Injuries

. . . The proper designation of radiation injuries is somewhat difficult. Probably the two most direct designations are radiation injury and gamma ray injury. The former term is not entirely suitable in that it does not define the type of radiation as ionizing and allows possible confusion with other types of radiation (e.g., infra-red). The objection to the latter term is that it limits the ionizing radiation to gamma rays, which were undoubtedly the most important; but the possible contribution of neutron and even beta rays to the biological effects cannot be entirely ignored. Radiation injury had the advantage of custom, since it is generally understood in medicine to refer to X-ray effect as distinguished from the effects of actinic radiation. Accordingly, radiation injury is used in this report to mean injury due only to ionizing radiation.

According to Japanese observations, the early symptoms in patients suffering from radiation injury closely resembled the symptoms observed in experimental animals receiving large doses of X-rays. The important symptoms reported by the Japanese and observed by American authorities were epilation (loss of hair), petechiae (bleeding into the skin), and other hemorrhagic manifestations, oropharyngeal lesions (inflammation of the mouth and throat), vomiting, diarrhea, and fever.

. . . Petechie and other hemorraghic manifestations were striking findings. Bleeding began usually from the gums and in the more seriously affected was soon evident from every possible source. Petechiae appeared on the limbs and on pressure points. Large ecchymoses (hemorrhages under the skin) developed about needle punctures, and wounds partially healed broke down and bled freely. Retinal hemorrhages occurred in many of the patients. The bleeding time and the coagulation time were prolonged. The platelets (coagulation of the blood) were characteristically reduced in numbers.

. . . Lesions of the gums, and the oral mucous membrane, and the throat were observed. The affected areas became deep red, then violacious in color; and in many instances ulcerations and necrosis (breakdown of tissue) followed. Blood counts done and recorded by the Japanese, as well as counts done by the Manhattan Engineer District Group, on such patients regularly showed leucopenia (low-white blood cell count).

> In extreme cases the white blood cell count was below 1,000 (normal count is around 7,000). In association with the leucopenia and the oropharyngeal lesions, a variety of other infective processes were seen. Wounds and burns which were healing adequately suppurated and serious necrosis occurred. At the same time, similar ulcerations were observed in the larynx, bowels, and in females, the gentalia [sic]. Fever usually accompanied these lesions.
>
> Eye injuries produced by the atomic bombings in both cities were the subject of special investigations. The usual types of mechanical injuries were seen. In addition, lesions consisting of retinal hemorrhage and exudation were observed and 75% of the patients showing them had other signs of radiation injury.

Notes

1 Quoted in Richard Rhodes, *The Making of the Atomic Bomb* (New York: Simon & Schuster, 1986), 672.
2 Ibid., 676.
3 Ibid., 678.
4 Gerald D. Nash, *The American West Transformed: The Impact of the Second World War* (Lincoln: University of Nebraska Press, 1985), 19.
5 Ibid., 30.
6 David Kennedy, *Freedom from Fear: The American People in Depression and War, 1929–1945* (New York: Oxford University Press, 1995), 668.
7 Rhodes, *The Making of the Atomic Bomb,* 497.
8 John Keegan, *The Second World War* (New York: Penguin Books, 2005), 218.
9 Kennedy, *Freedom from Fear,* 654, 655.
10 E. B. Sledge, *With the Old Breed: At Peleliu and Okinawa* (New York: Ballantine Books, 2007), 143.
11 Edmund Russell, *War and Nature: Fighting Humans and Insects with Chemicals from World War I to Silent Spring* (New York: Cambridge University Press, 2001), 124.
12 Ibid., 127.
13 Quoted in Ibid., 129.
14 Ibid.
15 John B. Dower, *War Without Mercy: Race and Power in the Pacific War* (New York: Pantheon Books, 1986), 90, 91.
16 Quoted in Rhodes, *The Making of the Atomic Bomb,* 510, 511.
17 Ibid., 511.
18 Russell, *War and Nature,* 131.
19 Ibid., 142.
20 Ibid.
21 Quoted in Thomas G Paterson, *On Every Front: The Making and Unmaking of the Cold War* (New York: W.W. Norton & Company, 1992), 17.

Further Reading

Dower, John W. *War Without Mercy: Race & Power in the Pacific War.* New York: Pantheon Books, 1986.
Keegan, John. *The Second World War.* New York: Penguin Books, 2005.
Kennedy, David. *Freedom from Fear: The American People in Depression and War, 1929–1945.* New York: Oxford University Press, 1999.

Murray, Williamson. *War in the Air: 1914–1945*. New York: Harper Paperbacks, 2005.
Nash, Gerald D. *The American West Transformed: The Impact of the Second World War.* Lincoln: University of Nebraska Press, 1985.
Overy, Richard. *Why the Allies Won*. New York: W.W. Norton, 1996.
Parker, Matthew. *Monte Cassino: The Hardest-Fought Battle of World War II*. New York: Doubleday, 2004.
Rhodes, Richard. *The Making of the Atomic Bomb.* New York: Simon & Schuster, 1986.
Russell, Edmund. *War and Nature: Fighting Humans and Insects with Chemicals from World War I to Silent Spring.* New York: Cambridge University Press, 2001.
Sledge, Eugene. *With the Old Breed: At Peleliu and Okinawa*. New York: Presidio Press, 2007.

Timeline

Employment Act	1946
Publication of *Sand County Almanac*	1948
Publication of *Road to Survival*	1948
Publication of *Fundamentals of Ecology*	1953
Ohio and Chesapeake Canal Hike	1954
The Federal-Aid Highway Act	1954
Brown v Board of Education Decision	1954
Defeat of the Echo Park Dam	1955
Montgomery Bus Boycott	1955–1956
Highway Revenue Act	1956
Olympic National Park Coast Hike	1958
Creation of the Greater St. Louis Committee for Nuclear Information	1958
Thalomide Scare	1961
Publication of *Silent Spring*	1962
Nuclear Test Ban Treaty	1963
Passage of the Wilderness Act	1964
Bureau of Reclamation attempt to dam the Grand Canyon	1966
Creation of the Environmental Defense Fund	1967
Banning of DDT	U.S., 1972; World, 2004 (except for malaria control)

Environmental Consensus in the Republic of Abundance 11

David Brower had been warned to not argue numbers in testimony against the Bureau of Reclamation in Washington, D.C. Brower, Director of the Sierra Club, was helping lead the opposition to the proposed Echo Park Dam in Dinosaur National Monument and had traveled to D.C. to challenge the Bureau of Reclamation. Reviewing numbers the night before his testimony, he discovered errors in the Bureau's evaporation calculations. The fight to protect Dinosaur had taken on some elements of the earlier Hetch Hetchy struggle. And, as before, proponents of the Colorado River Storage Project and the dam argued for "common sense" using scientific data against the abstract ideals of "nature lovers." Brower was determined to win that fight. In front of Congress the next day, he claimed to have found mathematical errors meaning that evaporation rate estimates were wrong and, if corrected, would argue against the Echo Park Dam. Following his testimony a Bureau of Reclamation employee presented to the sub-committee and attacked Brower's "ninth-grade arithmetic."[1] He presented an in-depth and likely confusing discussion of the role of trigonometry, calculus, and plane geometry in making these calculation estimates and, concluded, "you just cannot use ratios and run the old slide stick, and get any answer you want."[2] With Brower rebuffed and embarrassed, the Bureau stood strong behind its wall of mathematical formulas, calculated evaporation rates, and engineering expertise.

Her cancer had returned months earlier, and Rachel Carson had turned to deeply painful radiation therapy instead of further, futile surgery. Fighting through pain, nausea, and fear, she had finished *Silent Spring* and suffered through the attacks by the chemical industry, politicians, academic scientists, and others accusing her of communist intent, an anti-business attitude, and ineptness. Now, exhausted and battered, feeling pain in her back and fearing further metastases, she prepared for an appearance at the Women's National Press Club on December 5, 1962. In spite of her suffering, she came out swinging. Facing the audience and NBC and ABC television cameras,[3] she began with an account of negative responses to her book by two farm bureaus in Pennsylvania. In the article covering these responses, the reporter noted, and she quoted to her audience, "No one in either county farm office who was talked to today had read the book, but all disapproved of it heartily."[4] Noting this ignorance she then hammered at the violation of science by the pursuit of profits. Carson pointed out that multiple articles from a recent issue of the *Journal of Economic Entomology* had been funded by the chemical industry and that the American Medical Association referred physicians to a pesticide trade association to learn the health impacts of pesticides. "When the scientific organization speaks, whose voice do we hear, that of science or the sustaining industry?"[5]

Rachel Carson and David Brower represented and led two major fronts of environmental activism in the 1950s and early 1960s. Brower and the Sierra Club led a resurgent preservationism that would culminate in a string of victories for activists and nature. Carson represented a historically deeper tradition of environmental work seeking to clean the environment to protect human health. One of her important contributions was to extend that struggle into the natural world. The fight to cleanse air, water, soil, and the bodies of humans and animals of chemical and radiation contamination would require a national fight and significant change at the top levels of government. While pursuing seemingly different environmental goals, these groups shared tactics. Public information, committee hearings, letters to editors, articles, books, and letters to members of Congress were the arrows in these activists' quivers. The other integral strategy, and one that contributed to a growth of American democracy in this period, was scientific research, the study of mathematics, and the investigation of the assertions of scientists and engineers employed by federal agencies in order to better challenge them and the government agencies they represented. The insistence on public knowledge of these issues helped to widen participation in the environmental movement and build key support for other environmental issues. These measures strengthened a growing environmental consensus and, by expanding public knowledge and participation in government's management of environmental issues, made the United States a more democratic nation.

Following World War II Americans entered into an era of unprecedented prosperity. Exhausted and emotionally worn by the Great Depression and years of war, many Americans eagerly sought an improvement to their quality of life. Even as the Cold War introduced new stresses into American foreign policy and domestic life, the late 1940s through the early 1960s marked a period of noteworthy change in American society. The 1950s are regularly depicted as a decade of conformity, blandness, and cultural conservatism. There is much that is accurate in that appraisal, but it is also true that activists and cultural warriors began to demand and enact radical changes in this period. With the Brown v. Board of Education decision in 1954, Civil Rights activists were able to begin successful activism, particularly with a startling victory in the Montgomery Bus Boycott, forcing desegregation of busing in that city in 1956. Similarly, the beats were questioning central tenets of American culture, seeding the ground for protests and the counterculture that would emerge in the 1960s.

What is not well known to most Americans is that this seemingly staid, conservative decade was one of intense and largely successful environmental activism. Those successes and energy poured forward into the 1960s and helped create an environmental consensus that would not begin to unravel until the 1970s and 1980s. The fights to stop dams, protect forests, and create wilderness areas helped build the momentum and organizational skill and strategies that would lead to the even greater successes of the 1960s and early 1970s. The popularization of ecological thinking joined with increased information about radiation and chemical pollution made Americans more aware of environmental problems and the need for reform. With Barry Commoner and the Greater St. Louis Committee for Nuclear Information, Rachel Carson's *Silent Spring,* and other critical challenges to the status quo, a new environmental era was launched.

A Republic of Abundance

After World War II, America was a wealthy nation with a large and thriving middle class. As the nation provided an arsenal of democracy for allied powers and the American military, the economy expanded dramatically with full employment, high wages at all levels (with plenty of overtime), and generous corporate profits due to cost-plus contracts and other government incentives. It was not simply the war industries that thrived. Many ancillary industries such as logging, mining, agriculture, and others also benefitted from the wartime economy. Except for a short post-war recession, the prosperity persisted into the late 1940s and 1950s as Cold War policies such as the Marshall Plan[6] stimulated the American economy, and the development of a military-industrial complex guaranteed ongoing industrial war materiel production. Europe and Asia, physically and economically devastated by the war, purchased large quantities of American steel, lumber, agricultural goods, and manufactured items, thus helping drive economic growth while also causing further environmental damage in America.

Oddly enough, from our current perspective and understanding of the era, when the war was over Americans did not simply leap into mad-dash shopping, home building, car buying, and general habits of consumption. They hesitated, cautious to spend after the years of depression and the hard work, sacrifice, and horrors of World War II. It would require a new addition to the ideology of abundance and innovative, transformative federal policies and programs, to really get the supercharged economy of the 1950s underway.

In the post-war era, mass consumption emerged not only as a lifestyle and articulation of the American Dream but also as a central tenet of American ideology. Politicians and businessmen argued that Americans bore a responsibility to buy a new home, car, washing machine, clothes, and so forth to propel the economy on to greater prosperity for all Americans and as a bulwark against external and internal threats. In 1957 *Fortune* editor William H. Whyte declared, "thrift is now un-American."[7] A booming economy became patriotic, consumption a sword against communism, foreign and domestic. Nature's wealth, industrial manufacturing, and government policies made this possible, but this ideology of consumption came at great environmental and social costs.

Government policies made mass consumption possible. The federal government committed itself to fostering the strong growth of the gross domestic product and smoothing over tensions between labor and capital that had broken into open conflict numerous times during the 1930s. As historian Lizabeth Cohen writes, "Mass consumption, for the liberals particularly, provided a way of reconciling capitalist growth and domestic commitments, without endorsing too planned an economy or too powerful a welfare state."[8] For all of the economic development and natural resource exploitation of the previous three centuries, vast reservoirs of natural wealth remained to be tapped. Massive tracts of old growth timber in the West, energy created by dams, metals from national mining operations, and of course, industrial manufacturing would stoke the furnace of economic growth. The failure of society to pay for the costs of environmental degradation within the cost of a product subsidized the phenomenal economic growth from 1946 to 1973. The baby boom of 1946–1964, adding 75 million more Americans to the population provided another driver of economic growth.

The Employment Act of 1946 committed the government to supporting "maximum employment, production, and purchasing power," and the 1944 Servicemen's Readjustment Act required the government to increase its military defense spending and provide college money, small business loans, and home loans to returning veterans. Because of the G.I. Bill, by 1956 42 percent of veterans were homeowners. Programs such as these helped to pump money into the economy through jobs, education, and home purchases. As President Truman launched the Cold War in 1946, the military industrial complex geared up its production of bombers, fighters (later missiles), munitions, tanks, and so forth, becoming an integral component of the post-war American economy. Probably the strongest domestic engine of economic growth in the 1950s and 1960s was the rise of mass suburbanization. The building of the millions of homes, the tens of millions more cars necessary for living in such communities, and the forms of consumption associated with suburban living, as well as the numerous ancillary industries, kept the American economy humming for a couple of decades.

Suburbs existed prior to World War II, the first ones built with the expanding use of the car in the 1920s, but the decades following the war are known as the suburban era for a reason. The American landscape was transformed again as farms and fields were converted into suburbs, malls, and highway business strips. Home ownership doubled between 1940 and 1950. This is not simply the result of war-time and post-war prosperity but is a consequence of deliberate federal government policies. The Federal Housing Administration (FHA) enacted policies favoring new home construction in car-dependent areas, driving the rise of the suburbs and the decline of urban communities. FHA policies favored single-use residential zoning and single homes situated on large lots. This prevented factories, offices, stores, restaurants, and apartment complexes from being built in suburban neighborhoods, thereby increasing automotive dependence. These communities were not walkable; one could not commute by foot to work or stroll to the corner store to pick up bread or milk. Zoned away from the suburbs, residents were compelled to drive to accomplish basic tasks and get to work. Developers also made no effort to take public transportation into account when laying out and building these communities.

FHA rules promoted suburbs at the cost of cities and older homes. Mortgages were provided only for first mortgages and for purchasing or remodeling older homes. Racial bias also influenced the direction of development:

> Underwriting standards drew on appraisal techniques created by the Home Owners Loan Corporation (HOLC), which branded both older neighborhoods and racially and economically heterogeneous neighborhoods as "declining"—or too risky for investment. Known as "redlining" after the color used on HOLC maps to designate non-creditworthy areas, the standards automatically excluded neighborhoods with nonwhite residents from qualifying for mortgage insurance, creating an insidious federal endorsement of racially segregated housing—justified in the name of ensuring the long-term soundness of investments.[9]

Because of these policies, even new homes in many urban neighborhoods did not quality for FHA-insured mortgages. Many who would have chosen to buy a home in the city had no choice but to move to the suburbs.

Federal road policy also altered the urban and rural landscape, further incentivizing suburban living and increased automobile use. The Federal-Aid Highway Act of 1954 set aside $175 million specifically for the interstate system while dramatically increasing federal financing of highways. The Highway Act in 1956 and the Highway Revenue Act of 1956 further expanded funding and mandated high-quality standards for new highways and interstates. With this legislation the federal government created a self-perpetuating system of road growth because increasing gas consumption on the ever-expanding excellent road system meant more roads, increasing dependence on cars, and more gas sales generating more tax revenue for more roads. The very logic of an automotive society and an expanding capitalist economy built around it and by it was written through these federal policies.

Car sales increased dramatically after World War II, with 2.1 million sold in 1946 and more than doubling to 5.1 million in 1949. For the next 10 years new car sales averaged 5.9 million a year. More Americans owned cars, rising from half of American families in 1941 to almost 80 percent in 1960, with a great percentage of the car sales being new, replacement cars rather than first cars. Cars increased mobility, a necessity due to the rise of the suburbs and the decline of cities, and were integral to American leisure and conspicuous consumption. Even with road-building programs, the logic of suburbs was the primary reason for increased automotive use and car sales.

These series of acts created a federal planning and interstate system with dedicated funding for high-quality, direct interstate transportation routes across the nation regardless of local population and need. Rejecting local planning, engineers designed interstates on straight lines, promoting access to countryside and downtown cities alike. The roads were often built directly through existing neighborhoods, and the ones not destroyed declined in value because of overpasses, proximity of interstates and noise, and neighborhoods being divided by the roads. Decline drove businesses and jobs away, and cities turned from appealing locales to places to avoid or leave. Continuing urban decline and the logic of cheaper homes in the suburbs pulled even more urban residents out of the cities, causing a worsening downward spiral for urban areas. The flow of a whole generation of white affluent Americans from the city and, over time, increasing bright-flight, the migration of many successful African-Americans out of the cities, caused immediate damage to urban centers. The exodus meant the loss of tax income, spending money, and skills from communities that desperately needed those resources; this hastened urban economic decline or was the primary reason for it. These increasingly impoverished communities of African-American and Hispanic populations lost political representation, and without political and economic power they were easily targeted for creating garbage sites and dumping toxins.

Americans moved en masse to suburbs not only because of FHA policies and the growth of the highway and interstate system. Suburban homes were generally much less expensive than apartments in the city. More germane to American environmental history is the fact that many Americans sought a natural experience by relocating to the communities springing up in the countryside across the nation. Many developers presented suburban life as a way to reenter the long American pastoral tradition and be close to nature again away from the air, noise, and water pollution of the cities. Lawns, shrubs, and fruit trees were used to entice new residents, and some promoters emphasized wildlife such as songbirds and the beauty of the landscape. The description of suburbs as cookie-cutter, conformist communities that destroyed nature in their mindless sprawl has erased the fact that many Americans went there to get closer to nature, and their experiences with nature in suburbs would play a critical role in creating a generation of environmentalists in the 1960s and 1970s.

Construction and occupation of homes much larger than apartments (but still significantly smaller than a typical home today—average house size then was 1,300 square feet, and in 2014 it is approximately 3,000 square feet) meant that suburbs drove economic growth in jobs related to home construction, highways, and cars. The lumber industry, roofing industry, cement and construction jobs, electricians and plumbing work, and of course the white collar occupations in real estate, banking, and insurance all benefitted enormously from the housing boom. Similarly, jobs associated with the petroleum industry, which exploded along with increasing car and truck use, saw great increases. Businesses and workers in the steel industry, the synthetic rubber industry, glass manufacturing, road work, the fast food industry, motels, and so forth all benefitted from the new direction of the economy. The prosperity and jobs made it easy for Americans to become patriotic consumers and reinforced, even strengthened, the idea of America as a nation particularly blessed. Given the nation's recent victories in World War II and the swelling prosperity of the era, it is hard to blame Americans for their embrace of an ideology of abundance and making consumption a centerpiece of daily life. The blessings flowed, and they would be fools to not accept their share.

Suburban homes functioned as centers of consumptions. These young families moved into comparatively spacious homes and completed the homemaking project with durable goods that were becoming generally available and affordable in this era. Washing machines, refrigerators with freezer units, air conditioners, and televisions were all needed to complete the home. The manufacture and sale of these goods also benefitted the economy while raising the standard of living. Manufacturing, mining, the steel industry, the aluminum industry, and white collar jobs all saw an upsurge. The very existence of these machines in the home also transformed patterns of living and consumption. Refrigerators and freezers in millions of homes made it possible for food processing companies to invent chicken pot pie, fish sticks, orange juice from frozen concentrate, and the TV dinner, after many failed attempts such as lima bean and eggplant sticks, tuna pot pie, frozen camembert cheese, and "Papal approved" frozen whale steaks. The rise of the frozen food industry and processed canned and dry foods had serious implications for agriculture as well as family life. The combination of television, air conditioner, and refrigerator/freezer created a home environment in which it became much easier to spend greater amounts of time inside a home. Life within the domestic sphere and the creation of the nuclear family became the American ideal, one built on an ideology and reality of abundance. These features of convenience and greater prosperity could be quite isolating, also, as women saw their roles in society restricted to that of housewife and homemaker.

Another way Americans eat food found its genesis in the era of cars, suburbs, and prosperity. Ray Kroc began the transformation of world food culture with McDonalds in 1954. The exploding popularity of fast-food delivered to and consumed within the car appealed to Americans spending so much time traveling, commuting, recreating, and dating in Chevys, Oldsmobiles, and Fords. The low price of hamburgers, French fries, and milkshakes combined with Kroc's particular genius in creating foods that appealed to taste buds quickly made fast food very popular. The fact that the food was the same whether bought in Muncie, Indiana, or Truth or Consequences, New Mexico, contributed to the popularity of fast food franchises and fueled their national and international growth.

The combination of increased frozen and processed food consumption along with the rising popularity of fast food meant that Americans ate more food overall, and within that diet the proportion of meat and carbohydrates increased dramatically. The rising demand for meat, potatoes, dairy, and other products benefited farmers and food processors that could better make the change to much larger scales of production and processing. As this occurred Americans continued to lose track of how their food was made, where it came from, and farming's impact on the land.

Food Beyond All Measure

The mythology of America celebrates abundance, and the reality has often matched the narrative. In no other place is the nature of American abundance as clear as in post-WWII agriculture. That cornucopia of food has generated a set of environmental problems defying easy solutions both because they are complex and also because they are part and parcel of the expectation of "cheap food," an assumed benefit in the American standard of living. Historians refer to a "Green Revolution" in this era, a time of startling growth in crop yields resulting from the use of hybrid seeds, chemical fertilizers, herbicides, pesticides, and the ongoing and ever-increasing mechanization of agriculture. The "Green Revolution" also describes American exporting of this agricultural model and the introduction of market-oriented cash crops into third world countries in the same era. Unquestionably, farm production ascended to heights previously believed unattainable if even imagined.

The traditional American crop, corn, the key staple of native peoples and quickly adopted and popular food source for colonial Europeans and Americans, seemed to conquer the nation in the post-war years. The increase in corn yields is practically miraculous. From 25 bushels per acre in 1900 to 40 in 1950 and 80 in 1970, a more than threefold increase was seen in 70 years, less than two generations of farmers. By 2000 yields exceeded 120 bushels an acre, another 50 percent increase. Wheat production increased in similar measure, from 19 bushels per acre in 1950 to 36 per acre two decades later. As yields increased, the need for human labor declined. For wheat it slid from 147 hours of labor for 100 bushels in 1900 to only 6 hours of labor for the same amount in 1990. For corn the harvest hours were the same in 1900 as for wheat and then declined to 3 hours of labor for 100 bushels by 1990. The natural abundance of America first encountered by European settlers was now replaced by a number of innovations taking production to unseen levels.

The mechanization of agriculture helped achieve the perennial goal of farmers, that of increasing the land's productivity. But it worsened another perennial problem, the ongoing and expanding migration from rural America to urban communities. Gas-powered combines, corn harvesters, tomato pickers, cotton harvesting machinery, and many other petroleum-powered implements increased the speed of harvest while reducing labor needs.

As a result, many farms went out of business as increased yields drove down commodity prices. Agricultural historian Paul K. Conkin points out,

> In the 1930s around a million farmers grew cotton. By 1950, just as mechanical pickers began to have a wide impact, the number of cotton farms was down to just over 300,000. By 1974, when almost all cotton was picked by machine, the number was 80,000. In 2002 the number of cotton farms had shrunk to fewer than 25,000, with only 11,000 large farms producing 85 percent of the total.[10]

While the use of machines was essential to increasing yield while lowering manpower needs, chemical fertilizers were integral to this "Green Revolution." The origin of the chemical fertilizer industry of the post-WWII era is found in the TVA programs of the New Deal and after. In an effort to boost agriculture on the overworked, depleted soils of the Southeast, the government corporation produced stronger phosphate fertilizers called triple superphosphates. With the production of nitrates for munitions during World War II, surpluses developed and were sold as fertilizer to farmers. Farmers increased their dependency on nitrate fertilizers with steady growth of 4.5 percent a year from 1945 until 1980. The amount of nitrogen put onto and into soil rose from 2.7 million tons in 1960 to 11.4 million tons two decades later. Early stunning yields diminished over time. A million tons of fertilizer resulted in 10 million tons of grain in the 1950s, but this number shrank to 8.2 million tons of grain in the early 1960s. Yield continued to diminish with 7.2 million tons by the late 1960s and in the early 1970s approximately half of the 1950s numbers—5.8 million tons of grain for a million tons of fertilizer. Even though the return has declined steadily, American farmers still rely on chemicals to maintain productivity. Instead of learning to modify their practices and because there is no new land to expand onto, agriculturalists create a questionable abundance largely reliant on nitrate fertilizers that produce environmental consequences and costs not factored into the cost of food in stores.

With increased dependence on nitrate fertilizers, the nutrient value of soils themselves have declined steadily; dirt has become simply a way to hold roots and plants in place to absorb light, nitrates, and water. While food quality declines, the overuse of fertilizers also degrades streams, lakes, coastal estuaries, and the ocean. Farmers applying fertilizer at the wrong time or in too high a concentration cause runoffs of phosphates into local water supplies with negative and even toxic impacts on people. The excess is also carried into other bodies of water, causing algae blooms in lakes, streams, and oceans. The fertilizer and other sources of nutrients such as manure introduce nutrients into the water that algae feed on. They grow exponentially into a bloom and then quickly begin to die. With the algae deaths and decomposition the oxygen in the water is completely consumed; this process is known as eutrophication. It leaves the water without enough oxygen for other species, leading to the death of organisms such as fish, shrimp, coral, crabs, and numerous others. The dead zone in the Gulf of Mexico is one of the dire examples of the negative impact of fertilizer overreliance. Laying right off the mouth of the Mississippi River, where the runoff from a large portion of America's breadbasket runs into the Gulf, this dead zone averages 6,000 to 7,000 square miles in size. There are large dead zones off the coast of Oregon, in the Chesapeake Bay, and in Lake Erie as well.

Smaller eutrophication events occur regularly in lakes, streams, coastal estuaries, and bays. The obsession with productivity and unrealistic abundance based on an overreliance on chemicals means deterioration and a decline in natural abundance in other important areas such as the Gulf and in rivers, streams, and the Great Lakes. Those costs are never calculated or included in the "price" of corn, beef, pork, and wheat. The culture of abundance is predicated on a false economy in which the costs of environmental damage and declining human health are not part of the calculus. As throughout our history, nature continues to function as a sink to capture pollution and absorb the costs at the price of deteriorating oceans, soils, and atmosphere.

Farmers are not to blame for all of the fertilizer pollution. Lawn care businesses and homeowners have emulated the dependence on nitrate fertilizers and are, if anything, even more likely to apply too much of the nutrient at the wrong time. Whereas before World War II 1 pound of nitrogen per 1,000 square feet of lawn sufficed, the amount used on lawns grew to 8 pounds per 1,000 square feet by the 1970s. The phosphate runoff from millions of lawns and thousands of golf courses makes it difficult or impossible to control the pollution at the source and leads to steady degradation of all water environments. With so many source points, streams, creeks, rivers, wetlands, coastal estuaries, and the ocean all suffer from an ongoing, steady degradation of environmental quality and productivity due to the allures of plentitude and grassy green lawns.

Prosperity and innovations in agricultural production have enabled Americans in the post-WWII era to achieve a standard of living unprecedented by such a large population and such a large proportion of the whole population than at any previous time in history. A key feature of this high life is the regular presence of meat in meals. Before World War II the poor rarely ate beef, and it was not even a daily staple for the middle class. A revolution in meat production in the post-war years has changed that with the resounding answer to the question "what's for dinner?" often being beef. Antibiotics not only proved transformative in curing human illness following WWII, but it also radically changed animal husbandry. Many chronic diseases were now curable, and antibiotics made it possible to pack animals into disease- and epidemic-inducing environments. Chicken sheds holding 20,000 birds and livestock facilities packing in thousands of beef and pigs became common. It took 85 hours of work in 1929 to produce a half ton of chicken. Maybe no better example of the stunning impact of the industrialization of agriculture exists than the fact that the same amount of chicken required only 1 hour of labor a mere 51 years later, in 1980. The yields for hog production increased by 68 percent in the 1950s and more than 80 percent in the 1960s. The resulting glut of meat drove prices down and made beef, chicken, and pork a regular part of Americans' diet. But what seems a boon of cheap meat to consumers comes with other price tags.

Mixing antibiotics and hormones with animal feed became a standard practice in the 1950s, and almost all beef are now fed a steady mix of these, including anabolic steroids. This helped prevent disease outbreaks and provided pre-emptive and ongoing treatment for the infections always present in raising livestock but trebly so in the bog of mud, feces, and crowding that became the norm late in the 20th century. Hormones were implemented to increase the rate of growth. Moreover, because cattle do not eat grain (their stomachs are designed for digesting grasses), they automatically get sick on a corn diet; antibiotics were necessary so they could eat food inappropriate for them.

The Outdoor Life

The prosperity following the war meant many things for Americans. The opportunity to buy a new car for the first time in years as automobile factories converted from tanks, trucks, and jeeps to sedans was exciting, particularly as companies rolled out increasingly artistic and exciting models. Now it was possible to really commit to having a family, leading to the baby boom. The suburbs springing up across the nation were a place where money could be poured into a home, lawn, and middle-class comfort. Americans eagerly sought to enjoy their newfound prosperity and leisure time. Activities such as camping, hiking, and road trips neatly filled the bill. Americans took to the highways in station wagons, towing campers to visit the great outdoors. The more intrepid purchased military surplus rucksacks, boots, and tents and began the great backpacking craze that really took off in the 1960s. Visits to national forests increased from 10 million in 1945 to 27 million in a mere five years. The number reached 46 million in 1955 and approximately 92 million in 1960, marking a 900-percent increase during a time when population growth was only 35 percent. The national parks experienced a similar increase in visitation. Whereas Yosemite and Yellowstone National Parks averaged between 400,000 and 500,000 visitors annually before the war, in 1955 Yellowstone had 1.4 million visitors and Yosemite had 1 million. Rising incomes, increased savings, and shorter work weeks due to the successes of unions gave families more opportunity for travel and leisure.

Automobiles were more affordable, gasoline cheap, and the highway and interstate building boom following the war made it ever easier for Americans to venture to more distant locations for recreation and edification. Forests could be seen for their trees rather than their board-feet, mountains for their spires of rock, glittering streams and glaciers for more than the ore they might contain. With most Americans no longer in jobs directly reliant on nature's resources and living in a society that increasingly obscured the connection between natural resources, commodities, and people's lives, individual Americans increasingly were able to view nature in preservationist terms. As leisure time expanded with the resources and infrastructure for travel and vacation, enjoying nature became central to the American dream; this proved crucial to the rise of an environmental consensus. As Americans learned to enjoy and appreciate nature, the emerging new science of ecology gave them tools to better understand nature and the need to protect it.

Thinking Like Ecologists

Eugene and Howard Odum moved ecology into the mainstream and made Americans distinctly aware of their impact on nature as well as their own position within nature. This understanding helped drive the growth of American environmentalism as their ideas gained currency and the environmental movement increasingly adopted protection of ecosystems and species as part of its agenda. Both brothers published extensively in their careers, but it was the *Fundamentals of Ecology* by Eugene (the second edition was co-written with Howard) that exerted a profound influence on global ecological thinking, with multiple editions after the first edition in 1953 and translations into 20 different languages. In their theories and work they described a nature that moves toward balance and harmony over time. The central contribution of their work was the concept of the ecosystem, a community of organisms that also includes material cycles. Exploring the role of death and decomposition in enriching the ecosystem, they demonstrated the movement of nutrients through biota to soil and to biota again as well as the relationships between different ecosystems. According to their theory of ecology, the ecosystem sat atop of nature. While early natural systems were marked by struggle and conflict, as they matured species in a given ecosystem grew increasingly cooperative and "mutualistic," arriving finally at the ideal of the stable and harmonious natural community. Ecology research since then has shown that there is much unpredictability and flux in ecosystem development, and the idea of a fully realized harmonious state of climax in nature is inaccurate. But that very idea of ecosystems working toward a perfect state of nature proved very influential and powerful as ecology gained popularity and Americans increasingly saw themselves as agents of disruption in an otherwise ordered process.

Ardent environmentalists themselves, the Odums' conclusion, that nature moved from a tumultuous past to a stable present, threw the destructive activities of humans into stark contrast. Having described a nearly perfect nature, they now argued against deeply imperfect humans introducing chaos back into nature. They advocated family planning and population control, fought to stop the destruction of Georgia's coastal marshes, and argued for regional land-use planning as well as careful use and management of resources. Their work helped Americans see nature in more ecological terms and begin to understand the need to strike some balance between use and preservation, economic growth and environment-health, and they influenced numerous researchers, writers, and activists.

The Fight against Atomic Pollution

The ways in which environmentalism develops in the post-war era is rooted deeply not only in the nature of the economy and cultural practices but also from the very fact of the existence and use of nuclear weapons. The atomic age, springing from a particular way of war during World War II, increased fear and ecological awareness in Americans, bringing the threats to the environment home to their own bodies. Worster writes,

> the Age of Ecology opened on the New Mexican desert, near the town of Alamogordo, on July 16, 1945, with a dazzling fireball of light and a swelling mushroom cloud of radioactive gases . . . for the first time, there existed a technological force that seemed capable of destroying much of the life on the planet. As Oppenheimer warned, humans, through the work of the scientist, now knew sin. The implied question was whether they also knew the way to redemption.[11]

This newfound power introduced a cold fear into Americans and people throughout the world. This was certainly one factor in the rise of environmentalism in the second half of the 20th century. Fallout from atomic weapons testing at Bikini Atoll and above-ground tests in Nevada awakened Americans to a society gone desperately awry in its relationship to nature.

Figure 11.1 For several years following the end of World War II the United States conducted above-ground atomic bomb detonations. Some of these were conducted in Nevada with troops and officials in observance. Testing weapons was a way to conduct research on the increasing power of nuclear weapons but also a way to demonstrate military strength. In addition to the Nevada tests, bombs were detonated at the Bikini Atoll in the South Pacific, necessitating the removal of residents from local islands. Participants in above-ground detonations were struck by the blast wave and absorbed radiation from the detonation. Some of these observers later developed and died of radiation-related cancers. Courtesy of National Nuclear Security Administration / Nevada Site Office.

Above-ground nuclear weapons testing in the deserts of Nevada and Pacific sites were common occurrences with more than 200 detonations from 1945 to 1962. It would require committed scientist activists and an informed public citizenry to begin to bring the early excesses of the atomic era under control and thereby help launch the modern environmental age.

The U.S. government assured the public that there was nothing to fear as it proceeded with hundreds of tests over several years. Writer Terry Tempest Williams, who grew up in

Utah, writes of her memory of riding in the family car before dawn north of Nevada in 1957 and witnessing a nuclear blast:

> The September blast we drove through in 1957 was part of Operation Plumbbob, one of the most intensive series of bomb tests to be initiated. The flash of light in the night in the desert, which I had always thought was a dream, developed into a family nightmare. It took fourteen years, from 1957 to 1971, for cancer to manifest in my mother—the same time, Howard L. Andrews, an authority in radioactive fallout at the National Institutes of Health, says radiation cancer requires to become evident.[12]

Many scientists grew concerned over the massive scale of nuclear weapons testing and nuclear fallout pollution. Albert Schweitzer helped birth a rebuttal from the scientific community when he published an essay in 1957 denouncing the ongoing atmospheric testing of nuclear weapons. The philosopher and humanitarian, who was awarded the Nobel Peace Prize in 1953 for his medical and humanitarian work, sent the essay to the Nobel Prize Committee, and it was broadcast from Oslo to 50 countries in 1957. "A Declaration of Conscience" called for citizens of the world to insist on knowing more about the dangers of nuclear fallout and to protest and demand an end to testing. He countered the frequently deployed defense of nuclear weapons testing advocates that the radiation was minimal and not dangerous in trace amounts by explaining the accumulation of radiation poisoning as it moved up the food chain. To make this point he used evidence from the point of the Columbia River where the Hanford site conducted its processing. While noting that the water radioactivity was relatively low, he also wrote that

> the radioactivity of the river plankton was 2,000 times higher, that of ducks eating the plankton 40,000 times higher, that of fish 15,000 times higher. In young swallows fed on insects caught by their parents in the river, the radioactivity was 500,000 times higher and in the egg yolks of water birds more than 1,000,000 times higher.[13]

Schweitzer showed in no uncertain terms the process of accumulation of poison in various organisms, an important point that would be reiterated in *Silent Spring*. The declaration was not aired in the United States, and the *New York Times* declined to give it front-page coverage. But the rebuttal to the piece by the Atomic Energy Commissioner Willard Libby created a debate between the two men carried in the *New York Times, TIME,* and *The Saturday Review.* This ginned up greater interest in the topic and built momentum for activists concerned with the issue, such as Barry Commoner, Linus Pauling, and others.

It would take much more than Schweitzer's powerful plea to end the dangers of above-ground testing, so Barry Commoner and the Greater St. Louis Committee for Nuclear Information (CNI) rushed into the breach. The CNI was created by Barry Commoner, other Washington University (St. Louis, MO) scientists, and women reformers as a grassroots organization to challenge nuclear poisoning through education of the populace. Technically nonpartisan, a tactic that increased their credibility and success according to historian Michael Egan, they strived to undermine the facade of scientific objectivity that protected the Atomic Energy Commission (AEC) and other proponents of nuclear weapons and testing by providing citizens with accurate information. Using speakers, outreach programs, and media coverage, the CNI quickly captured the attention of the American people. Commoner earned his PhD in Biology at Harvard in 1941, and his early ecological awareness blossomed when he served in the South Pacific during World War II. Responsible for devising a method of dispersing DDT by plane to clear disease-carrying mosquitoes before the disembarkation of troops on beaches, he noticed that while flies were effectively eradicated, the DDT also killed large amounts of fish, which washed up on the beach and began rotting. The flies then swarmed back in even larger numbers. On that contested South Pacific beach, Commoner learned the dangers of disruptive technologies and ecological imbalance.

The AEC assured the American public of the safety of above-ground nuclear weapons testing and soothed the fears of those living downwind of test sites, but many scientists were discovering more evidence of unknown or unanticipated health risks from radiation exposure. They were able to demonstrate that people were exposed to dangerous amounts of Strontium-90 and Iodine-131. The half-life of Strontium-90 is 29 years, and it had been assumed that it would remain in the stratosphere for at least that length of time, diminishing the impact of tests on public health and the environment. This assumption turned out to be false, as was demonstrated when a surge of radioactivity occurred during a heavy rain event in New York state in 1953, 36 hours after tests in Nevada. Because AEC scientists were confident that Strontium-90 would dissolve in the stratosphere, it had not been carefully studied. Because of its chemical structure, which is similar to calcium, the radioactive material is carried by calcium-rich foods into the body and built into the very architecture of a person, in bones, bone marrow, and teeth. Its potential ubiquity in food and bodies and the corresponding threat of leukemia, bone cancer, and other cancers rendered this new invisible danger quite frightening. In 1956 Democratic presidential candidate Adlai Stevenson broached the topic of threats to human and environmental health and called for an end to testing, while President Eisenhower persisted in the government's position that the tests presented no public hazard.

In an effort to understand the threat to human health, the CNI, with strong support and assistance from the Washington University and St. Louis University dental programs, launched a baby tooth collection survey in 1958. While bone samples from children had been used to measure the accrual of Strontium-90 in young bodies, the sample set was small. The baby teeth survey offered the benefit of large samples from the key period when above-ground tests[14] increased dramatically with a corresponding surge of isotopes entering soils, plants, mother's milk, and the bones of their babies. Shooting for a goal of 50,000 teeth, the mostly women volunteers sending out questionnaires and collecting the forms and teeth had amassed 17,000 by the spring of 1960. Increased public support by the St. Louis mayor, school districts, and local dental and pharmaceutical associations triggered a flood of the little teeth, resulting in tens of thousands more within a few months. While the committee assured mothers that their milk was safe, the fact of the survey itself, along with the questionnaire, spoke otherwise. As Egan writes of this campaign,

> More effective than any advertising campaign, the Baby Tooth Survey served two purposes. First, it brought attention to the hazards of nuclear fallout to which the nation's children were particularly susceptible, and second, it required public participation by involving the public in the initial phase of the study and ensuring

> widespread interest in the committee's results. The overwhelming response to the requests for teeth, and the growing number of similar surveys around the country, suggested that Americans were becoming less willing to accept risk out of hand.[15]

Respondents displayed a general excitement for the scientific process, and parents and children evinced both anxiety and enthusiasm. One 11-year-old boy assured the survey managers that he normally put his tooth under his pillow for the tooth fairy's dime but was willing to absorb the loss for the advancement of science. The St. Louis mayor sent him a note thanking him and, expressing his desire he not suffer undue pecuniary pressure, gave him a dime. A note left for the tooth fairy by a young girl stated, "Dear Fairy, I would like to have a dime but do not take my tooth I am going to send it to siense [sic]."

Atomic anxieties were validated by the results of the study. Teeth from 1951–1952 contained 0.2 micromicrocuries per gram of Strontium-90. That number doubled by 1953 and quadrupled by 1954. The assurances of safety by the AEC and President Eisenhower were proven false. The following onslaught of letters to congressmen, public hearings, articles, and editorials with increasing media coverage popularized and energized the anti-testing movement. Contributing to this momentum was the increasing public awareness of numerous other radioactive poisons besides Strontium-90. Testimony in public hearings revealed that a number of isotopes comparable to Strontium-90 in the danger they posed. Strontium-89, Cesium-137, Barium-140, Iodine-131, and others had also not been closely studied by government scientists for their possible impact. The CNI pushed the AEC to move beyond simply considering direct impact on humans but to think ecologically. Considering both direct and indirect contamination would force the AEC to take threats to human health more seriously. The CNI pushed the argument by providing evidence of the impact of test fallout on people living downwind from test sites in Nevada, Idaho, and Utah, asserting that fallout levels directly endangered the health of children in these states.

As this debate on the science of fallout pollution proceeded, congressmen and senators were bombarded with letters from housewives and mothers calling for approval of a Test Ban Treaty. Their concerns and use of scientific evidence and arguments is a testament to the CNI's effective campaign of public education. Their efforts to stop the poisoning of the environment and their children reflects both the gendered environmental activism of women, the effort to extend care of the family into the world, and a sustained effort by women to clean up various forms of pollution extending back to the late 19th century. President Johnson's specific reference to the organization's studies in an address one year following passage of the Test Ban Treaty also speaks to the key role played by Commoner and the CNI in stopping above-ground nuclear weapons testing and helping launch the environmental era.

"The Repeated Refrains of Nature"

There is likely no other person in America so strongly emblematic of environmentalism as Rachel Carson. *Silent Spring* was so powerful in its language and argument as well as its consequent impacts it has been easy to see the moment of this book's publication as the launching of modern American environmentalism. Like *Uncle Tom's Cabin*'s impact on abolitionism in the pre–Civil War era, *Silent Spring* was transformative; Carson's influence was absolutely integral to the building of an environmental consensus in America. At the same time, it was because of the growing ecological understanding of the world and increasing concern about the overuse of a variety of chemical pesticides such as DDT, that Carson was able to even tackle the topic. The work of Barry Commoner and the Greater St. Louis Committee for Nuclear Information also helped seed the bed of American understanding, educating them on invisible dangers and cracking the facade of scientific objectivity and government protection. Carson gave voice to a growing community of dissenters for nature and brought on board the millions of Americans who to that point remained relatively oblivious to the threats to nature and health engendered by the explosion in chemical pesticide use. Carson transformed American environmentalism, playing the central role in the creation of an American environmental consensus that undergirded the reforms of the following 20 years.

As a child Carson displayed a great love of the natural world. Earning her Masters in zoology at John Hopkins University in 1932, she worked for the U.S. Bureau of Fisheries during the Great Depression then had a successful 15-year career for the U.S. Fish and Wildlife service as a scientist and editor, becoming editor-in-chief for all the agency's publications. Carson researched and wrote numerous pamphlets on conversation issues and the natural world, honing her knowledge and writing skills in the process. She published *The Sea Around Us* in 1957. She intended to teach readers about the vast and intricate life of the world's oceans, the role of the seas in creating and nourishing life, and to create a feeling of awe for the beauty and wonder of this environment. It was loved by readers and reviewers. In this same time period, Americans were growing alarmed at the threat to human health posed by the rampant aboveground nuclear weapons testing conducted by the United States and other countries. The CNI had seized the country's attention, and now Americans everywhere were experiencing and seeing the impacts of DDT and other chemical use.

In this era there was a confidence in the perceived benefit of DDT that reveals American trust in and acceptance of the benevolent authority of scientists and industry as well as an astonishing naiveté in their embrace of new technologies and chemicals (see Document 11.1). Poet and memoirist Mary Karr, in *The Liars Club,* writes of East Texas children in the 1960s following the DDT trucks:

> At dusk in the later summer in 1962, the mosquitos rose up from the bayous and drainage ditches. Kids fell ill with the sleeping sickness we called encephalitis. Marvalene Seesacque came out of a six-month coma that left her what we called half-a-bubble off plumb. Other kids weren't lucky enough even to wake up, and for the front page of the paper, Mother had taken a slew of funeral pictures with tiny coffins. A mosquito truck was dispatched . . . to smoke down the bad swarms. It puttered down the streets every evening trailing a long cloud of DDT from a hose as big around as a dinner plate. Our last game of the day that summer involved mounting our bikes and having a slow race behind the mosquito truck.[16]

Figure 11.2 Rachel Carson's work originated in a deep passion and love for nature. From her childhood to her last days, her best days were spent exploring nature. She particularly found tidepools, with their rich variety of species, a wonderful place to renew her passion for nature and the energy to sustain her important work. Courtesy of the Linda Lear Center for Special Collections and Archives, Connecticut College.

In this passage she captures the fact that DDT was being used effectively to reduce or eliminate diseases while also illuminating the innocence of American children's trust in those running society. Karr describes the nature of the "slow race":

> The trick was to pedal just fast enough to stay upright, but not fast enough to pull ahead of anybody. Add to this the wet white cloud of poison the mosquito truck pumped out to wrap around your sweaty body and send a sweet burn through your lungs, and you have just the kind of game we liked best—one where the winner got to vomit and faint.[17]

It wasn't only young children that placed too much faith in the chemical and industry, but adults as well. One man described an almost incomprehensible level of innocence:

> On a hunting trip in Northern British Columbia in the latter part of August 1957, we sprayed a tent for twenty-one nights with DDT. We did not sufficiently aerate the tent. When I got back home in September, my marrow and white and red corpuscles were terribly impaired.[18]

He was treated at the Mayo Clinic but died of leukemia in 1959.

In 1945 the United States produced 36 million pounds of DDT. By the late 1950s that number increased to 180 million pounds a year. It was used to destroy fire ants with no long-term success, and also to try and stop the spread of beetles causing Dutch elm disease and eradicate gypsy moths. Housewives used the chemical in their homes, spraying it on dishes and hanging wallpaper coated with the poison in children's rooms. Schoolchildren were sprayed with DDT in an effort to prevent polio before the development

Figure 11.3 DDT seemed to offer a perfect solution to pests' consumption of agricultural crops. The widespread, blanket use of the chemical in an effort to achieve total extermination introduced amounts of the toxin into the environment that were lethal to fish, birds, and other animals. At the same time insects evolved quickly, developing resistance to the chemical, reducing its efficacy. Getty Images.

of the polio vaccine. The chemical was so ubiquitous that by the late 1950s studies indicated that essentially every American carried DDT in their fat. Many entomologists and biologists grew concerned about the impact spraying had on other insects and species. While to most Americans Carson seems to have come out of nowhere with her startling revelation of the scale of industrial poisoning of the environment, in fact, it was organic farmers, women environmental activists, and government-employed scientists who provided her with critical information and data as well as crucial support for the research and writing of her manuscript. There was a community of concerned citizens and activists that provided her with evidence, documents, and moral support for the creation of this important book. Without their existing activism, knowledge, and encouragement, this book might not have been written. This community's passion for the issue and the book helped it gain rapid popularity.

But a great deal of credit for the book's success has to go Rachel Carson's skill as a writer. *Silent Spring* is not a simple treatise or evidence-driven argument. Rather, it is a manuscript of beautiful, powerful prose, driven by a passionate love of nature and fear for its future. This sense of a world poisoned and dying as a result of man's carelessness, hubris, and love of technology and science had to be at least partially informed by the cancer raging through her own body. Even as Carson wrote and researched one of the most important books in history, her body was a battleground between raging cancer and surgery and radiation treatments. Maybe even more important than her passion and powerful prose was the overwhelming and accumulating evidence she employed to show the careless use of DDT and other chemical pesticides as well as her explanation of the insidious infiltration of these chemicals into the bodies of bugs, fish, birds, and mammals, and the accumulation of poison in bodies up the food chain.

Carson provided abundant examples of fish, bird, wild animal, and domestic livestock die-offs from chemical use. In her writing she evoked the beauty and uniqueness of species in order to create empathy and even love for these creatures, much like the pro-nature essays and photos of the Sierra Club created a love for contested landscapes. Her rhetorical style is strongly evident in a passage about western grebes in Clear Lake, located in California:

> The following winter months brought the first intimation that other life was affected: the western grebes on the lake began to die, and soon more than a hundred of them were reported dead . . . It is a bird of spectacular appearance and beguiling habits, building its floating nests in shallow lakes of the western United States and Canada. It is called the "swan grebe" with reason, for it glides with scarcely a ripple across the lake surface, the body riding low, white neck and shining black head held high. The newly hatched chick is clothed in a soft gray down; in only a few hours it takes to the water and rides on the back of the father or mother, nestled under the parental wing coverts.[19]

In this passage Carson provided some species details and uses clear, evocative prose to create a picture of the western grebe and a baby, occupying its rightful place in nature. Moreover, she invoked the bonds of familial affection with her description of a downy,

Figure 11.4 This *Time* magazine picture of a woman in a bathing suit eating a hot dog and drinking a beverage while enshrouded in a cloud of DDT was fairly typical of the 1940s and 1950s. Other photographs in newspapers and magazines showed the spraying of crowds and neighborhoods, children at school, and people in swimming pools. Such images reveal an embrace of scientific innovation in chemical use as well as a level of innocence of the threat to their own health that is hard to understand in this day and age. Getty Images.

feathered chick protected and shielded by mother and father. What could be cuter and more fragile at the same time? Then arrives the deadly threat created by careless man.

> Following a third assault on the ever-resilient gnat population, in 1957, more grebes died. As has been true in 1954, no evidence of infectious disease could be discovered on examination of the dead birds. But when someone thought to analyze the fatty tissues of the grebes, they were found to be loaded with DDD [similar to DDT] in the extraordinary concentration of 1600 parts per million.[20]

Explaining the science and the accumulation up the food chain in clear and precise language, she wrote that plankton in the lake were found to have 5 parts of DDD per million, "plant-eating" fishes had 40–3,000 parts per million. Carnivorous fish carried the largest amounts of chemical poison in their bodies with one bullhead found to have a concentration of 2,500 parts per million. She summed up: "It was a house-that-Jack-built sequence, in which the large carnivores had eaten the smaller carnivores, that had eaten the herbivores, that had eaten the plankton, that had absorbed the poison from the water."[21] The chemical was no longer detectable in the water but "had merely gone into the fabric of life the lake supports."[22]

It was not only the overuse of pesticides that threatened the environment but industrial nonchalance as well. Carson described an incident in Austin, Texas, and downstream in 1961. A plant producing DDT, chlordane, toxaphene, and benzene hexachloride had allowed the run-off of waste into the Colorado River and Town Lake. The flushing of the city's storm-sewer system sent chemical waste residing in gravel and sand into the river system.

> As the lethal mass drifted down the Colorado it carried death before it. For 140 miles downstream from the lake the kill of fish must have been almost complete, for when seines were used later in an effort to discover whether any fish had escaped they came up empty.

She noted that 27 species of dead fish were found, and the kill amount was about 1,000 pounds per mile.

> There were channel cats, the chief game fish of the river. There were blue and flathead catfish, bullheads, four species of sunfish, shiners, dace, stone rollers, largemouth bass, carp, mullet, suckers. There were eels, gar, carp, river carpsuckers, gizzard shad, and buffalo. Among them were some of the patriarchs of the river, fish that by their size must have been of great age—many flathead catfish weighing over 25 pounds, some of 60 pounds reportedly picked up by local residents along the river, and a giant blue catfish officially recorded as weighing 84 pounds.[23]

This passage not only lays out the devastation of carelessness and toxic chemicals but also provides a list, or catalog, of the diversity of species in that river. To the reader it is not simply that fish were killed but the amount and variety of species and "patriarchs," fish that had survived everything else to be killed now and so quickly. The cumulative effect of the book was to show that nothing was safe from the irresponsible use of chemicals. It was a hit, sitting on *The New York Times* bestseller list for 31 weeks, and mobilized Americans to begin demanding more information and controls on chemical use.

A Great Sound and Fury

Industry-sponsored scientists responded angrily, even hysterically, to *Silent Spring*, questioning Carson's scientific credentials and her competence as a woman to accurately conduct or evaluate scientific research and empirical evidence. They even made comments about her appearance and sexuality and lobbed accusations of socialism at her. The attack by male scientists, chemical trade leaders, and others used the assumptions of gender against her, accusing her of hysteria, softness, lack of clear thinking, and an inability to really understand and use science. Biographer Linda Lear captures the tone of criticism quite effectively when describing William J. Darby's response to *Silent Spring.* Darby was the chair of the Department of Biochemistry and director of the Division of Nutrition at Vanderbilt University School of Medicine. He

> exhibited his lack of objectivity when he titled his review in *Chemical & Engineering News,* "Silence! Miss Carson." Referring to Carson's list of scientific sources, Darby sarcastically castigated her supporters as "organic gardeners, the antifluoride leaguers, the worshippers of 'natural foods' and those who cling to the philosophy of a vital principle, and other pseudo-scientists and faddists."[24]

He also questioned her use of scientific sources and summarized by stating that "in view of her scientific qualifications in contrast to those of our distinguished scientific leaders and statesmen, this book should be ignored."[25]

The book, media coverage, and the growing public outcry caught the attention President Kennedy, who appointed a scientific advisory committee to review the issue. Although it took 10 years, DDT was banned for agricultural use in America in 1972. The Stockholm Convention in 2004 banned worldwide agricultural use. It is still occasionally used to fight disease-carrying mosquitoes in epidemic areas.

The fight for nature and ecosystem health was waged on two broad fronts. One area of conflict was over the poisoning of nature and human bodies. Activists worked to limit the impacts of chemical pesticides and radiation fallout on all species, including humans. Women's activism and voice was integral to this effort; their work represents a continuation from the cleaning of the cities in the late 19th and early 20th centuries to smog abatement efforts in the mid-20th century to the organizing in this era. At the same time that activists sought to end radiation and chemical poisoning, activist groups such as the Sierra Club and the Wilderness Society ramped up their efforts to preserve valuable, beautiful pieces of nature from the river-harnessing dreams of the Bureau of Reclamation, and built on momentum from those struggles to create federal legislation to protect some ecosystems forever.

No Hetch Hetchy Repeat

Flush with two decades of success damming rivers in the West and seeing water control as the key to further economic development in the region, the Bureau of Reclamation proposed the Colorado River Storage Project (CRSP), a plan to build 10 dams on the Colorado River system for water storage, flood control, and economic development. The proposed Echo Park Dam, like the earlier Hetch Hetchy Dam, would violate the sanctity of land set aside for the enjoyment of Americans, the preservation of dinosaur fossils and dig sites, and Indian petroglyphs and pictographs. The reservoir would flood the striking river valleys of the Yampa and Green Rivers and drown healthy river habitat. In desert ecosystems like this, river bottoms or riparian zones are areas of rich diversity and heavy population. Losing that to the reservoirs would remove a key engine of ecosystem health in the monument. Fierce opposition arose against the plan, but the Bureau likely did not feel much concern at the beginning stages of a fight that would be so consequential in the development of American environmentalism. Echo Park activists emphasized the beauty and wildness of the Dinosaur landscape. In doing so they created widespread support for their cause and helped bring the "Go-Go Era" of dam construction to an end.

The Sierra Club and the Wilderness Society, under the leadership of David Brower and Howard Zahniser, organized a well-orchestrated attack on the Bureau of Reclamation's plan. With the support of such prolific and prodigious writers as Wallace Stegner and Bernard DeVoto, both devoted westerners who used their writing to articulate a sense of place, environmentalists garnered widespread, national support for their efforts to block the dam. They employed national media coverage, a special photo and essay collection titled *This*

Figure 11.5 Echo Park is a site of stunning beauty where the Yampa and Green Rivers come together inside Dinosaur National Monument. Bureau of Reclamation plans to build a dam there ignited a strong opposition movement that mobilized environmentalism in the post-war years to even greater accomplishments. Library of Congress.

is Dinosaur, and a nationwide publicity campaign to mobilize Americans against the plan. Even Eleanor Roosevelt wrote a column against the proposed project.

In addition to building broad national popular and political support, the activists were able to show that many of the claims of the Reclamation engineers were exaggerated or false. Particularly important were Brower's efforts at getting the evaporation numbers right. Recovering from his embarrassment in Congressional testimony and comments about "9th grade arithmetic," he was put in touch with Ric Bradley, a physics professor at Rutgers University. Bradley conducted a close study of the science of evaporation as well as the Bureau's methodologies and studies, and found numerous problems in the agency's numbers and confident evaporation predictions. The Bureau was forced to revise its numbers. As it turned out, employees of the agency had known that their evaporation rate estimates were flawed. Brower's testimony was validated when the House subcommittee that had mocked his critique was notified by the Bureau of its errors. With this publically established, the Echo Park Dam opponents argued that a higher Glen Canyon dam would render the dam in Dinosaur unnecessary. This proved instrumental to protecting Echo Park. Of more historical importance, the willingness and ability to take on the engineers and the Bureau of Reclamation, much like Rachel Carson had taken on scientists, industry, and the Department of Agriculture, marked the increasing sophistication and growth of the environmental movement. These activists and organizations democratized the United States as they forced greater government accountability and responsiveness. Moreover, showing the dishonesty and mistakes by industry and government specialists emboldened environmentalists to continue their challenges and push for further reforms.

But to close the deal they had to agree to not oppose the damming of the Colorado at Glen Canyon, a compromise Brower and others would deeply regret when the beautiful watershed began filling up. Regardless, this victory marks a major moment in the history of American environmentalism. This was the moment in which the environmental groups managed to assert themselves as powerful protectors of nature opposed to dramatic transformations of the landscape. When the Bureau announced plans to build two dams in the Grand Canyon in 1966, environmentalists quickly shut the project down.

Creating Wilderness

But the dominant ethos of environmental thought among activists in the 1950s and 1960s was preservation, and this is best demonstrated in the next major national effort. The ongoing tension over the best uses of nature is best exemplified in the struggles of activists to create and protect wilderness and pass wilderness legislation. The image of wilderness in most people's minds is one of streams frothing through canyons, wolves loping across tundra, and untouched forest stretching for mile upon endless mile. But fights over smaller, less grandiose pieces of wilderness helped build support for the concept of wilderness, and for setting aside vast, important swathes of relatively healthy habitat and providing them permanent and inviolable protection.

Supreme Court Justice William O. Douglas helped popularize the idea of wilderness and led efforts to create wilderness areas across the nation. As the longest serving member of the Supreme Court, he employed an influential bully pulpit to help popularize environmentalism. His numerous editorials, essays and books, public protests, and judicial activism as well as his Chesapeake & Ohio Canal and Olympic Beach hikes helped to limit development and create wilderness areas by popularizing and generating public sympathy for preservation efforts. His first major public activism occurred close to where he served the judicial branch of government. Seeing the rise in automobile-based tourism, and still adhering at least partially to their original mission of expanding access to beautiful and inspiring landscapes to Americans, the National Park Service looked to create a parkway through the historic Chesapeake & Ohio Canal site in 1954. This was to be built on the model of the Blue Ridge Parkway and other popular road projects. But like the earlier Blue Ridge work by the Civilian Conservation Corps, the National Park proposal and a parkway bill triggered powerful resistance from environmental groups like the Izaak Walton League, the Audubon Society, and others. They sought to prevent the loss of this natural oasis so close to the buzzing hub of national politics and also wanted to preserve important habitat for birds and other species. The opposition remained in the minority and was making little headway, so the leaders of this effort, including Justice Douglas, decided a well-publicized hike would build support and also educate the media and the public to the benefits of preserving this habitat. Historian Adam Sowards, the chronicler of Douglas's important role in American environmentalism in the post-war era, describes the gathering the night before the epic hike:

> The group ate steaks at the Cumberland Country Club and heard various politicians pontificate. Republican Senator J. Glenn Beall of Maryland forecasted the group's defection to the parkway cause, and former state Senator William A. Gunter put the hike's stakes in sharp relief: "the question is 'whether this long strip of land should be converted into a runway for goldplated cadillacs in a Machine Age, or to be reserved as a foot path for the pleasure of nature-loving pedestrians of the Paleolithic Age,' such as Douglas and his fellow nature lovers."[26]

There was a general assumption the hike would fail. Accepting the gauntlet, Douglas and the others reeled off 22 miles that first day, the justice carrying his own 40-pound pack while other hikers' supplies were hauled by car. They successfully completed the trek, and the attendant media attention generated sufficient positive publicity to convince the National Park Service to abandon its plan and protect the area.

Another road out west caught the Judge's attention. Businessmen and chambers of commerce on the Olympic Peninsula of Washington State looked to build a coastal road through the Olympic National Park along the rugged shoreline with its views of sea stacks, crashing waves, and abundant wildlife. They hoped to attract the same kind of lucrative tourist business seen on Oregon's coastal highway. Environmentalists worried about the impact of a road on coastal habitat as well as its ruining what Douglas referred to from his backpacking experience "a place of haunting beauty, of deep solitude."[27] Inspired also by the recent victory in the Echo Park fight, conservationists locally and nationally, while enduring accusations of elitism, colonialism, and disregard for local economic issues, insisted on protecting the coastal wilderness. In 1958 they organized a local hike to the coast in order to publicize and educate Americans on the benefits of the coastal wilderness and the need for permanent protection. Douglas helped plan and participated in the hike. This hike created more local disgruntlement than the East Coast version. One local woman stated, "If I ever saw a place without a road that needed a road, this is it. As far as I'm concerned they can build it down the cotton-pickin' beach."[28] A prominent peninsula businessmen met the hikers at their finish with four large signs stating,

> WE OWN THIS PARK, TOO. WE WANT A SHORELINE ROAD
> FIFTY MILLION U.S. AUTO OWNERS AND THEIR FAMILIES LIKE SCENERY, TOO!
> SUPER HIGHWAYS FOR 47 STATES BUT PRIMITIVE AREAS FOR US
> BIRD WATCHERS GO HOME

Pressed by a reporter that protecting wilderness was reminiscent of Alexander Hamilton's elitist ideas about government and its relationship to its citizens, Douglas emphasized the duty of government to protect minority rights and those of individuals. With wilderness lovers constituting a minority population, their rights deserved protection as well. He also made the point to the local businessman with the signs that everything else in America was developed. "We'll give you 99 percent of the U.S. but give us the other 1 percent please."[29] This hike and Douglas's participation raised the profile of the effort, preventing the road from being built. In making his argument Douglas captures the essential tension between development and preservation. The exploitation of nature and transformation of the landscape drove American growth, power, and prosperity, and in this era resolved political and class tensions. But, for many Americans it was becoming manifestly clear this came at great ecological cost, and it became necessary to stop further destruction when and where possible.

These isolated local efforts were significant, but environmentalists sought much stronger measures that could provide nature and habitat with widespread, permanent protection. Buoyed by the momentum created in the campaign and victory against the Echo Park Dam, and anger over the Forest Service's decision to reduce the Three Sisters Primitive Area in Oregon by 53,000 acres, Howard Zahniser and the Wilderness Society, the Sierra Club under David Brower's leadership, and numerous other organizations launched a campaign lasting eight years, from 1956 to 1964, to create a federal wilderness system. Increased timber harvests in the national forests in the post-war era mobilized wilderness advocates in a manner similar to the organizing against the CCC in the 1930s. The timber cut in the Willamette National Forest of Oregon quadrupled from 1945 to 1955. This was part of a broader trend. National Forest Timber Sales (almost entirely in the West and mostly in the Pacific Northwest and Northern Rockies) had increased from 1.3 billion board feet (bbf) in 1939 to 3.1 bbf in 1945. The numbers continued to ramp up in the post-war years. From 2.7 bbf in 1946, the cut increased to 6.3 bbf in 1955 and 9.4 bbf a mere five years later. As the Forest Service reclassified prior wilderness areas for logging and assisted the logging companies in destroying old growth forests and ecosystems, environmentalists scrambled to create wilderness legislation in order to protect, or permanently zone off the areas, in the words of historian Kevin Marsh. Hence, wilderness advocates were motivated to create a national wilderness system because of the frailty and impermanence of the wilderness designations already in use.

An old conservationist provided some of the ideas crucial to the wilderness movement. Aldo Leopold revealed how far he had traveled intellectually with the release of his essay collection *Sand County Almanac* in 1948. Leopold broke strongly from his own conservationist heritage and background, particularly in his rejection of a simply instrumentalist view of nature and asserting the inherent rights of nature. In what reads as a series of critiques of land use, of waste, of consumption, of the valuing of economic arguments over esthetic, spiritual, and scientific arguments, there resides a radical idea that was not popular at the time nor is even today. Rejected by multiple presses the collection sold poorly and received generally negative reviews early in the book's life. The land ethic would not really grab hold until the 1960s with the growing popularity of the environmental movement and the use of the book to argue for wilderness legislation. Leopold argued that man would never change its destructive practices without some significant and dramatic change in thinking, a new ethical system; the "land ethic" must be constructed to protect the rights of nature as a system.

The desire to set aside large expanses of relatively healthy landscape and ecosystems certainly marked a new high-water mark for environmental activists in America. While the idea of wilderness had been bouncing around since Leopold and Marshall's early efforts, increasing outdoor recreation in the post-war era helped create a wider constituency for the concept. Popular literature such as Edward Abbey's *Desert Solitaire* created both a tone of resistance to the administered life while articulating a view of nature that emphasized the wild and rugged experience, and criticized "industrial tourism." In one of many memorable and vivid passages, Abbey wrote that it was necessary to get the rangers out of their offices to "work off a little office fat" and help the numerous Americans wandering the back country in his vision of an ideal use of national parks.

> Once we outlaw the motors and stop the road-building and force the multitudes back on their feet, the people will need leaders. A venturesome minority will always be eager to set off on their own, and no obstacles should be placed in their path; let them take risks, for Godsake, let them get lost, sunburnt, stranded, drowned, eaten by bears, buried alive under avalanches—that is the right and privilege of any free American.[30]

Having played such a critical role in building opposition to the Echo Park Dam, Wallace Stegner also provided cultural arguments for the importance of wilderness. In his "Wilderness Letter," which was used to introduce the legislation, he captured the passion for wilderness landscapes felt by many activists, tying it to American exceptionalism.

> Something will have gone out of us as a people if we ever let the remaining wilderness be destroyed; if we permit the last virgin forests to be turned into comic books and plastic cigarette cases; if we drive the few remaining members of the wild species into zoos or to extinction; if we pollute the last clear air and dirty the last clean streams and push our paved roads through the last of the silence, so that never again will Americans be free in their own country from the noise, the exhausts, the stinks of human and automotive waste.[31]

Moving from the critique of destruction and petty, consumptive waste, he transitioned to talking about humans as citizens within nature: "that never again can we have the chance to see ourselves single, separate, vertical and individual in the world, part of the environment of trees and rocks and soil, brother to the other animals, part of the natural world and competent to belong in it."[32] Pulling out all the stops, Stegner stated,

> Without any remaining wilderness we are committed wholly, without chance for even momentary reflection and rest, to a headlong drive into our technological termite-life, the Brave New World of a completely man-controlled environment. We need wilderness preserved . . . because it was the challenge against which our character as a people was formed. The reminder and the reassurance that it is still there is good for our spiritual health even if we never once in ten years set foot in it. It is good for us when we are young, because of the incomparable sanity it can bring briefly, as vacation and rest, into our insane lives. It is important to us when we are old simply because it is there—important, that is, simply as an idea.[33]

Frederick Jackson Turner and Theodore Roosevelt's belief that a rugged landscape toughened Americans and built a particular American personality and national exceptionalism echo in Stegner's plea for wilderness. These ideas have lost much of their currency in recent decades. But he also offers a profound critique of consumption and meaningless activity that echoes more radical thinkers and anticipates contemporary criticism. Stegner also makes the case for the right of nature to exist for its own sake, a position that would have seemed alien to Roosevelt and Turner.

The benefits of the Wilderness Act are not only protected habitat and beautiful places to hike, camp, and escape the stresses of daily life. Historian Kevin Marsh states that this federal legislation effectively expanded American democracy. While he does not privilege wilderness activism over Civil Rights organizing and protest from the New Left, Marsh does effectively argue that

> although the wilderness movement is often characterized as elitist . . . the movement to protect de facto wilderness and the increasingly organized political efforts on the part of both conservationists and resource industry coalitions and corporations changed the debate to a far more grassroots political discussion. The end result was to make wilderness politics more open to public participation through the political arena of Congress rather than the previously closed confines of Forest Service administrative decision making.[34]

Whereas in the 1950s only a limited few Americans had any influence or say in the decisions made by the Department of Agriculture, "by the 1970s, all citizens had some form of direct influence through their congressional representatives."[35] Fighting to protect nature made America a more democratic society.

Over the course of the 1960s a powerful environmental consensus developed as Americans mobilized to fight to preserve parks, rivers, and wildlife and to stop or diminish the poisoning of the environment through the use or overuse of above-ground nuclear weapons testing and chemicals. The movement marched on to further successes and an increasingly litigious approach to fighting environmental damage. The plan by New York State to build a power plant on Mt. Storm King on the Hudson River opened the era of environmental law. The Sierra Club joined in a suit to stop the plant, and the court took the unprecedented step of finding for the environmental organization's non-economic interest in the case and ordered a halt to the project. The use of a lawsuit a year later to stop aerial spraying of DDT on Long Island strengthened this new strategy. The man who filed the suit, Victor Yannacone, joined Charles Wurster and other scientists at the State University of New York at Stonybrook to create a new organization, the Environmental Defense Fund (EDF), in 1967.

They believed that letters to the editor, testimony before politicians, and photo books would no longer suffice; only in a court of law could they cut through political manipulation and bureaucratic strategies to receive a fair hearing and ruling. An expanding regulatory framework gave them the basis to do this as did the growing environmental consensus and ecological knowledge in society. The Sierra Club Legal Defense Fund and the Natural Resources Defense Council were quickly created and joined the EDF in using legal strategies as their primary tool. Environmental law took off in the 1970s, and environmental groups put an increasing amount of resources into lawyers, technical experts, and lawsuits.

The 1950s and early 1960s were a time when environmental thinking jumped forward dramatically in response to rapid and destructive innovations in science and technology and because of ground-breaking work in ecological science, and the influence of books such as *Silent Spring* and *Sand County Almanac.* The popularity of nature and recreation helped boost support for environmental agendas at the same time that many writers engaged in a critique of American patterns of consumption and commitment to growth that threatened the health nature. Environmental organizations grew dramatically in membership and developed effective strategies for ending or limiting chemical pesticide use, blocking dams, and creating wilderness areas. The successes of these groups, along with their ability to convince most Americans that the natural world required protection, laid the foundation for the stunning achievements of the next several years.

Document 11.1 DDT Advertising

Americans were amazed by the remarkable ability of DDT and other chemicals to kill insects, increase crop yields, and potentially reduce the possibilities of illness. As noted already, they were very trusting of scientists and the chemical industry and, because of this trust and a lack of critical information about problems with the chemicals, were reckless in their use of the chemical.

1. Analyzing the following advertisement and World War II propaganda poster, what promises are made about the benefits of DDT? How is popular culture and gender used to promote and sell the product? How is the domestic sphere and children's health employed to manipulate consumers to make emotional decisions to purchase the products?
2. Studies of the response to *Silent Spring* emphasize the revelations of the book and its powerful prose in convincing Americans to oppose and end DDT production and spraying. To what degree might the introducing of this poison into their children's lives have informed American parents' fear and anger and mobilized their support for controls on chemical pesticides as well as more general environmental reforms?

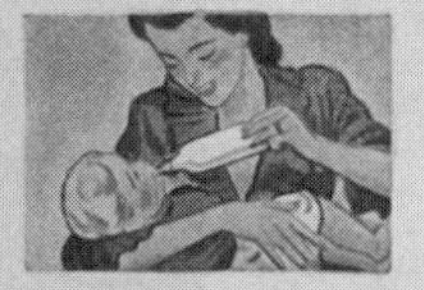

Figure 11.6 Crossett Library, Bennington College

Figure 11.7 National Archives

Document 11.2 *Agricultural Chemicals, Robert H. White-Stevens,* 1962 (Reprinted by permission of CropLife)

In the essay from which this was excerpted, the author decries the failure of the modern agricultural industry to communicate its successes and importance to the increasingly urban American population. He opens with a long section essentially summarizing the success of the Green Revolution. At the end of the piece he paints a picture of future starvation and misery due to the jeremiads of Carson and others.

1. Based on the following passage, what are the author's assumptions about the relationship of science, industry, and agricultural production?

2. How does he view the perspective of critics and the urban American audience? To what degree does this reveal a certain set of assumptions on the part of scientific experts?

. . . Our USDA and land-grant college system, now celebrating its centennial, stands witness to the tremendous flow of education, research, and published information which has emerged over the past one hundred years. How is it possible that so much of this essential and valuable information has not reached the majority of our people?

The answer is simple. We in agriculture have shrunk to a distinct minority during the past twenty-five years, and have become so by dint of our own competence. We have, in effect, researched ourselves into obscurity both politically and socially. In addition, we have engendered the ire of the urbanite by producing a so-called surplus of food and fiber, which, it is claimed, cost the urban taxpayer several billion dollars per year to support.

This gross misunderstanding, punctuated all too frequently by crisis and sensationalism, is the target at which we need to fire our arsenal of scientific truth. We have been so preoccupied in agricultural research, rural education, and extension that we have essentially ignored the urban peoples in our communications, and have been talking largely to ourselves.

We need now to tell the urban peoples in a thousand places and a thousand ways what scientific agriculture, including agricultural chemistry, has meant to their health, their welfare, and their standard of living. We should make it clear in schools, service clubs, in church meetings, and in the hundreds of other groups to which our people attach themselves:

> That the entire cost of agricultural research by federal, state, and industry is less than the savings it brings in cost of food alone to the American people each year;
>
> that DDT alone has saved as many lives over the past fifteen years as all the wonder drugs combined;
>
> that insecticides have been credited with extending the prospective life span in at least one Asiatic country from thirty-two to forty-seven years;
>
> that our so-called surplus of food is really no more of a surplus than a healthy reserve in the bank which represents our margin of safety in a hungry and envious world, and which ensures a steady course of food and prices in place of wide seasonal fluctuations which could cost our consumers several fold the expense of support programs;
>
> that our cheap, nutritious and wide array of foodstuffs in America today costs us less in terms of hours of work than it did any other nation since time began;
>
> that agricultural science with all its disciplines working in close collaboration in the lab, in the college, in the field, in the factory, and out on the farm, has for the first time in man's long struggle against want procured the means to banish hunger from the earth in our time;
>
> that our knowledge and control of the chemistry and function of the pesticides and additives we use vastly exceed that of the natural compounds which invariably contaminate our food supply when it is unprotected;

that based on this cheap, safe, prolific, and varied food supply is man's ability to turn his surplus time and energy into research, education, and culture, for by DeGraff's law the progress of a people is inversely proportional to the time and effort required to produce the necessaries of existence.

This is what we must tell our urban peoples in all its brilliance and luster, and then rely upon their good sense to decide whether the alarums of the Longgoods, the Bicknells, and the Carsons are valid or whether the patient researches of scientific agriculture published from thousands of laboratories over the past century can be reliably accepted with confidence . . .

Miss Rachel Carson has done it from the opposite side in her book *Silent Spring*, for she is a writer on biological subjects with an extraordinary, vivid touch and elegance of expression. She paints a nostalgic picture of Elysian life in an imaginary American village of former years, where all was in harmonious balance with Nature and happiness and contentment reigned interminably, until sickness, death, and corruption was spread over the face of the landscape in the form of insecticides and other agricultural chemicals.

But the picture she paints is illusory, and she as a biologist must know that the rural Utopia she describes was rudely punctuated by a longevity among its residents of perhaps thirty-five years, by an infant mortality of upwards of twenty children dead by the age of five for every 100 born, by mothers dead in their twenties from childbed fever and tuberculosis, by frequent famines crushing the isolated peoples through long dark, frozen winters following the failure of a basic crop the previous summer, by vermin and filth infesting their homes, their stored foods and their bodies, both inside and out.

Document 11.3 "The Wilderness Act," 1964

In 1964 environmental activists won legislation that would have been thought impossible a few years earlier. The Wilderness Act was the result of eight years of activism, lobbying, congressional committee meetings, and a nationwide effort by environmental groups, particularly the Sierra Club and the Wilderness Society, but with the support of many others, to educate Americans as to the benefits of wilderness and the reasons some nature must be given total protection from overuse, development, and destruction. Over this time Howard Zahniser's original bill went through 66 revisions. Passage of this legislation marks the culmination of several decades of developing thought on the benefits of preserving large sections of ecosystem along with the increasing popularity of the belief that nature has inherent rights of its own. From the original 9 million acres protected as wilderness in 1964, the amount of public land designated as wilderness has expanded to 109 million acres.

1. According to the language of this legislation, how is wilderness defined and characterized? What are the benefits of wilderness?
2. How are the assumptions of the wilderness legislation different from earlier attitudes about the best uses of nature? To what degree is this a codification of earlier environmental thought such as the ideas of Muir, Edge, and Leopold? What is

problematic about this approach? Some historians such as William Cronon have argued that creating wilderness makes nature a place where people are not and thereby contributes to the continued degradation of natural places outside of wilderness. Is this a legitimate critique, and what are possible responses to this?

Wilderness System Established Statement of Policy

Sec. 2. (a) In order to assure that an increasing population, accompanied by expanding settlement and growing mechanization, does not occupy and modify all areas within the United States and its possessions, leaving no lands designated for preservation and protection in their natural condition, it is hereby declared to be the policy of the Congress to secure for the American people of present and future generations the benefits of an enduring resource of wilderness. For this purpose there is hereby established a National Wilderness Preservation System to be composed of federally owned areas designated by Congress as "wilderness areas," and these shall be administered for the use and enjoyment of the American people in such manner as will leave them unimpaired for future use as wilderness, and so as to provide for the protection of these areas, the preservation of their wilderness character, and for the gathering and dissemination of information regarding their use and enjoyment as wilderness; and no Federal lands shall be designated as "wilderness areas" except as provided for in this Act or by a subsequent Act.

(b) The inclusion of an area in the National Wilderness Preservation System notwithstanding, the area shall continue to be managed by the Department and agency having jurisdiction thereover immediately before its inclusion in the National Wilderness Preservation System unless otherwise provided by Act of Congress. No appropriation shall be available for the payment of expenses or salaries for the administration of the National Wilderness Preservation System as a separate unit nor shall any appropriations be available for additional personnel stated as being required solely for the purpose of managing or administering areas solely because they are included within the National Wilderness Preservation System.

Definition of Wilderness

(c) A wilderness, in contrast with those areas where man and his own works dominate the landscape, is hereby recognized as an area where the earth and its community of life are untrammeled by man, where man himself is a visitor who does not remain. An area of wilderness is further defined to mean in this Act an area of undeveloped Federal land retaining its primeval character and influence, without permanent improvements or human habitation, which is protected and managed so as to preserve its natural conditions and which (1) generally appears to have been affected primarily by the forces of nature, with the imprint of man's work substantially unnoticeable; (2) has outstanding opportunities for solitude or a primitive and unconfined type of recreation; (3) has at least five thousand acres of land or is of sufficient size as to make practicable its preservation and use in an unimpaired condition; and (4) may also contain ecological, geological, or other features of scientific, educational, scenic, or historical value.

Notes

1 Quoted in Mark Harvey, *A Symbol of Wilderness: Echo Park and the American Conservation Movement* (Seattle: University of Washington Press, 1994), 195.
2 Ibid.
3 There were only the three networks at that time, so this represented two-thirds of television programming.
4 Quoted in Linda Lear, *Rachel Carson: Witness for Nature* (New York: Henry Holt and Company, 1997), 426.
5 Ibid.
6 The U.S. government provided $15 billion in direct aid and technical assistance to western European nations in order to rebuild their infrastructures and get their war-torn economies moving again. This program was created as a way to contain the expansion of communism by rebuilding thriving capitalist economies. Operating from 1947 to 1951, the program proved integral to rebuilding those economies and also helped the American economy as those nations purchased equipment and resources and, once their economies started to grow, increasing amounts of American manufactured goods.
7 Quoted in Lizabeth Cohen, *A Consumers' Republic: The Politics of Mass Consumption in Postwar America* (New York: Vintage Books, 2004), 121.
8 Ibid., 116.
9 Christopher W. Wells, *Car Country: An Environmental History* (Seattle: University of Washington Press, 2012), 259, 260.
10 Paul K. Conkin, *A Revolution Down on the Farm: The Transformation of American Agriculture since 1929* (Lexington: The University Press of Kentucky, 2008), 106.
11 Donald Worster, *Nature's Economy: A History of Ecological Ideas* (New York: Cambridge University Press, 1994), 342.
12 Terry Tempest Williams, *Refuge: An Unnatural History of Family and Place* (New York: Vintage Books, 1991), 286.
13 Quoted in Michael Egan, *Barry Commoner and the Science of Survival: The Remaking of American Environmentalism* (Cambridge, MA: The MIT Press, 2007), 38.
14 Ibid., 70.
15 Ibid., 72.
16 Mary Karr, *The Liars' Club: A Memoir* (New York: Viking, 1995), 40.
17 Ibid., 41.
18 Quoted in Lear, *Rachel Carson,* 357.
19 Rachel Carson, *Silent Spring* (Boston, MA: Houghton Mifflin Company, 1962), 47.
20 Ibid.
21 Ibid., 48.
22 Ibid.
23 Ibid., 145, 146.
24 Lear, *Rachel Carson,* 433.
25 Quoted in Ibid.
26 Quoted in Adam Sowards, *The Environmental Justice: William O. Douglas and American Conservation* (Corvallis: Oregon State University Press, 2009), 41.
27 Ibid., 50.
28 Ibid., 54.
29 Ibid.
30 Edward Abbey, *Desert Solitaire: A Season in the Wilderness, A Celebration of the Beauty of Living in a Harsh and Hostile Land* (New York: Ballantine Books, 1968), 63, 64.
31 Wallace Stegner, "Wilderness Letter," 1960. The Wilderness Society website, wilderness.org/bios/former-council-members/wallace-stegner. Accessed 10 August 2014.
32 Ibid.
33 Ibid.

34 Ibid.
35 Kevin Marsh, *Drawing Lines in the Forest: Creating Wilderness Areas in the Pacific Northwest* (Seattle: University of Washington Press, 2007), 97, 98.

Further Reading

Carson, Rachel. *Silent Spring.* Boston, MA: Houghton Mifflin Company, 1962.

Cohen, Lizabeth. *A Consumers' Republic: The Politics of Mass Consumption in Postwar America.* New York: Vintage Books, 2004.

Egan, Michael. *Barry Commoner and the Science of Survival: The Remaking of American Environmentalism.* Cambridge, MA: The MIT Press, 2007.

Harvey, Mark. *A Symbol of Wilderness: Echo Park and the American Conservation Movement.* Seattle: University of Washington Press, 1994.

Hays, Samuel. *Beauty, Health, and Permanence: Environmental Politics in the United States, 1955–1985.* New York: Cambridge University Press, 1987.

Hirt, Paul. *A Conspiracy of Optimism: Management of the National Forests Since World War II.* Lincoln: University of Nebraska Press, 1996.

Lear, Linda. *Rachel Carson: Witness for Nature.* New York: Henry Holt and Company, 1997.

Marsh, Kevin. *Drawing Lines in the Forest: Creating Wilderness Areas in the Pacific Northwest.* Seattle: University of Washington Press, 2007.

Nash, Roderick. *The Rights of Nature: A History of Environmental Ethics.* Madison: The University of Wisconsin Press, 1989.

Oelschlaeger, Max. *The Idea of Wilderness: From Prehistory to the Age of Ecology.* New Haven, CT: Yale University Press, 1993.

Pasternak, Judy. *Yellow Dirt: An American Story of a Poisoned Land and a People Betrayed.* New York: Simon & Schuster, Inc., 2010.

Rome, Adam. *The Bulldozer in the Countryside: Suburban Sprawl and the Rise of American Environmentalism.* New York: Cambridge University Press, 2001.

Sellers, Christopher C. *Crabgrass Crucible: Suburban Nature and the Rise of Environmentalism in Twentieth Century America.* Chapel Hill: The University of North Carolina Press, 2012.

Shapiro, Laura. *Something from the Oven: Reinventing Dinner in 1950s America.* New York: Penguin Books, 2004.

Souder, William. *On a Farther Shore: The Life and Legacy of Rachel Carson, Author of Silent Spring.* New York: Crown Press, 2012.

Sowards, Adam M. *The Environmental Justice: William O. Douglas and American Conservation.* Corvallis: Oregon State University Press, 2009.

Wells, Christopher W. *Car Country: An Environmental History.* Seattle: The University of Washington Press, 2012.

Worster, Donald. *Nature's Economy: A History of Ecological Ideas.* New York: Cambridge University Press, 1994.

Timeline

Spraying of Agent Orange in Vietnam	1962–1971
Santa Barbara Oil Spill	1969
Burning of the Cuyahoga River	1969
Earth Day	1970
National Environmental Policy Act	1970
Creation of the Environmental Protection Agency	1970
Revised Clean Air Act	1970
Marine Mammal Protection Act	1972
Passage of the Clean Water Act	1972
Creation of the Endangered Species Act	1973
Oil Crisis	1973
Oregon Land Use Law	1973
Safe Drinking Water Act	1974
Love Canal	1978
Creation of Earth First!	1979
Creation of the Superfund Program	1980
Warren County Protests	1982
Listing of the Spotted Owl as an Endangered Species	1988
Exxon Valdez Oil Spill	1989
Elwha Restoration Act	1992
End of Chlorofluorocarbon Production and Use	1996

Environmental Reform and Schism

12

George Alexander was a sawmill worker for the Louisiana Pacific Lumber Company in 1987. Twenty-three years old and just married with a young wife who was three months pregnant, Alexander worked one of the most dangerous, high-skill jobs in the mill. Using a steel blade of 10 inches to cut through large old-growth trees, he wore a heavy face mask and paid careful attention for odd sounds that might warn of danger and the blade jumping off knots or metal debris such as old fencing or a choker chain. The bandsaw had developed cracks and was wobbly. He had asked his supervisor for a replacement but was advised to keep using the damaged blade until new ones arrived. One day in May 1987, Alexander was halfway through a 20-foot log when the blade hit a 60-penny nail. The saw blade broke off, striking him in the face and wrapping around his body. Cutting through his mask and tearing into his jugular vein, the saw also broke his jaw in multiple places and knocked out a dozen teeth, leaving him on the edge of death. While co-workers used a blowtorch to unwrap the saw from his body, a friend held his veins together so George did not bleed to death before the ambulance arrived.

As he recovered in the hospital, Louisiana Pacific went on the offensive, using the incident to accuse eco-saboteurs of terrorism. The press ran with the story, and George Alexander's maiming became a cautionary tale in the conflict between advocates of logging and those attempting to protect the ancient forests through extreme measures.

During the late 1960s and 1970s, American environmental crises spilled and flamed into the American consciousness. Events such as the Santa Barbara Oil Spill, the burning of the Cuyahoga River, and the health crisis from toxic pollutants at Love Canal frightened Americans and mobilized them to support environmental reforms and greater regulation of waste disposal, resource extraction, and pollution. In this era, the nation made major strides in managing a series of environmental problems while creating mechanisms and regulations for ongoing enforcement and environmental repair. The relationship of society and nature was radically transformed in a few short years with a raft of legislation during the Nixon administration that greatly expanded the federal government's power and responsibility for environmental protection; this was largely possible because of a still growing environmental consensus.

In a broader context, Americans perceived a culture spinning out of control. Environmental activism and reforms occurred against a backdrop of great tumult and conflict in American society. Student protests against the Vietnam War, Civil Rights activism, and the Women's rights movement gripped the nation. The rise of a countercultural protest challenged the central tenets of American life, infused the environmental movement with greater dynamism and energy, and contributed to more reforms. As a result, Americans pushed the environmental consensus further to achieve unprecedented reforms through the 1970s. At the same time, this era also saw the high-water mark of environmentalism, and the environmental consensus began to fragment as the robust economy of the 1950s and 1960s began to fade and as business and, increasingly the Republican Party began pushing back against environmental reforms.

Disasters and Environmentalism

Even as Americans began to think more ecologically and push for laws to protect nature and themselves, environmental disasters provided potent reminders of the need for immediate and drastic change. The Santa Barbara Oil Spill, the burning of the Cuyahoga River, the Love Canal toxic pollutants crisis, and other incidents transfixed the nation and helped drive environmental reform. The 1969 Santa Barbara, California, oil spill was the largest spill in U.S. waters up to that time. Before the well was plugged, oil poured out for 10 days, discharging approximately 100,000 barrels of oil that coated the waters and beaches and killed thousands of birds and numerous elephant seals, sea lions, and dolphins. Coverage of the spill by local and national media centered the issue in the American consciousness. Historian J. Brooks Flippen characterizes the coverage in sarcastic tones consistent with his argument that the media exaggerated the environmental impact of the event:

> The disaster was, in fact, no greater than several oil tanker spills the world had suffered but, with the scenic beauty of the Californian coast as a backdrop, it still made for great television. Birds covered with sticky oil struggled for life; dead seals floated ashore; enraged Santa Barbara housewives cried for the cameras.[1]

Santa Barbara government officials were enraged by the lackadaisical response of the federal government and oil company officials.

The coverage of the disaster tapped into increasing distrust of corporations and questioning of government's commitment to protecting the environment and provoked a powerful national outcry. Anger at the spill reflected the ecological moment—the environmental consensus shared by a large majority of Americans. This is evident when compared to the response to the Deepwater Horizon spill of 2010. Whereas strong protest erupted following Santa Barbara, the response to the Deepwater spill, which was much more environmentally destructive, was muted in comparison and resulted in no changes in regulations or oil drilling practices. In contrast, the 800-square-mile oil slick off the coast of Santa Barbara and the attendant media coverage proved decisive for the environmental movement. As Park Service director George Hartzog stated, "Ecology has finally gained currency."[2] The Sierra Club and other environmental groups seized on the issue. President Nixon ordered a halt to drilling in federal waters off the California coast, and the catastrophe and response lent weight to several regulatory moves that expanded the power and scope of the federal government on environmental issues. A report by the President's Council on Environmental Quality several years later assessed the impact:

> the event dramatized what many people saw as thoughtless insensitivity and lack of concern on the part of government and business . . . It brought home to a great many

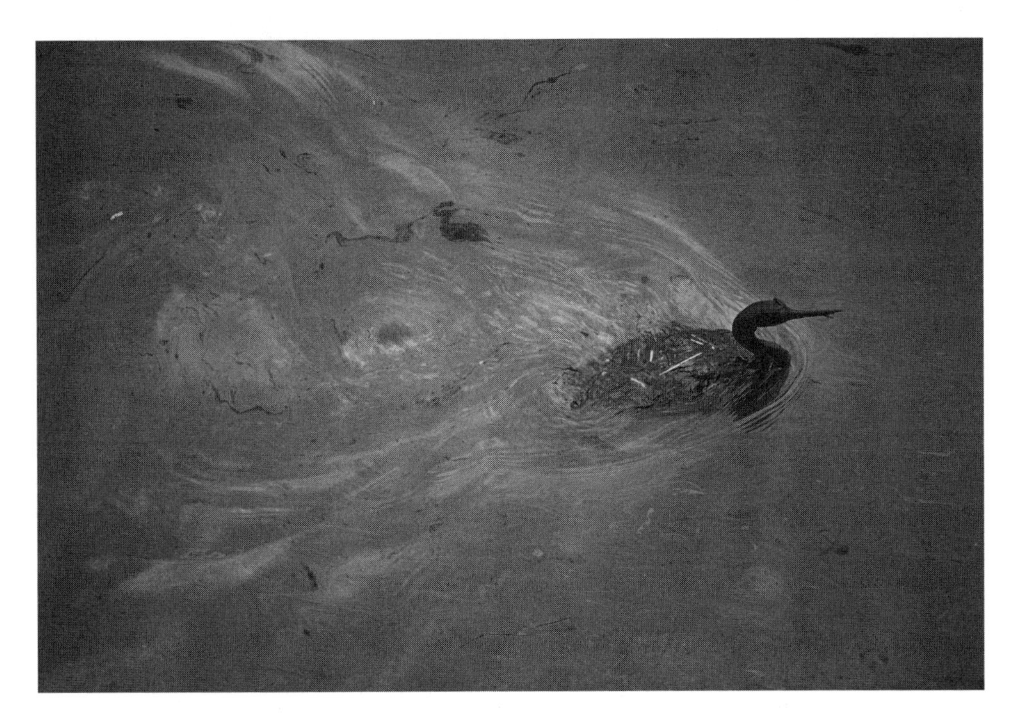

Figure 12.1 The Santa Barbara oil spill released 100,000 barrels of oil over 10 days. The disaster gained national attention and triggered protests and demands for greater government control of environmental issues. Getty images.

> Americans a feeling that protection of their environment would not simply happen, but required their active support and involvement.[3]

The passage of the Clean Air Act, the creation of the Environmental Protection Agency, and the National Environmental Policy Act are at least partly the result of American concern over what the oil spill represented. Disasters such as this reinforced the message coming from books such as *Silent Spring* and from environmentalists that nature was under severe threat and required reforms providing protection.

When water wasn't tarred with oil, it was burning. The unnatural conflagration on the Cuyahoga River in 1969 is viewed as a pivotal moment in the history of American environmentalism. In fact, this river running through downtown Cleveland and draining into Lake Erie had caught fire previously at least 10 times, if not more, before the famous event of 1969.[4] Ironically, the fire of 1969 was small in comparison to earlier conflagrations and originally garnered little media interest. But the growing ecological consciousness and consensus, and the media's interest in environmental stories meant that the 1969 fire was exaggerated, hyped, and misunderstood. The media version and oft-repeated story is that a massive conflagration caught the city and nation by surprise, triggering the first real efforts to clean the river. In fact, Cleveland had already begun implementation of a river and downtown cleanup in 1968. But the fire happened at a time when Americans were highly tuned in to environmental crises and shared a sense of the natural world spinning out of control.[5] The environmental consensus was strengthened and justified by just such events. As environmental groups worked to push through reforms and legislation and the media provided constant coverage of ecological disasters, Americans flocked to environmental organizations. The Audubon Society more than doubled its membership from 1962 to 1970, and the Sierra Club grew from 20,000 in 1959 to 113,000 in 1970. The National Wildlife Federation doubled in four short years from 1966 to 1970. The disasters, media coverage, and environmental organizations' use of these events to build support expanded the environmental consensus, making it broad and strong enough to radically transform the relationship between government, business, and nature over the next several years.

While extensive media coverage of crises captured the nation's attention and increased support for reforms, ongoing environmental problems also reinforced calls for change. The regular reportage of the decline and impending extinction of iconic species such as the bald eagle and others such as the American kestrel, alligators, sea otters, sea lions, and so forth reminded Americans that unchecked environmental degradation could result in extinctions of loved and necessary animals. Regular smog events were a constant reminder as well. In Los Angeles in July 1969, radio and television stations ran the following public service announcement: "The children of Los Angeles are not allowed to run, skip, or jump inside or outside on smog alert days by order of the Los Angeles Board of Education and the County Medical Association."

The Earth Gets a Day

U.S. Wisconsin senator Gaylord Nelson came up with the idea of Earth Day after watching the Santa Barbara oil spill play out. Designed as a national teach-in on environmental issues, he hoped to bring the tactics and energy of the anti-war movement to environmentalism. Part of the "genius of Earth Day," to borrow historian Adam Rome's phrase and book title, was the decision to leave organizing events to local communities, public schools, colleges, and universities. Up to 20 million Americans participated in some way, and universities and colleges as well as high schools organized protests, teach-ins, workshops, and other programs to teach about and act against pollution, toxins, wilderness destruction, and other environmental issues. While the media focused on protests— there were 250,000 marchers on Fifth Avenue in New York City, and in Omaha, Nebraska, high-school kids bought out the local supply of gas masks to theatrically protest air pollution—in reality most of the events were educational and organizational in nature. At more than 12,000 celebrations and events across the nation, activists embraced this opportunity. These events emphasized local issues, and this day was in many cases the starting point for important local reforms and education. Many speakers were neither professional nor experienced, but their amateurism and excitement energized the movement from the grassroots up as participants learned how to take local action to improve the environment. Also, activists from the anti-war movement and other areas of protest were attracted to environmentalism, as were many Americans who were concerned about problems in the nation but worried about the radicalism of other opposition movements.

The event provided opportunities and education to a diverse array of Americans. Rome captures the eclectic nature of participants in describing a protest: "some were hippies. Many wore suits."[6] Thousands of housewives, stultified by their suburban housewife role and worried about the environment, threw themselves into the organizing of Earth Day.

Figure 12.2 Earth Day has lost much of its relevance in recent years with corporations using the day as an opportunity to "greenwash," that is, appear more environmentally friendly than they really are or groups and agencies passing out information on water use reduction and alternative energy opportunities. But the first Earth Day was a radical, transformative moment with long-term consequences for environmentalism and American society. While protests did occur across the country, much of the focus of Earth Day and the week surrounding it was on the study of local environmental issues, teach-ins, and strategizing for local action. Getty Images.

Many sustained their environmentalism after the big day. School children represented the largest proportion of participants by age as they planned and participated in educational events, staged protests, or watched documentary films about environmental disasters. Whether this is a truly "Green Generation," as Rome asserts, is open to debate. Regardless, contrary to the contemporary view of Earth Day as a largely corporate-sponsored event at which companies and universities can establish their green bona fides, this first Earth Day provided a strong push to the environmental movement, expanding the ranks of activists, garnering widespread media attention, and creating a permanent eco-infrastructure of lobbying organizations, environmental studies programs, and community ecology centers. In many cases, local environmental reforms followed events that explored with local problems. There is no question that these events and this day strengthened the environmental consensus and lent momentum to state and federal environmental reform policies.

Land-Use Laws and a Land Ethic

In the beginning of this period the environmental consensus was still strong enough to culminate in what historian Adam Rome terms a "Quiet Revolution" in land-use laws and ideas on property rights. Hawaii in 1961, then several other states between 1963 and 1970, passed laws protecting farmlands and wetlands and seeking to limit suburban sprawl and unplanned growth. Vermont and Maine passed strong land-use laws and also began requiring permitting for environmental conditions. In 1972 Florida gave state agencies the power to control land use. Twenty-three states had passed land-use laws of some sort by 1975. The most famous and long-lasting of the land-use laws was passed in the Pacific Northwest. Employing powerful rhetoric, and tapping into deep-seated concerns while benefitting from a strong environmental consensus, Oregon Governor Tom McCall and other activists achieved significant land-use reform. In 1973 he declared, "There is a shameless threat to our environment and to the whole quality of life—the unfettered despoiling of the land."[7] Criticizing "sagebrush subdivisions" and "coastal condomania" as well as general sprawl from the cities, he stated, "The interests of Oregon for today and in the future must be protected from the grasping wastrels of the land."[8] That land-use law is still in place and has effectively contained sprawl to protect farmlands and habitat.

Citizens in California tried to move legislation through three times to protect the state's long coastline; they were rebuffed each time. Turning to a ballot initiative, they passed the Coastal Conservation Act in 1976, providing safeguards for 1,100 miles of shoreline and coast. Rome argues that frustration over local governments' inability or unwillingness to deal with environmental issues or manage land use meant that, across the nation, Americans were willing to accept increased state power at the expense of local government in order to protect forests, coasts, farmland, and clean drinking water.

The environmental consensus and the land-use movement at the state level influenced legal cases and law review articles with developing interpretations of property rights that included community concerns and incorporated ecological concepts. These suggested that in some cases the traditional emphases on private land ownership were incompatible with ecosystem integrity and community rights to a healthy landscape. Revolutionary court cases in the early 1970s built on the ecological view of property, and something like a land ethic came close to being institutionalized at the state level. The California Court of Appeals in 1970 provided a clear example of this with their confirmation of the power of the San Francisco Bay Commission. The court wrote,

> the legislature has determined that the bay is the most valuable single natural resource of the entire region and changes in one part of the bay may affect all other parts; that the present uncoordinated, haphazard way in which the bay is being filled threatens the bay itself and is therefore inimical to the welfare of both present and future residents of the bay area; and that a regional approach is necessary to protect the public interest in the bay.[9]

In *Just v. Marinette County* in 1972, the Wisconsin State Supreme Court challenged the traditional view of property rights giving the land owner great latitude in use regardless of community and environmental needs. The court used ecological language to assert "the interrelationship of the wetlands, the swamps and the natural environment of the shorelands to the purity of the water and to such natural resources as navigation, fishing, and scenic beauty"[10] and pointed out the ecological importance of wetlands.

With the widespread passage of state land-use protection and planning laws it was logical that Washington State Democratic Senator Henry "Scoop" Jackson proposed the National Land Use Policy Act in 1970. He hoped that this legislation would provide a balance between environmental and development interests. The Nixon administration introduced a competing bill; both emphasized federal aid to state governments and avoided punitive regulations or required actions. Crafting compromise legislation, they introduced it with bipartisan support in 1972. The primary target of the proposed law was suburban sprawl, and while environmentalists sought modifications to the bill to require states to protect farmland, steep hillsides, wetlands, and other important habitat and vulnerable sites, bill sponsors believed that including such requirements would prevent passage and that federal aid and guidance was enough to protect vulnerable ecosystems.

The assumptions and beliefs inherent in advocates' endorsement of the bill reveal the sea change in attitudes about the land. Jackson's adviser, Lynton Caldwell, invoked Aldo Leopold, arguing for

> a land ethic that is consistent with our best knowledge of the natural world, including the full range of human needs in land and environment. The National Land Use Policy Act, as law would be a major step toward the development of a popular land ethic.[11]

Others decried the frontier ethic and expansion that had encouraged wasteful and destructive practices throughout American history. What Rome refers to as "one of the most influential reform documents of the early 1970s,"[12] *The Use of Land* was published in 1973 and argued for a radical change in views of land use and ownership, specifically quoting Leopold to the effect that treating land as a simple commodity was automatically destructive. These debates and discussion garnered widespread press coverage with the media emphasizing needed changes in legal views of property rights and the "Takings Clause." While environmentalists could not know it, this effort marked the high water mark of the national environmental movement.

The nation hadn't changed as much as land-use control advocates hoped. Legislators and property rights advocates, concerned about an erosion of private-property rights and the expansion of government power at the state and federal level, began publishing opposition pieces and organizing protests against the proposed law. Trade groups tied to development worried about increased costs. The Chamber of Commerce played an instrumental role in resisting the legislation, organizing a lobbying coalition, and generating activism at the grassroots level through its approximately 2,500 state and local chambers. In building a strong wall of opposition they emphasized the loss of private property rights and economic self-determination. The minority statement in the 1973 Senate Committee report put it in stark, ideological terms. The Land Use Act would "jeopardize the one single characteristic of American life, the right of private ownership of property, that so distinguishes our lives from those of people in other countries."[13] While meaningful reforms would still be passed in 1973 and upcoming years, and the clear counterthrust to environmentalism would not become truly apparent until the Reagan Administration in the 1980s, a schism was now opened that would spread further as economic limits or constraints ran up against environmental reforms.

An Environmental Government

Environmental disasters and the growing environmental consensus explain the pinnacle of environmental reforms during the Nixon administration. Unlike the earlier Republican president who did so much for the environment, Theodore Roosevelt, President Nixon was not deeply interested in conservation or environmental issues. But the momentum for environmental reform was so powerful and pervasive in this era he could hardly resist. It was a bipartisan moment, and Nixon was determined to capture the broad middle of American society in order to sustain Republican power. And while he was not an active environmentalist, the consensus was so ubiquitous in this era that Nixon likely shared the concerns of many Americans. The legislation passed and agencies created were common sense solutions to both ongoing problems and crises. The power of the Democratic Party in Congress also meant that a presidential veto would likely not stand. The upshot is that a major transformation of Americans and the American government's relationship with nature was enacted during this Republican administration. The Clean Air Act was passed in 1970, and the Water Pollution Control Act; Federal Insecticide, Rodenticide, and Fungicide Act; the Marine Mammals Protection Act; and the Safe Drinking Water Act all gained approval in 1972. One of the most powerful agencies to emerge from this era of environmental regulation and reform was, appropriately enough, the Environmental Protection Agency. It was part of the National Environmental Policy Act of 1970, which also required environmental impact statements for federally funded projects. Starting with a budget of $455 million in 1971, by the beginning of the Reagan administration, it more than doubled from 6,000 to 13,000 employees, and its budget climbed to more than $5.6 billion. In addition to these reforms, the Endangered Species Act was enacted in 1973.

The Endangered Species Act was unprecedented in its comprehensive protection of nature. All species except bacteria, viruses, and insect pests that presented a threat to humans were acknowledged to have a right to exist. This law brought the full weight of the federal government behind that presumption. "Because Congress invariably justifies its actions in terms of the public interest,"[14] the law constructed nature and the law in instrumentalist terms, emphasizing the benefits to humanity. This would allow later challenges and modifications to the law that would weaken it. Even so, the Endangered Species Act came remarkably close to codifying a land ethic. Because of its acknowledgement of the role of "critical habitat," it almost stated that ecosystems had a right to existence as well. The federal government now had the power and responsibility to some degree to interfere with or limit private land use to protect threatened or endangered species. Citizens were given the right to petition for the listing of species, marking a further democratization of America arising from environmental issues and laws. Environmental reforms further democratized the American political system while they also expanded the power and size of the government. Lobbying by environmental activists, the clear ecological crises in numerous areas, and the unifying and powerful environmental consensus propelled the federal government to expand its ability and power to manage complex environmental issues and redefine its relationship with both society and nature. Government now increasingly bore the role of nature's protector.

The Limits of Consensus

The gushing flow of petroleum after World War II was not simply a consequence of innovative entrepreneurship; federal policy created an environment of petroleum abundance. Because of concerns that national security would be compromised by an overreliance on imported oil, President Eisenhower's administration in 1959 set a limit of 12.2 percent of oil consumed in the United States coming from foreign sources. This triggered a rush to drill more wells, build more refineries, and fully tap the nation's vast petroleum resource. The tsunami-sized wave of oil that flooded the American economy underwrote the nation's growth in this era. "Cheap gas" kept other costs down and made suburbs, recreational travel, and shipping of goods cross-nationally, as well as long-distance wars in Korea and Vietnam, possible. The booming economy and American way of war was predicated on cheap petroleum, and for a while this helped Americans achieve a level of prosperity undreamt of in the wildest imaginings of Alexander Hamilton. Besides generating

Figure 12.3 For a nation deeply dependent on petroleum, the Oil Crisis of 1973 was an abrupt jolt to the economy and American society. Long lines at gas stations, tanks running dry, and increasing fuel prices created conflict and concern. While some Americans turned to riding bikes as transportation and carpooling, the crisis engendered no long-term changes in economic behavior. National Archives.

ever-escalating amounts of consumption, increased air pollution and smog, water and soil pollution from oil wells, and the damage to the health of children from leaded gasoline, the petroleum foundation of this post-war economy created economic expectations that were never realistic but that inform many Americans' dissent against environmental reforms.

By the early 1970s domestic oil production dropped off precipitously, and Americans' thirst had to be quenched with oil from around the world. A great deal of this petroleum was provided by Persian Gulf Arab States, leaders and members of the Organization of Petroleum Exporting Countries (OPEC) created in 1960 to reduce drilling of oil in order to raise prices. OPEC was also well positioned to control the spigot of oil if desired. In 1973, the same year that Nixon lifted Eisenhower's import quota, Syria and Egypt attacked Israel in the Yom Kippur War. When the United States rushed military supplies to Israel, OPEC responded by raising prices then closing the spigot completely. The embargo was a knife to the heart of the American economy as oil prices almost quadrupled by year's end.

Americans were forced to confront their dependency during the oil crisis of 1973–1974 when prices climbed dramatically, and low oil supplies meant gas pumps running dry and stations rationing fuel. They did not always handle this situation gracefully. Fights at long gas lines, offers of sex for gas, hijacking of fuel trucks, and people cutting into long gas station lines stretching for several blocks are some of the examples that illustrate the stress created by the shortage; these stories don't reveal the low-grade anxiety of the majority of Americans who managed to behave themselves.

State governments and the federal government were largely ineffectual in resolving this crisis. But there was not much that could be done given such a rapid cut-off and the degree of American dependency. Limited efforts to fix prices and ration gas failed. The creation of a national speed limit of 55 miles per hour did significantly reduce gas usage, but in later years when Americans forgot the stress of the crisis, many interpreted the speed limit as classic government overreach and lobbied for its repeal. For a short time Montana removed speed limits on interstates (then reinstituted them when there were too many accidents) and the speed limit on Interstate-10 in West Texas is 85 miles per hour as of 2014. Environmentalists thought this event might provide impetus for increased energy conservation, and there were efforts to create car pools and cooperatives. Many Americans began using the bike as a transportation tool and ridership on public transit increased in urban areas. But driving habits did not change and neither did the American dependence on petroleum. As prices stabilized and even after a second energy crisis of 1979, there would be no radical rethinking of oil dependence. Our reliance has only deepened in the decades since.

High Water Mark

The Endangered Species Act became law in 1973, the same year that the National Land Use Policy Act almost passed. In retrospect this represented the high-water mark of the environmental movement's passage of laws and regulations and creation of agencies. The defeat of the land-use bill meant the end of the effort to institutionalize the land ethic. There were continuing state reforms, but comprehensive land-use reform came to a shuddering halt. An economic downturn, the oil crisis with corresponding increases in gas prices, and a severe constriction in home sales and construction in 1973 and 1974 helped turn opinion against this bill and increasingly against environmental reforms because of the costs associated with them. Clean water and air legislation passed with little resistance because they targeted industry and required a reduction in pollution, a relatively straightforward action. Land-use reform was infinitely more complex and involved a much wider swath of American society. Homeowners, construction workers, farmers, and many others would be affected in numerous ways economically, not just corporations. This high water mark of the environmental movement also generated the key concern and rhetorical strategy of environmental opponents ever since; simply, the threat to property rights and economic growth.

The first major test for the Endangered Species Act occurred when controversy erupted over a small, 3-inch fish called the snail darter. It was listed as endangered and stopped the construction of the Tellico Dam on the Little Tennessee River of Tennessee. Media coverage and public debate focused on the "ridiculousness" of such a small, non-appealing fish halting economic development and costing jobs. This conflict reached the U.S. Supreme Court, and in *Tennessee Valley Authority v. Hill* in 1978 the Supreme Court found in favor of the Endangered Species Act and protecting the snail darter. However, where economic concerns arose, the land ethic ended, and the limits of the Endangered Species Act were reached. The first revisions to the law came as a result of this standoff, with the creation of a committee of seven high-ranking federal officials (unofficially known as the "God Squad") authorized to grant exemptions to the law. Ironically, they unanimously voted to preserve the fish over dam construction, so Tennessee congressmen engaged in sneaky political maneuvering to get the bill through Congress and onto President Jimmy Carter's desk. Under intense public scrutiny and pressure, he did not veto the bill, and the dam was completed in 1979. As it turned out, resident populations of the fish were found in other streams, so the species has not been extirpated.

Toxic Bodies

Even with the new regulatory framework and increased ecological knowledge of Americans, the long history of sloppy and thoughtless disposal of wastes meant that the American landscape was littered with thousands of sites filled with toxic poisons; many of them forgotten, some just ignored. The Love Canal crisis brought this issue to the forefront of the American consciousness. Hooker Chemical had used the never completed Love Canal to dump a variety of dangerous industrial byproducts such as solvents, chlorinated hydrocarbons, caustics, fatty acids, alkalis, and other toxic materials. Filling the canal and covering the waste with a layer of dirt, the company then sold the property to the local school board for one dollar. While the sale contract did include information about the buried materials, this was either forgotten or ignored. Above this unregulated, untreated, and unmanaged stew of chemical manufacturing waste arose a bustling suburban community of affordable homes, lawns, and parks; directly over the toxins a new school was built. In the late 1970s, illness spread among the Love Canal residents as poisons leaked from deteriorating metal barrels and entered the water supply, soils, and their bodies. A 1979 EPA study found elevated white blood cell counts among residents, and 33 percent of residents showed chromosomal damage. Other studies were inconclusive with some not showing

Figure 12.4 The Love Canal neighborhood. Following the removal of residents from the polluted site, the EPA and Hooker Chemicals spent more than $400 million remediating the area to make it habitable. Residents moving back into the area for the inexpensive home prices after 2004 have complained of health problems. Love Canal Images, University Archives, University at Buffalo, The State University of New York.

health impacts and carcinogenic affects, while others did. Residents did note a number of birth defects such as cleft palate, a child born with an extra row of teeth, and cognitive development issues.

An EPA administrator described a visit to the site:

> I visited the canal at that time. Corroding waste-disposal drums could be seen breaking up through the grounds of backyards. Trees and gardens were turning black and dying. One entire swimming pool had been popped up from its foundation, afloat now on a small sea of chemicals. Puddles of noxious substances were pointed out to me by the residents. Some of these puddles were in their yards, some were in their basements, others yet were on the school grounds. Everywhere the air had a faint, choking smell. Children returned from play with burns on their hands and faces.[15]

As community members experienced and learned of various health problems, their concerns mounted. Housewife Lois Gibbs tried to transfer her son from the polluted school site to a safer one. Denied, she and other women from the neighborhood organized a high-profile protest campaign employing marches, rallies, letter-writing, and other tactics. They quickly gained national attention as they appeared on the Donahue show and, frustrated at the slow federal and state responses, burned government officials in effigy. In desperation they even took two EPA officials hostage for a few hours.

President Carter finally stepped in, using federal emergency funds for the first time in U.S. history to deal with a non-natural disaster. Drainage lines were built to move toxins away from homes, and eventually, more than 900 families were relocated away from the site. It cost $27 million to evacuate and purchase the polluted properties. Americans' awareness of the threat of dumped toxics was reinforced with the discovery of 100,000 barrels leaking hazardous waste in West Point, Kentucky. Love Canal propelled Congress to pass the Comprehensive Environmental Response, Compensation, and Liability Act (CERCLA), otherwise known as the Superfund Act. Managed by the EPA, this legislation created a means for distributing funds and applying expertise to cleaning and remediating the very worst polluted sites in America. Because so many companies had dumped toxic wastes and moved or gone out of business, the U.S. government was forced to pick up the costs of cleanup. An EPA report produced in 1979 estimated the existence of between 32,000 to 50,000 hazardous waste sites across the nation, with approximately 2,000 of them posing serious health risks. Lois Gibbs went on to found the Citizens Clearinghouse for Hazardous Wastes in 1981. By 1986, 1,300 groups were part of this network, and the number grew to 7,000 by 1990. Their staunch defense of homes and communities has compelled governments to more carefully plan, educate, and protect populations when creating new landfill and toxic waste sites. While women were not the only ones involved in these efforts, they played a prominent role, continuing a trend from the urban activism of the Progressive Era.

Carter and Environment

The administration of President Jimmy Carter, from 1976 to 1980, did not break radical new ground on environmental issues and struck a centrist tone. He had inherited an economic mess with unemployment at 11 percent, inflation around 12 percent, and high gas prices from the first oil embargo and crisis of 1973. The second oil embargo and energy crisis of 1979 contributed to a growing backlash against environmentalism as the price of gas tripled to a dollar a gallon. The Carter administration sought to implement reforms to reduce fossil fuel energy consumption and dependency, but the proposed reforms, such as mileage standards for automobiles, generally failed in Congress. His request in a national address that Americans change their lifestyles, reduce consumption, and turn down their heaters to reduce fuel use was generally met with derision and marks the last time an American president asked this nation's citizens to change their consumer habits and make sacrifices for the common good.

The impacts of the oil embargo and the resulting economic doldrums convinced President Carter to provide government support for the development of alternative energy sources. However, Carter lent more support for Nuclear power over solar; frustrated environmentalists turned to Congress for increased funding for solar power in the 1979 Energy Act. His support for and signing of the Surface Mining Control and Reclamation Act of 1977, and his shepherding through a stronger Clean Air Act to expand the EPA's role in managing air-pollution problems arising from the coal industry, were positive contributions to air and water pollution issues. Moreover, the EPA began hosting brown bag lunches with environmental groups in order to inform them of actions and policies, gain their input, and build constituency support for the agency. President Carter did have 32 solar panels installed on the White House roof to heat water. Seven years later, in 1986, President Reagan had them removed, reflecting his gutting of funding for solar power and his elimination of tax breaks for wind turbine power.

Earth Day, ongoing awareness of environmental pollution issues, and positive actions on the part of government to reform pollution and waste disposal practices contributed to a still growing environmental consensus through the 1970s, even as schism was beginning. Environmental organizations were front and center in the efforts to expand government's role in managing environmental issues. The five top environmental organizations grew from a membership base of 841,000 in 1970 to almost 1.5 million a decade later, reflecting more than 76 percent growth. The groups emphasizing pollution issues grew much more quickly than those still focused on wilderness and preservationist agendas. At the same time the make-up of these groups was shifting. The increased use of legal tactics meant the hiring of more lawyers and the commitment of more resources to protracted and expensive lawsuits. Also, the increasing complexity of environmental problems, particularly dealing with toxins and in negotiating the regulatory system and various agencies, meant that environmental groups had to dedicate more of their finances to technical expertise. With these changes, increased money because of growing membership rolls, and their victories of the previous two decades, environmental groups oriented increasingly toward Washington, D.C. They became powerful players in that political arena, providing expert testimony, lobbying congressmen, and helping to write bills providing environmental protections but also including compromises with business interests. They were in a strong position to influence legislation and environmental agencies, but to many activists they were losing touch with their base and showing too great a willingness to compromise.

Backlash

The fragmentation of the environmental consensus really gained pace under the sustained efforts of President Ronald Reagan and members of his administration. Elected in 1980, promising to bring "morning" to America and claiming to represent a silent moral majority,[16] the Reagan Administration sought to roll back the environmental laws, regulations, and agencies created in the previous two decades. One of the leaders of the attack against environmentalism was newly appointed Secretary of the Interior James Watt. He came from a background of challenging environmental regulations as director of the Mountain States Legal Foundation, an organization created by business interests to use the courts to protect business from environmental laws and regulations.

Watt struck quickly on several fronts. He sought to cease the expansion of national parks because he believed that the park system was too large already and unfairly blocked economic use of land and resources. The Interior Secretary also wanted to privatize some park administrative work and use Land and Water Conservation Fund monies (designated for land acquisition) for park upkeep rather than continuing the funding of that work from revenue and appropriations; this effort was stopped by Congress. Watt and Reagan both advocated selling off public lands in the West to pay against the ever-increasing deficits of the Reagan administration and to build support among western constituencies, particularly the "wise-use"[17] movement and "sage-brush rebellion" of this era. Watt's courting of these radical movements generated strong criticism and contributed to his eventual removal from that position.

John Crowell had been a lawyer for the Louisiana Pacific Corporation, a logging company. When President Reagan made him Assistant Secretary of Agriculture and for the Environment, Crowell issued instructions to double timber company cuts on federal lands. His increased harvest goals were pushed down to forest managers forced to dramatically increase sales to meet the high numbers. Other federal agencies took their cue from government-hostile and business-friendly leadership. The EPA was slow in cleaning hazardous waste sites, and the Office of Surface Mining did not sufficiently enforce the Surface Mining Control & Reclamation Act of 1977. As Samuel Hays notes, "on any number of occasions the resolution of issues in the EPA involved only agency and industry personnel; those who represented the viewpoints of affected citizens or public-health advocates were not represented."[18] This was deliberate and in direct contradiction to practices during the Carter administration. With increasing evidence that lackadaisical enforcement was tied to the desires of business, environmental organizations and individuals filed hundreds of lawsuits to compel agency compliance with their mandates and the law.

The Reagan administration's management of environmental issues was deeply cynical. Knowing that regulation and enforcement was predicated on scientific evidence and analysis, the administration replaced scientific managers with those who embraced skepticism and supported business interests. They also sought to create a nearly insurmountable evidence barrier to regulation, requiring a much higher threshold of evidence for policy changes. This strategy would bear fruit for Republicans in their later opposition to climate change legislation or agreements. Mirroring their efforts against social welfare programs, the Reagan administration worked to weaken environmental enforcement and regulation by slashing agency funding. Under Reagan, Anne Gorusch Bedford set about weakening the EPA, reducing the budget by $200 million and staff by 23 percent. She restricted the role of experts in determining policy and running the agency; veterans left in droves. As a result, the agency became demoralized and noticeably less effective.

The democratization that accompanied environmental reforms was also undermined by this administration. Many of the public information and increased access victories of the 1960s and 1970s were rolled back as the administration moved a good deal of environmental policy making into the Office of Management and Budget. Presidents Nixon and Carter had shifted the office from its primary job of producing budgets to one that centralized and expanded executive power by reducing congressional, court, and public oversight. Public information programs were cut or reduced, and the Reagan administration supported industry efforts to conceal information about chemical products to protect their trade interests. Therefore, there was a general, sharp cutback in citizen involvement in agency decision-making in the Reagan years.

Although the Reagan administration was not particularly effective at dismantling government or severely reducing the operations of environmental agencies within government, it accomplished two things. Their use of demonizing rhetoric against environmentalists as anti-growth and anti-jobs effectively pushed the movement left in Americans' minds as increasing numbers of Republicans grew uninterested in environmental activism. The same rhetoric and threats against environmental organizations and agencies revitalized environmental organization as activists embraced more radical measures: membership numbers spiked as Americans worried about the removal of key features of American life that protected nature and made the nation more democratic. In contradiction to the environmental consensus built in the 1950s and 1960s, a cultural bifurcation was occurring in the 1980s with environmentalism depicted as a democratic, left political issue and Republicans abandoning environment as inconsistent with their political agenda.

Figure 12.5 Clearcuts of private and public land were common in the West into the 1980s when new land policies were put into place to protect stream banks and leave patches of forest intact. This photo of a clearcut along the Satsop River of Washington State in 1973 shows the immediate damage. It does not take much imagination to predict the following deterioration. Habitat for wildlife was lost, but with the steep slopes in that drainage, the fall and winter rains would cause heavy erosion, soil loss, and deposition of dirt in streams. It was just such common practices like these that environmentalists opposed in the 1970s and again during the Reagan Administration. Environmental Protection Agency.

Environmental groups gained increased membership because of the attacks by the administration. Radical groups such as Earth First!, Greenpeace, and Sea Shepherd grew in popularity, contributing to the perceived radicalism of the overall environmental movement by Americans. Earth First! enacted a campaign of political and direct action based on the idea of nature having its own inherent rights requiring protection. A central leader in the effort, Dave Foreman, grew up in a traditional fashion earning his Eagle badge in the Boy Scouts, serving in the Marine Corps, and even campaigning at one point for conservative Republican Barry Goldwater. He worked as a lobbyist for the Wilderness Society but grew disillusioned at what he perceived as the failures of the modern environmental movement. In the first publications of the *Earth First! Journal* he declared the equality of nature with man and the need for activists to come to the natural world's defense, turning to the activism and courage of Civil Rights activists for both inspiration and strategy. The early publications also invoked Henry David Thoreau's assertion that unethical laws must be defied by acts of civil disobedience. From the perspective of Foreman and many others that joined Earth First! or acted on their own, the environmental organizations were too timid, having grown entirely too corporate and fond of compromise. The group grew to 75 chapters in 24 states by 1990. Since there is no official membership in Earth First!, it is impossible to know the total number of participants in this group. Adding to the complexity of assessing their popularity and impact is the fact that many who sympathize with the group's philosophy and program do not attend public events in order to preserve their anonymity and protect themselves from prosecution for acts of eco-sabotage.

Critics of Earth First! and later groups such as the Earth Liberation Front have been quick to label them as terrorist, but the primary strategy of these activists has been and is to inflict property damage or block development or resource extraction in a manner that renders project costs prohibitive and unprofitable. As *Desert Solitaire* contributed to an increasing passion for wilderness, Edward Abbey's *The Monkeywrench Gang* helped drive the rise of this movement as Foreman and others used the novel and its manifesto as a guide, while the novel popularized the idea of eco-sabotage.

Earth First!ers also articulated a deep ecological vision for protecting nature for its own sake. In a 1983 "wilderness preserve system" proposal the group argued that 50 preserves of 716 million acres of healthy ecosystems be made off-limits to "industrial human civilization as preserves for the free-flow of natural processes."[19] They believed that large, complex swaths of wilderness were necessary to protect a wide diversity of habitats and species. In this proposal one can see the culmination of the wilderness ideas initiated by Leopold and Marshall early in the century as well as a logical conclusion of the research showing the need for large functional ecosystems for maximum viability of diverse species. Regardless, the idea of setting this much land off limits from logging, mining, suburban sprawl, hunting, four-wheeling, and snowmobiling was far beyond what the vast majority of Americans were or are ready for.

Bold proposals aside, Earth First!ers had their hands full defending what was supposed to be protected. A fight over a healthy but poorly protected wilderness in southwestern Oregon raised the ante between the activists, economic interests, and the American judicial system. Mike Roselle, co-founder of Earth First!, and three others blockaded a road into the North Kalmiopsis area in April 1983 because they believed the road and ensuing cuts violated Forest Service regulations. The activists hoped to delay logging long enough for a court injunction to stop the cut. The stand-off grew increasingly confrontational as more protestors joined the resistance, and the logging company could not get to its timber, loggers could not work and pay their bills, and the Forest Service was unable to provide its promised services to the logging economy and local community. Dave Foreman was intentionally hit and drug by a truck he was trying to block, suffering permanent knee damage. While the driver avoided arrest, 24 Earth First!ers did not. A court injunction was finally granted, but the 1984 Oregon Wilderness Act opened the area to logging anyway.

In 1987 George Alexander, a sawmill worker, was maimed by a blade striking a nail in a tree. He worked for Louisiana Pacific Lumber, one of the companies aggressively clear-cutting old growth forest from public lands under the guidance of Crowell. While the company and press condemned Earth First!ers, there is strong evidence against the tree being spiked by a member of that group. As it turned out, a local survivalist spiked trees along the border of his property because he believed that the logging companies would consistently cut across property boundaries into private held land. He was being watched by the FBI, and had made public announcements that trees on his property were spiked in order to prevent their cutting and just such an accident. Earth First! noted that an essential part of tree-spiking strategy was to inform the logging companies where trees had been spiked and to flag the areas. Therefore, this could not be the result of their eco-sabotage. While well-meaning, the idea that Earth First! could trust corporate leadership and management to warn their workers of potential danger flies directly in the face of their own rhetoric asserting deep irresponsibility and greed of those very companies. In the end, used as a martyr for the lumber industry, George Alexander grew increasingly angry against the company. He disagreed with their clear-cutting of forests, and when he filed a damage claim against Louisiana Pacific for his injuries, they fought him in court; Alexander finally received only $9,000 and an involuntary transfer to the night shift.

In the 1980s Pacific Lumber Company was taken over by Maxxam, a Texas-based corporation. Pacific Lumber had been a northern California operation for decades and practiced sustained cut harvests. The debt accrued by Maxxam in its takeover of the company necessitated a dramatic increase in the pace of cutting down old-growth, redwood forests. Judi Bari, a union organizer by trade, was infuriated by the impact on human lives as well as on the ancient groves. Her anger and activism propelled Bari into a leadership position within Earth First!. Forced overtime undercut family health and meant more accidents and injuries while working already dangerous jobs. What hurt forests also damaged logging communities in Northern California because when the trees were liquidated by the Texas conglomerate, the logging jobs would disappear. This is an old, oft-repeated story in the West but Earth First! and some union activists fought to prevent the assumed, foregone conclusion. Bari worked hard to unite activists and lumber union members together in opposition to the destruction of the redwood forest. The creation of the Industrial Workers of the World (IWW)-Earth First! Local 1 signaled a unique relationship and potentially dramatic step forward in the relations of resource extraction workers and radical environmentalists. Two years of sustained organizing built local support, and at a critical juncture, Bari and other Earth First!ers renounced tree-spiking. Other chapters of Earth First! in northern California and Oregon also renounced this strategy in an effort to support this significant moment of coalition building.

The ongoing destruction of the redwood forests prompted Bari to organize a "Redwood Summer" for 1990. Based on the Mississippi Summer of 1964, when Civil Rights activists used voter organizing and protests to demand civil liberties and desegregation in the Magnolia State, Bari and other organizers hoped to similarly use protest and civil disobedience to end the cut. In the midst of planning the summer actions, a pipe bomb exploded in Bari's car as she and another activist, Darryl Cherney, drove to an organizing event in Santa Cruz, California. Placed under Bari's seat, the pipe bomb drove fragments into her lower abdomen and injured Cherney's left eye. Oakland police and the FBI immediately arrested them on suspicion of making the bomb. Making no effort to pursue or even investigate other possible suspects, while proffering flimsy evidence that was easily shown suspect, the law enforcement agencies publically stated that Bari and Cherney had made the bomb. Environmental groups rose to their defense and demanded that the California House and Senate judiciary committees, as well as the state attorney general, examine the bombing investigations. Charges were quickly dropped, but no other suspects were pursued.

The spotted owl debate inflamed the timber wars even further. In 1988 biologists discovered that the small and elusive spotted owl required large acreages of old growth forest to reproduce. Environmentalists sought to protect the bird but also viewed its habitat protection as a means to preserving the wide stretches of old growth forest aggressively harvested under the Reagan Administration. The U.S. Fish and Wildlife Service originally refused to list the species as endangered in 1988 but, under increased pressure from environmental groups, reversed course, listing it in 1991. Accordingly, the federal government instituted stricter limits on harvests on national forest lands. Lumber firms seized upon the restrictions to lay the blame for economic malaise in logging communities at the feet of the government and environmentalists. The rhetoric grew red hot. "Save a logger, kill a spotted owl" stated one common bumper sticker. One northern Olympic Peninsula restaurant offered "spotted owl burgers" to travelers. While a sound and fury was targeted at the owls and their advocates, much of the economic crisis was in fact created by logging company practices and Crowell's increased cuts, which pushed a glut of lumber into the economy. Also, automation in mills reduced the need for employees. Even more important is the fact that the majority of Northwest lumber was shipped to Japan with no milling or preparation for the market, costing even more American jobs.

In 1989 and 1990 the FBI arrested several Earth First! members in Arizona and Montana. In Arizona there had been a number of cut power lines from a nuclear power plant, power line poles were knocked over, and chairlifts at a ski area built on Native American sacred land had been vandalized. Allegations of tree spiking in the Clearwater National Forest of Idaho triggered the Montana arrests. The FBI used a number of tools such as agent infiltration, phone tapping, and bugging of homes to gather information and try to break up the group. Internal stresses over tree spiking, political debates within the organization, and the increased media spotlight brought pressure on the group and weakened it for a time.

Moderation and New Crises

Following the end of President Reagan's term, his vice-president, George H. W. Bush, won the election to the presidency in 1988. While running for office, he billed himself as an "environmental president" in an implicit acknowledgement that the anti-environmental rhetoric and policies of his boss had inflicted damage on the Republican Party. While generally in agreement with the party's position that environmentalism limited economic growth, his administration was less ideologically aggressive than Reagan's, and some environmental successes were achieved during his four years in office. His mixed record also reveals the growing tension between economic growth and environmental issues. Unlike Reagan, Bush appointed a noted environmentalist to run the EPA. He helped move the 1990 revisions to the Clean Air Act out of Congress, although they were greatly weakened in the process. His administration and a democratically controlled Congress also cooperated to fund cleanup of nuclear power plants and weapons facilities. President Bush also continued to fund cleanup of toxic military sites, costs which will likely eventually reach at least $400 billion. With more than 18,000 such sites across the nation, this is a persistent and ongoing environmental challenge. Finally, he also signed an unprecedented environmental law, the Elwha Restoration Act, which called for the removal of two dams on the Elwha River in Washington State if required for restoration of the river and the salmon fisheries. A unanimous, bipartisan congressional delegation from Washington State sponsored the legislation. Given the change in rhetoric and tone on environmental issues in the 21st century, this seems a remarkable moment in recent history.

Americans could see environmental progress even as new crises emerged. Two decades after the passage of the Clean Air Act of 1970, less than 10 million tons of emissions were released into the air compared to 25 million tons in 1969. The campaign to remove lead from gas, paint, and other products resulted in a reduction of lead emissions from more than 200,000 tons a year to less than 10,000. But incidents such as the *Exxon Valdez* oil spill in 1989 served as a stark reminder of how quickly the environment could suffer severe damage as a result of human carelessness or mistakes. Run by an exhausted crew working long shifts, and with ship officers making critical navigational errors, the single-hull tanker *Exxon Valdez* struck a reef in Prince William Sound in Alaska. A pristine and rich marine environment of herring, salmon, orca, bald eagles, seals, and sea otters, the sound was devastated by the flow of up to 32 million gallons of oil, the largest to that time in U.S. history and not eclipsed until the Deepwater Horizon spill of 2010. The remote location and limited local resources made response difficult, and the oil quickly spread, eventually coating 1,300 miles of beach and covering 11,000 square miles of ocean. While estimates of animal deaths are difficult, approximately 1000,000–250,000 seabirds (puffins, murres, and others), close to 3,000 sea otters, 22 orca, and hundreds of bald eagles were killed by the oil spill. Salmon, herring, and other fishery populations were heavily damaged as well, and the herring have not yet rebounded. Criticism settled on Exxon Mobil for its use of single-hull tankers when double-hull tankers were available, and for the reliance on an overworked crew and the navigation mistakes. Extensive legal action has provided money for remediation and to offset losses to the local fishing and tourist economy, albeit not without some problems. A 1990 law was quickly passed requiring oil companies to use double-hull tankers, but a phase-in schedule has allowed the companies to continue to use single-hull tankers; as of 2013, Exxon Mobil and others still regularly use these more vulnerable tankers.

Americans were also reminded that human health could be threatened by environmental deterioration, when, in 1991, an EPA report provided incontrovertible evidence that the ozone layer was being depleted by chlorofluorocarbons (CFCs) at a much higher rate than originally predicted in 1986. CFCs were used in air conditioning, as refrigerator refrigerants, and as a propellant in hair spray, furniture polish, deodorizers, and other products. CFCs and other ozone-depleting substances chemically altered the ozone layer in the

Figure 12.6 The *Exxon Valdez* oil spill in Prince William Sound in 1989 was ecologically and economically devastating. The man in this picture is holding a dead sea otter. It is believed that approximately 3,000 sea otters and close to a quarter million birds died in this spill, which could have been prevented with the use of double-hulled tankers. The impacts of the spill still linger. Herring populations have not recovered, and it is still easy to find oil under rocks on beaches. U.S. Fish and Wildlife Service.

stratosphere, and the result was a dramatic thinning of the ozone layer on an annual basis and the yearly appearance and expansion of ozone holes above the Antarctic, the Arctic, and northern North America. With this thinning and disappearance of the ozone layers, more harmful ultraviolet light was able to penetrate to the earth's surface, causing increased skin cancer, cataracts, mutations to frogs and plants, a reduction in plankton populations, and even skin damage to some whale populations. The potential impact was the possibility of ending life as we know it, if the process was not stopped and reversed. Fortunately, politicians acted relatively quickly to end this crisis. The Montreal Protocol, signed in 1987 by the United States and 23 other countries, scheduled the end of CFCs by 1999. With the unequivocal evidence of the disappearing stratospheric protective layer, President Bush agreed with and supported efforts to move that deadline up to 1996.

Environmental Injustice

Even when the nation shared a strong environmental consensus, the overwhelmingly white, middle-class and upper-class nature of its membership meant that impoverished, ethnic communities were neglected or ignored by environmental organizations. Because of their lack of representation and economic and political power as well as persistent racism, it was often those communities where garbage sites were automatically located and toxins dumped. White flight from middle-class communities that African-Americans moved into and the rise of the petro-chemical and other chemical industries worsened this trend. In 2000 the University of Texas at Dallas and the *Dallas Morning News* conducted a study that found almost half of the housing for the poor of Dallas, most of whom are minorities, sit within a mile of toxin-emitting factories. Many sites are also rural, located close to petro-chemical facilities like in Diamond, Louisiana, or close to military bases, some of the most polluted landscapes in America.

Chicago was one of the major cities that saw significant white flight. Government housing projects worsened the plight of impoverished African-Americans remaining in the city. Altgerd Gardens was opened by the Chicago Housing Authority in 1945. Built to house low-wage African-American residents and war veterans, this public housing sustained the color line in the Windy City as the metropolis declined and more prosperous residents migrated to the suburbs. In addition to numerous other problems besetting this housing facility, in the early 1980s it became clear that chemical pollutants were sickening and killing residents of Altgerd. Following the deaths of four little girls to cancer, community leader Hazel Johnson was determined to find answers. Seeing linkages between the deaths and a spate of illnesses, she organized a community campaign that flooded the Illinois Environmental Protection Agency with thousands of complaints. The resulting study revealed the extent to which Altgerd was dangerously polluted. The site had originally been used for sewage, then PCBs had been illegally dumped there. The project was also encircled by at least 50 landfills and more than 250 leaking storage tanks. Johnson and other community activists demanded further investigation. The results were shocking. Half of the pregnancies examined in the study ended in stillbirths, miscarriages, birth defects, or premature births. It also revealed that a large number of residents suffered from a variety of maladies; most of those who moved out immediately recovered.

But most who lived there were either trapped by poverty or loved their community and refused to be forced out by industrial pollution. They fought for and won safe sewer and water lines; Johnson, because of her key role, was given the title of "Mother of the Environmental Justice Movement" at the First National People of Color Environmental Justice Summit in 1991.

The only major American city without zoning laws, Houston, experienced robust economic growth during World War II, and cyclically after that, with an economy primarily based on the petro-chemical industry. It is also one of the most segregated cities in the nation. African-Americans have represented from 21 to 28 percent of the population from

1960 to 2000. In 1980, 82 percent of this group lived in city blocks that were predominantly black. The lack of zoning alongside persistent community segregation made it easy for white city officials to disproportionately locate dumps, landfills, and incinerators in multiple black neighborhoods and one Hispanic neighborhood. In addition to polluting their community and reducing their quality of life, such practices made middle-class and poor neighborhoods even poorer. As sociologist Robert Bullard notes, "Five decades of this discriminatory practice lowered residents' property values, accelerated the physical deterioration of Houston's black neighborhoods, and increased disinvestment in these neighborhoods."[20] He also adds that "stigmatizing" black communities in such a manner led to even more dumping and operations such as chop shops, salvage yards, and recycling operations that undercut middle-class life.

Black Houstonians frequently opposed the siting of garbage waste facilities in their neighborhoods long before the events in North Carolina that are typically seen as launching the environmental justice movement. One site was protested in 1967 after a young girl drowned there. Anger from that protest helped fuel the Civil Rights riot at Texas Southern University soon afterwards. Demands to shut down a dump in the Trinity Gardens neighborhood in 1971 almost turned ugly, and six months of protest and demonstration led to the closure of that site. An effort in the late 1970s to block the placement of yet another landfill in a black neighborhood, in this case the predominantly middle-class Northwood Manor community, led to an important lawsuit, *Bean v. Southwestern Waste Management Corp.* It was the first case to use the Civil Rights Act to charge discrimination in landfill siting on the basis of race. Bullard was hired to conduct research and interviews for the prosecution and argues the importance of this case in the development of the environmental justice movement.

> Significantly, the 1979 Houston case developed and tested a new methodology and legal theory of environmental discrimination without the benefit of any regional or national studies on the subject. The case predated some important landmark studies and events. These include the 1983 study of off-site commercial hazardous waste landfills in the South by the U.S. General Accounting Office; the 1987 release of the report *Toxic Wastes and Race,* authored by the Commission for Racial Justice; the letters written in 1990 by the Gulf Coast Tenants Organization and SouthWest Organizing Project to the "Big Ten" environmental groups accusing them of elitism and racism.[21]

The 1991 First National People of Color Leadership Environmental Leadership Summit and President Clinton's signing of Executive Order 12898 were at least partially mobilized by events in and documents stemming from the Chicago pollution and Houston cases. Although they lost the Houston case in court, it helped to mobilize activists against garbage siting in black neighborhoods and provided education on the discrimination inherent in creating those sites.

In the summer of 1978 the Ward Transformer Company of Raleigh, North Carolina, hired men to illegally dispose of large amounts of PCB-contaminated liquids. For two weeks, along the shoulders of 240 miles of country roads winding through 13 rural counties, they drove at night, surreptitiously dumping the poison in violation of the Toxic

Substance Control Act. The company owner sought to avoid the costs of complying with EPA regulations. The state was forced to remediate this pollution and decided to transfer the contaminated soil to a landfill to be built in Warren County, North Carolina, a poor, rural area populated primarily by poor blacks. White and black activists joined together in 1982, incorporating recently employed Civil Rights strategies to use non-violent resistance, vocal and visible rallies with the capture of media attention, and civil disobedience to stop this potential threat to their own health. They brought lawsuits against the state, sought out other sites, built national support, and employed a direct action campaign, even laying down on the road at one point to block the delivery of contaminated soil to the landfill. There were more than 550 arrests during the protest, and the nation was riveted by the events in Warren County. Even though they failed to stop the polluting of their community, their movement protest did manage to launch the environmental justice movement. The protests led the Congressional Black Caucus to request that the U.S. General Accounting Office study hazardous waste siting and race, and the United Church of Christ Commission for Racial Justice was compelled by events to research and publish its important *Toxic Wastes and Race* report in 1987. Because of the series of toxic pollution and garbage dumping events in poor, predominantly black communities, Americans began to comprehend the racial overtones of dumping, and environmentalists started to address the issue as community activists not normally interested in environmental activism demanded greater protections for their families and homes.

Greater knowledge of race and dumping of waste, along with increased environmental justice activism in the years after Warren County, helped prevent another such environmental injustice in Louisiana. In an egregious case that starkly illuminates how race and poverty intersected to increase vulnerability to discriminatory pollution, a site selection process by the U.S. Nuclear Regulatory Commission and Louisiana Energy Services (LES) honed in on their ideal location for what was planned as the first privately owned uranium enrichment facility in the nation. They progressively narrowed the sites down based on race. According to Bullard,

> the aggregate average percentage of blacks in the population within a one-mile radius of the seventy-eight sites examined (in sixteen parishes) is 28.35 percent. When LES completed its initial site cuts and reduced the list to thirty-seven sites within nine parishes, the aggregate percentage of blacks in the population rose to 36.78 percent. When LES then further limited its focus to six sites in Clairborne Parish, the aggregate average rose again, to 64.74 percent. The final site selected, the LeSage site, has a 97.10 percentage of blacks in the population within a one-mile radius.[22]

This community earned a per capita average of $5,800 a year, less than half the national average at the time, and the two black communities dating back to the end of the Civil War were left unmentioned in the draft environmental impact statement. Like the native peoples of Yellowstone National Park, these impoverished rural blacks were literally written out of the historical record. But this protest accomplished what the Warren County activists could not. Local activists were able to organize and effectively used Executive Order 12898, an order signed by President Bill Clinton in 1994 requiring federal agencies consider environmental justice issues in planning, and the National Environmental Policy Act to defeat the planned facility.

The two decades of previous environmental justice activism with the national legislation passed under President Clinton increased the power of the federal government to protect these vulnerable populations in some cases.

Military posts nationwide and overseas suffer from such extensive chemical, petroleum, and biological weapons pollution that many of them are de facto toxic waste cleanup sites. Much of this is a legacy of the Cold War and ongoing conflicts, the military industrial complex, and the overreliance on chemical and biological weapons. Moreover, the materials needed to maintain and prepare a military machine so dependent on petroleum, steel, and numerous chemicals have contaminated military sites across the nation and in U.S. territories. Kelly Air Force Base in San Antonio, Texas, was in use from 1916 until it was closed in 2001, and it was the source of dangerous pollution for the surrounding Hispanic community of working-class families. The primary function of the facility was maintenance of weapons systems and airplanes during World War II and the Cold War. Millions of gallons of various chemicals, jet fuels, volatile organic compounds, acids, and lead were handled and disposed of at the base. Included among these were dangerous materials such as dichloroethylene, tetraehloroethylene, polychlorinated biphenyls, vinyl chloride, benzene, and others. The handling of these materials was lackadaisical at best with some of these toxics dumped directly into Leon Creek, pits, and sometimes on open ground. One base employee noted that he was told annually to empty vats of chemicals directly onto the ground over the Christmas holidays when military posts function on skeleton crews and it would escape detection. The occasional hard rains, as much as a dozen inches in a day, and flash flooding that are common in the San Antonio area, meant that the toxic chemicals dumped into pits were occasionally flushed into creeks and sewers and into the streets and lawns of East Kelly.

As a result, the plume of poison spread through the shallow water aquifer that lay under approximately 20,000 houses in this community. Residents were endangered by flooding waters dropping contaminants in their lawns and gardens and by their use of water from shallow wells tapping the poisoned aquifer a few feet below the ground. A test well on one homeowner's property near the base showed the presence of TCE and tertrachloroethylene. These can cause damage to the liver, nervous system, and kidneys. After breaking down, these chemicals also release active carcinogens into the environment. To worsen matters, a redacted Air Force report from 1972 indicated that radioactive waste was dumped on the post as well. Reports of high miscarriage rates, birth defects, anemia, and asthma predominated. There were also a reported 120 incidents of Lou Gehrigs disease in the affected community. One study indicated the existence of liver cancer clusters in East Kelly, and a 2006–2007 study concluded that dozens of cancer cases were the results of toxins in the environment. However, an Air Force commissioned study failed to find a link. As of 2012, multiple sites tested along Leon Creek still exceed state limits for pesticides, heavy metals, semi-volatile organic compounds, and polychlorinated biphenyls. The failure to declare Kelly a superfund site precluded citizens from securing outside experts via EPA grants and technical assistance. In 1999 local activists, the Southwest Workers Union, and the Texas chapter of the Sierra Club petitioned Texas Governor George W. Bush and the EPA for a superfund site designation, but Bush and state officials blocked the designation. As a result, the information on health impacts remains incomplete. By 2012, the Air Force had spent more than $325 million cleaning up the mess. The U.S. government will have spent approximately

\$450 million dollars by the time cleanup is complete, but toxins still remain in the shallow watershed and in the lawns of community homes. Achieving true protection, remediation, and health care has remained difficult for these communities as responsible environmental agencies have grown weaker over time and these issues fail to capture the national interest.

Environmentalism peaked in the 1970s with several key successes during the Nixon administration. The tide began to turn with push back against land-use laws and the rising conservative ascendancy of the 1970s and 1980s. The Reagan Administration launched a determined Republican assault against the achievements of the previous 20 years but ironically triggered an environmental resurgence and the rise of more radical groups such as Earth First! The coarsening of rhetoric and debate, as well as Republican efforts to paint environmentalists as radicals opposed to economic growth, and occupying the American fringe, meant that environmentalism became less a centrist issue and one Americans increasingly associated with the Democratic Party and leftist politics. Mainstream environmentalism came under attack also from people of color living in polluted communities. They demanded protections for their health and received some support from the federal government and environmental groups. More importantly, environmental justice activists organized themselves to protect their communities and families. In so doing, they helped expand the definition and agenda of environmentalism. Even though strong attempts were made to derail much of the environmental accomplishments of the 1960s and 1970s, environmentalists and Democrats were able to fight those efforts and even see continued successes. The ongoing creation of wilderness areas, the protection of barrier islands from development, the listing and protection of the spotted owl, the banning of CFC production, and legislation to remove major dams in the Pacific Northwest to restore salmon indicate the persistence of a shared belief in the need for ongoing and increased environmental protections and reforms.

Document 12.1 Introduction to "The Endangered Species Act," 1973

Over the 20th century the federal government expanded its power over wildlife management through a series of laws. The Lacy Act of 1900 was designed to prevent extinction of birds due primarily to the millinery trade and market hunting. Following the extinction of the passenger pigeon in 1914, Congress passed the Migratory Bird Treaty Act in 1918. This required a federal permit for hunting migratory birds. It was challenged by the state of Missouri, but the Supreme Court found in favor of the federal government. The passage of the Migratory Bird Conservation Act in 1929 and the 1934 Fish and Wildlife Coordination Act further strengthened the federal government's role in, and commitment to, protecting and managing wildlife. The creation of the Bald Eagle Protection Act in 1940 required the federal government to preserve a single species from extinction, the national bird. While *Silent Spring* led to federal action banning chemicals dangerous to wildlife, creation of the National Wilderness Preservation System in 1964 greatly strengthened the federal government's ability to protect wildlife via preservation of wilderness landscapes and habitat. At the same time as this legislation was passed, the Bureau of Sport Fisheries and Wildlife, later known as the U.S. Fish and Wildlife Service, organized a committee of nine biologists who published a list of 63 species likely to go extinct without federal intervention. This was quickly followed by the Endangered Species Preservation Act of 1966. While this was the first comprehensive federal effort to prevent extinctions caused by humans, it was flawed—agency cooperation was voluntary, and the act did not extend protection to plants. The passage of this bill in 1973 sought to correct these weaknesses and provide substantive federal government oversight and management to the protection of species that faced extinction.

1. What are the assumptions about the value of nature that are built into this document? As you read the different statements, what are the origins of these ideas historically, ecologically, and philosophically?
2. What power does the federal government give itself to protect endangered or threatened species? Does this constitute a legitimate expansion of federal power? What justifications can be made for this?

(a) Findings

The Congress finds and declares that—

(1) various species of fish, wildlife, and plants in the United States have been rendered extinct as a consequence of economic growth and development untempered by adequate concern and conservation;

(2) other species of fish, wildlife, and plants have been so depleted in numbers that they are in danger of or threatened with extinction;

(3) these species of fish, wildlife, and plants are of esthetic, ecological, educational, historical, recreational, and scientific value to the Nation and its people;

(4) the United States has pledged itself as a sovereign state in the international community to conserve to the extent practicable the various species of fish or wildlife and plants facing extinction, pursuant to—

   (A) migratory bird treaties with Canada and Mexico;
   (B) the Migratory and Endangered Bird Treaty with Japan;
   (C) the Convention on Nature Protection and Wildlife Preservation in the Western Hemisphere;
   (D) the International Convention for the Northwest Atlantic Fisheries;
   (E) the International Convention for the High Seas Fisheries of the North Pacific Ocean;
   (F) the Convention on International Trade in Endangered Species of Wild Fauna and Flora; and
   (G) other international agreements; and

(5) encouraging the States and other interested parties, through Federal financial assistance and a system of incentives, to develop and maintain conservation programs which meet national and international standards is a key to meeting the Nation's international commitments and to better safeguarding, for the benefit of all citizens, the Nation's heritage in fish, wildlife, and plants.

(b) Purposes

The purposes of this chapter are to provide a means whereby the ecosystems upon which endangered species and threatened species depend may be conserved, to provide a program for the conservation of such endangered species and threatened species, and to take such steps as may be appropriate to achieve the purposes of the treaties and conventions set forth in subsection (a) of this section.

(c) Policy

(1) It is further declared to be the policy of Congress that all Federal departments and agencies shall seek to conserve endangered species and threatened species and shall utilize their authorities in furtherance of the purposes of this chapter.

(2) It is further declared to be the policy of Congress that Federal agencies shall cooperate with State and local agencies to resolve water resource issues in concert with conservation of endangered species . . .

- (1) The Secretary shall develop and implement plans (hereinafter in this subsection referred to as "recovery plans") for the conservation and survival of endangered species and threatened species listed pursuant to this section, unless the finds that such a plan will not promote the conservation of the species. The Secretary, in developing and implementing recovery plans, shall, to the maximum extent practicable—
  - (A) give priority to those endangered species or threatened species, without regard to taxonomic classification, that are most likely to benefit from such plans, particularly those species that are, or may be, in conflict with construction or other development projects or other forms of economic activity;
  - (B) incorporate in each plan—
    - (i) a description of such site-specific management actions as may be necessary to achieve the plan's goal for the conservation and survival of the species;
    - (ii) objective, measurable criteria which, when met, would result in a determination, in accordance with the provisions of this section, that the species be removed from the list; and
    - (iii) estimates of the timer required and the cost to carry out those measures needed to achieve the plan's goal and to achieve intermediate steps toward that goal.

Document 12.2 "Principles of Environmental Justice," The First National People of Color Environmental Leadership Summit, 1991

The environmental justice movement began in the 1970s with efforts to fight chemical and municipal pollution in Texas, North Carolina, and other locations. The central assertion of environmental justice activists is that traditional environmentalism is elitist, driven by the concerns of prosperous, mostly white Americans seeking to protect their own lifestyles and freedom to enjoy nature while ignoring the plight of those suffering from water, air, and industrial pollution. The First National People of Color Environmental Leadership Summit was held in Washington, D.C. on October 24–27, 1991. They drafted the following documents, defining environmental justice through 17 principles.

1. How do the concerns articulated in the following document correspond with or differ from the traditional conservation and environmental issues through the 20th century?
2. To what degree do the issues and points in this document reflect the previous work by the Committee for Nuclear Information, Rachel Carson, ecologists like the Odum brothers, and others? Is it accurate that the environmental movement largely ignore(d) the particular problems of ecology and health threatening impoverished communities?

WE, THE PEOPLE OF COLOR, gathered together at this multinational People of Color Environmental Leadership Summit, to begin to build a national and international movement of all peoples of color to fight the destruction and taking of our lands and communities, do hereby re-establish our spiritual interdependence to the sacredness of our Mother Earth; to respect and celebrate each of our cultures, languages and beliefs about the natural world and our roles in healing ourselves; to ensure environmental justice; to promote economic alternatives which would contribute to the development of environmentally safe livelihoods; and, to secure our political, economic and cultural liberation that has been denied for over 500 years of colonization and oppression, resulting in the poisoning of our communities and land and the genocide of our peoples, do affirm and adopt these Principles of Environmental Justice:

1) **Environmental Justice** affirms the sacredness of Mother Earth, ecological unity and the interdependence of all species, and the right to be free from ecological destruction.
2) **Environmental Justice** demands that public policy be based on mutual respect and justice for all peoples, free from any form of discrimination or bias.
3) **Environmental Justice** mandates the right to ethical, balanced and responsible uses of land and renewable resources in the interest of a sustainable planet for humans and other living things.
4) **Environmental Justice** calls for universal protection from nuclear testing, extraction, production and disposal of toxic/hazardous wastes and poisons and nuclear testing that threaten the fundamental right to clean air, land, water, and food.
5) **Environmental Justice** affirms the fundamental right to political, economic, cultural and environmental self-determination of all peoples.
6) **Environmental Justice** demands the cessation of the production of all toxins, hazardous wastes, and radioactive materials, and that all past and current producers be held strictly accountable to the people for detoxification and the containment at the point of production.

7) **Environmental Justice** demands the right to participate as equal partners at every level of decision-making, including needs assessment, planning, implementation, enforcement and evaluation.
8) **Environmental Justice** affirms the right of all workers to a safe and healthy work environment without being forced to choose between an unsafe livelihood and unemployment. It also affirms the right of those who work at home to be free from environmental hazards.
9) **Environmental Justice** protects the right of victims of environmental injustice to receive full compensation and reparations for damages as well as quality health care.
10) **Environmental Justice** considers governmental acts of environmental injustice a violation of international law, the Universal Declaration of Human Rights, and the United Nations Convention on Genocide.
11) **Environmental Justice** must recognize a special legal and natural relationship of Native Peoples to the U.S. government through treaties, agreements, compacts, and covenants affirming sovereignty and self-determination.
12) **Environmental Justice** affirms the need for urban and rural ecological policies to clean up and rebuild our cities and rural areas in balance with nature, honoring the cultural integrity of all our communities, and provided fair access for all to the full range of resources.
13) **Environmental Justice** calls for the strict enforcement of principles of informed consent, and a halt to the testing of experimental reproductive and medical procedures and vaccinations on people of color.
14) **Environmental Justice** opposes the destructive operations of multi-national corporations.
15) **Environmental Justice** opposes military occupation, repression and exploitation of lands, peoples and cultures, and other life forms.
16) **Environmental Justice** calls for the education of present and future generations which emphasizes social and environmental issues, based on our experience and an appreciation of our diverse cultural perspectives.
17) **Environmental Justice** requires that we, as individuals, make personal and consumer choices to consume as little of Mother Earth's resources and to produce as little waste as possible; and make the conscious decision to challenge and reprioritize our lifestyles to ensure the health of the natural world for present and future generations.

Document 12.3 From *Ecodefense: A Field Guide to Monkeywrenching*, *Dave Foreman*, 1993

Earth First! was created in 1979 by Dave Foreman, Mike Roselle, and others. Foreman, a former marine and environmental lobbyist, felt that America's long history of environmental destruction joined with the major national environmental organization's embrace of cooperation and compromise necessitated a stronger, principled stand employing civil disobedience and acts of property damage in order to protect the land. The creation of Earth First! reflects the concern over pushback against environmentalism and what some saw as the increasing corporatization of environmental groups. Also, the creation of this group indicates the increasing commitment by some environmentalists to the tenets of deep ecology as well as their willingness to embrace radical measures in defense of nature.

1. What are the arguments and justifications provided by Foreman for taking a more aggressive stand against environmental destruction? Is there merit in his arguments? Does he make a convincing case for more radical action?
2. In evaluating this piece, what influences do you see from past environmental thinkers and activists? To what degree is a tradition developing, or does this position mark a radical departure from the trend in environmental thought and action over the course of the 20th century?

In the space of a few generations we have laid waste to paradise. The tall grass prairie has been transformed into a corn factory where wildlife means the exotic pheasant. The short grass prairie is a grid of carefully fenced cow pastures and wheat fields. The passenger pigeon is no more. The last died in the Cincinnati Zoo in 1914. The endless forests of the East are tame woodlots. The only virgin deciduous forest there is in tiny museum pieces of hundreds of acres. Six hundred grizzlies remain and they are going fast. There are only three condors left in the wild and they are scheduled for capture and imprisonment in the Los Angeles Zoo. Except in northern Minnesota and Isle Royale, wolves are known merely as scattered individuals drifting across the Canadian and Mexican borders (a pack has recently formed in Glacier National Park). Four percent of the peerless Redwood Forest remains and the monumental old growth forest cathedrals of Oregon are all but gone . . . Domestic cattle have grazed bare and radically altered the composition of the grassland communities of the West, displacing elk, moose, bighorn sheep and pronghorn and leading to the virtual extermination of grizzly, wolf, cougar, bobcat and other "varmints." Dams choke the rivers and streams of the land.

. . . In January of 1979, the Forest Service announced the results of RARE II: of the 80 million acres of undeveloped lands on the National Forests, only 15 million acres were recommended for protection against logging, road building and other "developments." In the big tree state of Oregon, for example, only 370,000 acres were proposed for Wilderness protection out of 4.5 million acres of roadless, uncut forest lands. Of the areas nationally slated for protection, most were too high, too dry, too cold, too steep to offer much in the way of "resources" . . . Important grizzly habitat in the Northern Rockies was tossed to the oil industry and the loggers. Off-road-vehicle fanatics and the landed gentry of the livestock industry won out in the Southwest and Great Basin.

During the early 1980s, the Forest Service developed its DARN (Development Activities in Roadless Non-Selected) list outlining specific projects in specific roadless areas. The implication of DARN is staggering. It is evidence that the leadership of the

United States Forest Service consciously and deliberately sat down and asked themselves, "How can we keep from being plagued by conservationist and their damned wilderness proposals? How can we insure that we'll never have to do another RARE?" Their solution was simple and brilliant: get rid of the roadless areas. DARN outlines *nine thousand* miles of road, one and a half million acres of timber cuts, seven million acres of oil and gas leases in National Forest RARE II areas by 1987. In most cases, the damaged acreage will be far greater than the acreage stated because roads are designed to split areas in half and timber sales are engineered to take place in the center of roadless areas, thereby devastating the biological integrity of the entire area. The great roadless areas so critical to the maintenance of natural diversity will soon be gone. Species dependent upon old growth and large wild areas will be shoved to the brink of extinction.

. . . It is time for women and men, individually and in small groups, to act heroically and admittedly illegally in defense of the wild, to put a monkeywrench into the gears of the machine destroying natural diversity. This strategic monkeywrenching can be safe, it can be easy, it can be fun, and—most importantly—effective in stopping timber cutting, road building, overgrazing, oil & gas exploration, mining, dam building, powerline construction, off-road-vehicle use, trapping, ski-area development and other forms of destruction of the wilderness, as well as cancerous suburban sprawl.

But it must be strategic, it must be thoughtful, it must be deliberate in order to succeed. Such a campaign of resistance would follow these principles.

Monkeywrenching Is Non-Violent

Monkeywrenching is non-violent resistance to the destruction of natural diversity and wilderness. It is not directed toward human beings or other forms of life. It is aimed at inanimate machines and tools. Care is always taken to minimize any possible threat to other people (and to the monkeywrenchers themselves) . . .

Monkeywrenching Is Individual

Monkeywrenching is done by individuals or very small groups of people who have known each other for years. There is trust and a good working relationship in such groups. The more people involved, the greater are the dangers of infiltration or a loose mouth. Earth defenders avoid working with people they haven't known for a long time, those who can't keep their mouths closed, and those with grandiose or violent ideas (they may be police agents or dangerous crackpots).

Monkeywrenching Is Targeted

Ecodefenders pick their targets. Mindless, erratic vandalism is counterproductive. Monkeywrenchers know that they do not stop a specific logging sale by destroying any piece of logging equipment which they come across. They make sure it belongs to the proper culprit. They ask themselves what is the most vulnerable point of a wilderness-destroying project and strike there. Senseless vandalism leads to loss of popular sympathy.

Monkeywrenching Is Timely

There is a proper time and place for monkeywrenching. There are also times when monkeywrenching may be counterproductive. Monkeywrenchers generally should not act when there is a non-violent civil disobedience action (a blockade, et.) taking place against the proposed project. Monkeywrenching may cloud the issue of direct action and the blockaders could be blamed for the ecotage and be put in danger from the work crew or police. Blockades and monkeywrenching usually do not mix. Monkeywrenching may also not be appropriate when delicate political negotiations are taking place for the protection of a certain area. There are, of course, exceptions to this rule. The Earth warrior always thinks: Will monkeywrenching help or hinder the protection of this place?

Notes

1 J. Brooks Flippen, *Nixon and the Environment* (Albuquerque: University of New Mexico Press, 2000), 25.
2 Quoted in Ibid.
3 Quoted in Kirkpatrick Sale, *The Green Revolution: The American Environmental Movement, 1962–1992* (New York: Hill and Wang, 1993), 19.
4 David Stradling and Richard Stradling, "Perceptions of the Burning River: Deindustrialization and Cleveland's Cuyahoga River," *Environmental History* Vol. 13, No. 3, July, 2008, 515–535. According to the authors: "The first conflagration was likely the one in 1868 when an oil slick caught fire. The Rockefeller refineries were at fault, and the *Plain Dealer* made an impassioned call for a cleanup of the oil waste from the river to protect the city and the lumber yards on the banks from future flames." They describe a fire fifteen years following this (the advice of the papermen clearly being ignored). "During a dramatic late-winter flood in 1883 a spectacular fire raced across the high waters of Kingsbury Run, a creek that ran past the Standard Oil Refinery, before joining the Cuyahoga at the Great Lakes Towing Company boat repair yard, just south of downtown. Leaking oil from a still at the Thurmer and Teagle refinery was ignited by a boiler house standing in the rising water. The *New York Times* described the horror of burning water moving downstream toward Standard Oil's massive refinery. Although the heroic efforts of firemen and employees saved much of the plant, several Standard tanks exploded and buildings burned. Men jumped into the high water to dam up the culvert that separated Kingsbury Run from the Cuyahoga, successfully keeping the fire from the flooded flats along the river."
5 Another large fire three decades later killed three men.
6 Quoted in Benjamin Kline, *First Along the River: A Brief History of the U.S. Environmental Movement* (New York: Rowman & Littlefield, 2007), 79.
7 Quoted in Adam Rome, *The Genius of Earth Day: How a 1970 Teach-In Unexpectedly Made the First Green Generation* (New York: Hill and Wang, 2013), 121.
8 Adam Rome, *The Bulldozer in the Countryside: Suburban Sprawl and the Rise of American Environmentalism* (New York: Cambridge University Press, 2001), 228.
9 Quoted in Ibid., 234.

10 Ibid.
11 Ibid., 240.
12 Ibid.
13 Ibid., 246.
14 Roderick Nash, *The Rights of Nature: A History of Environmental Ethics* (Madison: The University of Wisconsin Press, 1989), 176.
15 Eckardt C. Beck, "The Love Canal Tragedy," *EPA Journal* 1079. Accessed at http://www.epa.gov/history/topics/lovecanal/01.html, 7 July 2013.
16 The phrase "the moral majority" goes back to Evangelical leader Jerry Falwell and Richard Nixon. They claimed that a silent majority of Americans did not agree with the New Leftists of the 1960s and 1970s or with recent changes in American society. The argument further went that they retained traditional Christian values, were conservative, and needed better representation in government. President Ronald Reagan employed this idea rhetorically and supported legislation and policies that would appeal to this group of Americans.
17 The wise-use movement gained prominence in the 1980s and was located primarily in western states. Members of the group argued that federal forests, parks, and monuments "locked-up" land and unfairly blocked use by locals. They sought greater recreational and economic access to federal lands and even advocated selling off some of this property.
18 Samuel Hays, *Beauty, Health, and Permanence: Environmental Politics in the United States, 1955–1985* (New York: Cambridge University Press, 1987), 496.
19 Quoted in Rik Scarz, *Eco-Warriors: Understanding the Radical Environmental Movement* (Walnut Creek, CA: Left Coast Press, 2006), 66.
20 Robert D. Bullard, *The Quest for Environmental Justice: Human Rights and the Politics of Pollution* (San Francisco, CA: Sierra Club Books, 2005), 45.
21 Ibid., 55.
22 Ibid., 9.

Further Reading

Blum, Elizabeth D. *Love Canal Revisited: Race, Class, and Gender in Environmental Activism.* Lawrence: University Press of Kansas, 2008.

Bullard, Robert. *Dumping in Dixie: Race, Class, and Environmental Quality.* Boulder, CO: Westview Press, 2000.

———. *The Quest for Environmental Justice: Human Rights and the Politics of Pollution.* San Francisco, CA: Sierra Club Books, 2005.

Dowie, Mark. *Losing Ground: American Environmentalism at the Close of the Twentieth Century.* Cambridge, MA: The MIT Press, 1996.

Fiege, Mark. *The Republic of Nature: An Environmental History of the United States.* Seattle: University of Washington Press, 2012.

Flippen, J. Brooks. *Nixon and the Environment.* Albuquerque: University of New Mexico Press, 2000.

Hays, Samuel. *Beauty, Health, and Permanence: Environmental Politics in the United States, 1955–1985.* New York: Cambridge University Press, 1987.

Hurley, Andrew. *Environmental Inequalities: Class, Race, and Industrial Pollution in Gary, Indiana, 1945–1980.* Chapel Hill: The University of North Carolina Press, 1995.

Kline, Benjamin. *First Along the River: A Brief History of the U.S. Environmental Movement.* New York: Rowman & Littlefield, 2007.

Lerner, Steve. *Sacrifice Zones: The Front Lines of Toxic Chemical Exposure in the United States.* Cambridge, MA: The MIT Press, 2012.

———. *Diamond: A Struggle for Environmental Justice in Louisiana's Chemical Corridor.* Cambridge, MA: The MIT Press, 2005.

Markowitz, Gerald and David Rosner. *Lead Wars: The Politics of Science and the Fate of America's Children.* Berkeley: University of California Press, 2013.
McGurty, Eileen. *Transforming Environmentalism: Warren County, PCBs, and the Origins of Environmental Justice.* New Brunswick, NJ: Rutgers University Press, 2009.
Nash, Roderick. *The Rights of Nature: A History of Environmental Ethics.* Madison: The University of Wisconsin Press, 1989.
Neer, Robert M. *Napalm: An American Biography.* Cambridge, MA: Harvard University Press, 2013.
Pulido, Laura. *Environmentalism and Economic Justice: Two Chicano Struggles in the Southwest.* Tucson: The University of Arizona Press, 1996.
Rome, Adam. *The Bulldozer in the Countryside: Suburban Sprawl and the Rise of American Environmentalism.* New York: Cambridge University Press, 2001.
———. *The Genius of Earth Day: How a 1970 Teach-In Unexpectedly Made the First Green Generation.* New York: Hill and Wang, 2013.
Scarz, Rik. *Eco-Warriors: Understanding the Radical Environmental Movement.* Walnut Creek, CA: Left Coast Press, 2006.
Steinberg, Ted. *Acts of God: The Unnatural History of Natural Disaster in America.* New York: Oxford University Press, 2000.
Unger, Nancy C. *Beyond Nature's Housekeepers: American Women in Environmental History.* New York: Oxford University Press, 2011.

Timeline

Passage of the Elwha Restoration Act	1992
Passage of the California Desert Protection Act	1994
Reintroduction of Wolves into Yellowstone National Park	1995
Removal of the Quaker Neck Dam (Neuse River)	1997, 1998
Signing of the Kyoto Protocol	1998
Removal of the Edwards Dam	1999
Release of *An Inconvenient Truth*	2006
Delisting of Wolves from the Endangered Species List	2009
Copenhagen Conference on Climate Change	2009
Deepwater Horizon Oil Spill	2010
Removal of the Elwha River Dams	2011–2014
Hurricane Sandy	2012

A Time of Environmental Contradictions

13

The 2009 Copenhagen Conference on Climate Change was originally intended to measure progress on the reduction of greenhouse emissions since the Kyoto Accord and enact further measures to slow the process of global warming. American resistance to cap and trade legislation, rejection of the earlier agreement by many Americans, and anti-global warming activism in the United States put a damper on the upcoming talks, reducing the odds of meaningful progress. Thousands of climate change protestors descended on the city to demand action.

Senator James Inhofe of Oklahoma, an aggressive climate change denier who received more than $300,000 in oil contributions during the 110th Congress (2007, 2008) and more than $600,000 from 2000–2008, as well as large sums from the coal industry, was the Ranking Member of the Senate Committee on Environment and Public Works. He traveled to Copenhagen with the sole intention of announcing his opposition to global warming agreements, determined to block any restrictions on greenhouse emissions that might emerge from the conference. In a speech in Copenhagen he declared, "China and India have pledged to reduce the rate of growth, or intensity of their emissions. But that's not acceptable to the U.S. Senate."[1] Denouncing climate change science as incorrect, he quickly proceeded to his real concern, the economic impact of restrictions on greenhouse gases:

> According to WEFA [Wharton Economic Forecasting Associates] economists, Kyoto would cost 2.4 million U.S. jobs and reduce GDP [Gross Domestic Product] by 3.2% or about $300 billion annually, an amount greater than the total expenditure on primary and secondary education. Because of Kyoto, American consumers would face higher food, medical, and housing costs—for food, an increase of 11%, medicine, an increase of 14%, and housing, an increase of 7%. At the same time an average household of four would see its real income drop by $2,700 in 2010, and each year thereafter.[2]

Inhofe continued to discuss the economic impacts of potential cap and trade legislation on Americans and returned to the issue of emissions by countries not bound by earlier agreements, particularly China and India. He wrapped up his speech by stating,

> My stated reason for attending Copenhagen was to make certain the 191 countries attending . . . would not be deceived into thinking the U.S. would pass cap-and-trade legislation. That won't happen. And for the sake of the American people, and the economic well-being of America, that's a good thing.[3]

A short two years later and thousands of miles away, 400 guests stood at the Elwha Dam on a beautiful September day in 2011. They were there to celebrate the tearing down of this dam that was so integral to the early economy of the town of Port Angeles, Washington, and that had also stripped the Lower Elwha Klallam Indians of a key resource, the salmon. The dam had almost destroyed the once prodigious salmon and steelhead runs, and 1992 federal legislation called for removal of this dam and the Glines Canyon Dam above it, in order to restore the fisheries. After 20 years of fighting for appropriations, the dams' end was near. Representative Norm Dicks, one of the sponsors of the original legislation, and Senator Maria Cantwell were there for the celebration, as was Washington Governor Christine Gregoire and Assistant Secretary of Indian Affairs Larry EchoHawk. Among the vast crowd were Elwha river restoration activists, members of the business community, Port Angeles City government representatives, and numerous Olympic National Park employees. Also present were many members of the tribe that had lost the river's wealth and were instrumental in returning the salmon to the river.

Before the crowd stood Ben Charles Sr., an elder of the Lower Elwha Klallam and a descendent of those who had seen their livelihood destroyed. In the spirit of reconciliation he spoke of the efforts to restore the river and, noting the crowd of supporters, quoted Hebrews 12:1: "We are encompassed about with a great cloud of witnesses." The Elwha Dance Group performed and the Port Angeles High School chamber orchestra and jazz ensemble played music as the community celebrated the spirit and promise of ecological restoration.[4]

These two moments capture the interesting juxtaposition of environmental attitudes in America today. There is enough knowledge and understanding of the importance of ecosystems to generate widespread support for a variety of restoration projects such as dam removal, wolf reintroduction, oyster reef construction, and many others. At the same time, Americans have largely resisted acknowledging or taking action on climate change because of the threat to the American standard of living and the confusion created by organizations opposing reforms.

The close of the 20th century and the beginning of the 21st is a confusing time of both remarkable environmental successes alongside continuing and, in some cases, worsening environmental pollution. The fragmentation of the environmental consensus of the middle of the 20th century is nowhere more stark than in the public dialogues on climate change. It is also likely that many Americans' interest in environmental issues waned because of the clear evidence of success. Recovering eagle populations across the country, wolves running wild in the northern Rockies, cleaner rivers, and restored fisheries from the Great Lakes and their tributaries to salmon returns on the Sacramento River are clear indications of environmental progress. The height and health of second-, third-, and fourth-growth forests supporting exponentially growing deer populations also signify to many Americans that the environment is flourishing. Where the Cuyahoga once burned, fishermen now cast for bass. Whereas the Kennebec River of Maine was once an industrial slough, it is now a popular recreational site and healthy fishery. The sheared landscape of East Texas, like so many parts of the country, is now covered again with healthy forests and has seen the return of deer, cougars, and many other birds and animals. The sense of crisis that saturated America in the late 1960s and early 1970s has been replaced with a feeling of success and a sense of accomplishment.

President Clinton

The election of President Bill Clinton in 1992 elicited great hope among environmentalists. His advocacy of environmental issues during the election and Vice-President Al Gore's publication of *Earth in the Balance* in 1992 convinced many that a Clinton administration would embrace environmental issues. That hope was soon sorely tested by the moderate democratic president's efforts to foster relationships with the business community by backing away from and even weakening environmental regulations. The vice-president went mute on environmental issues for the most part. Clinton did give an Earth Day speech in 1993 announcing the creation of a new biological survey with a budget of $180 million. Its mission was to provide scientific advice to the federal government and to conduct a survey of and monitor biological resources on federal lands. But this was quickly followed by efforts to make environmental regulation more amenable to business. The EPA's budget was cut from $6.9 billion to less than $6.4 billion. The pushback of the business community against environmental regulations led Clinton to emphasize cooperation with business, and he created a more relaxed regulatory system. This enabled states, businesses, and local communities to create more inexpensive, less punitive ways of reaching pollution limits.

Even while making regulation more lenient, Clinton supported expanding ecosystem preservation. The California Desert Protection Act was passed in 1994 and set aside 69 areas with wilderness designation and also included the creation of Death Valley and Joshua Tree National Parks. The Mojave National Preserve was designated with more than 1.4 million acres of land. The majority of land protected under this legislation was already federally owned, but this legislation provided geologic sites, native lands, and ecosystems greater protection under the law.

House Republicans won a majority in the mid-term 1994 elections largely on the basis of their "Contract with America," advocating term limits and a reduction of government size and power. The anti-environmental extremism of the 1994 Republican class is exemplified by Congresswoman Helen Chenoweth of Idaho who spoke openly and aggressively against all types of government regulations and laws and famously hosted an "endangered salmon bake." Another congressman mocked worries over threatened and endangered salmon because of the abundance of the fish he viewed in grocery stores. While sad or funny, these comments revealed anger at any sacrifice of human comfort or expenditure of costs for nature as well as a fundamental ignorance of wild salmon versus hatchery and farmed salmon. The ideology of abundance is made manifest here as threats to prosperity and consumption lead to rage and rejection of environmental laws and even the determination of an increasing number of Republican officials and Americans to reject the legitimacy of the Endangered Species Act and other environmental reforms.

A major challenge to environmental laws was launched by a House Republican group in 1995, as they proposed a slate of bills designed to weaken the federal government's regulatory power. Their bills proposed a variety of tactics, from forcing government to pay for any costs to property owners arising from environmental laws to making the process of regulating pollution and ordering cleanups more difficult. The House also attempted to dramatically weaken water regulation with a 1995 Clean Water Act; President Clinton referred to it as the "Dirty Water Act." All of these efforts were defeated.

Radical defenders of nature regained their stride during this period, revitalized by aggressive Republican attacks on the environment and angered at President Clinton's centrist position. Earth First! gained national prominence again with their opposition to a planned cut in the Cove Mallard district of the Nez Perce National Forest in Idaho in 1994. More than 6,000 acres of forest were slated for removal, but the overall impact of the cut, with a planned 145 miles of access roads, meant a greater overall effect than the timber harvest

Figure 13.1 Members of Earth First! employed a number of strategies to combat destruction of the environment and to draw media attention. Dressing up like endangered species, rolling cloth cracks down the face of dams, and protest were ways of protecting the environment short of sabotage. Tree-sitting, like chaining themselves to a tree or together in a logging road, is a highly effective form of civil disobedience. It demonstrates the commitment of the activist, and it is difficult to remove tree-sitters without injuring them. Therefore it slows down resource extraction activities or road building and drives up the cost. Getty Images.

alone. This largest remaining roadless area in the lower 48 states functioned as de facto wilderness. Roads mean ecosystem fragmentation, increased erosion into streams and rivers hosting trout and threatened and endangered salmon and steelhead, and increased human pressure on animal populations. Roads would also facilitate greater access by hunters, atv drivers, and snowmobilers with a corresponding impact on habitat and wildlife. The activists set up a base of operations with tents, school buses, and tepees. With 40 protestors camped there, another 60 distributed themselves tactically throughout the forest. They used road blocking and tree sitting to try and stop the cut.

These actions created anger in local logging communities reliant on timber jobs and money and resulted in limited scuffling and physical assaults on some protestors. Some Earth First!ers employed tree spiking and vandalized road construction equipment; the very livelihood of loggers and other occupations related to logging were threatened by these actions. From the perspective of environmentalists and Earth First!ers this cut was egregious in its threat to ecosystem health and its violation of a vast, beautiful, and ecologically healthy region. Economic needs and values drove the conflict and emotions ran high. The Forest Service and local law enforcement eventually conducted a round up, arrested protestors, and broke up the camp.

In the year before the next presidential election Clinton went green again. He vetoed Republican driven budget bills that would have opened the Arctic National Wildlife Refuge to oil drilling and blocked an attempt to approve clear-cutting in Alaska's Tongass National Forest. Republican efforts to cut EPA funding and open access to the California Mojave National Preserve were also fought back. Republican attempts to roll back environmental programs and protections revealed the breakdown of the earlier environmental consensus and a schism that politicized environmentalism. While this split had been developing for two decades, it was abundantly clear by the mid-1990s that the Democratic Party was now considered the environmental party and the Republican Party avowedly and openly opposed policies and programs designed to protect and improve the health of nature. This dichotomy has strong implications for the debate about environment and contributes to a stasis that makes resolving crises much more difficult.

Clinton made positive contributions to environmental improvement as well, using the 1906 Antiquities Act to create the Grand Staircase-Escalante National Monument in southern Utah and to set aside 1.7 million acres of wilderness within its borders. Even as President Clinton tried to find that middle ground between environmental protection and economic prosperity, he also recognized the need for the United States to play a role in curbing greenhouse gases. Clinton expanded U.S. participation in negotiations at the Kyoto Conference of 1997; the Kyoto Protocol was established that same year, setting limits on greenhouse emissions by 38 industrialized nations. They agreed to reduce greenhouse gases by varying amounts from the 1990 levels to occur between 2008 and 2012. The United States agreed to reduce emissions by 7 percent, while the European Union agreed to 8 percent and Japan 6 percent. Because developing countries were exempted from the protocol, global greenhouse gases would continue to increase in spite of the agreed-upon cuts. Therefore, President Clinton required an agreement by leading developing countries that they would try to control emissions. After receiving verbal agreement, he signed the protocol in late 1998 but never forwarded it to the Senate for ratification. Clinton's concerns about economic impact and the responsibilities of developing economies would be echoed by President George W. Bush and Senator Inhofe; the protocol was abandoned by the Bush administration.

President George W. Bush

The presidency of President George W. Bush echoed Reagan's in its approach to environmental issues. Aggressively attacking environmentalists as anti-business and anti-growth, this administration refused to acknowledge or respond to climate change, abandoning the Kyoto Protocol completely and instructing the EPA to not address climate change in its policy documents. The Bush administration strived for eight years to open the Arctic National Wildlife Refuge to oil drilling. His inability to do so speaks to a base consensus on the benefit of public land protections as well as proving that wilderness designation is a valuable strategy for habitat protection. Bush threw open public lands in the West for increased natural gas exploration, drilling, and mining. One of the major points of conflict during this administration was over the best uses of the national parks. The National Park Service had initiated major changes in the 1990s designed to highlight, facilitate, and implement research on species and ecosystem protection and restoration within the parks. The reintroduction of wolves to Yellowstone National Park is but one example. Increased funding had also been provided to improve water and air quality. However, like two decades earlier, the entry of a President opposed to environmental initiatives and determined to provide greater opportunities for business interests meant another round of attacks on environmental programs, regulations, and parks. The tension between the parks' expanding ecological stewardship and conservatives' desire to restrict the size and power of government was made explicit in a contestation of the use of national parks during this administration.

Gale Norton took over as Secretary of the Interior and shared much in common with her Reagan administration predecessor James Watt. She had opposed government regulation of resource extraction industries and maintained close ties to companies in that economy. Like President Bush, she rejected climate change. Norton also sought to commercialize the national parks and showed little interest in preserving their ecological integrity. Her "National Outdoor Recreation Policy" was an effort to open as much protected, environmentally sensitive land to motorized, off-road activities as possible. With input from lobbyists for recreation commercial interests and mostly excluding park officials and experts, the plan emphasized increased use and access by snowmobile, sport aviation, jet ski, and motor boat enthusiasts. Over the second half of the 20th century the Park Service had increasingly supported low-impact activities such as hiking, backpacking, cross-country skiing, and canoeing, while also protecting and restoring habitat and ecosystems. Now, under the aegis of Norton's plan, noisy, air-polluting activities would interfere with and undermine the quality of low-impact recreation as well as damage habitat. Also, these increased motorboats, all-terrain-vehicles, and snowmobiles, ranging over a greater portion of park territory, would generate more air and noise pollution. The speed, noisiness, and reach of gas-fueled vehicles frighten and endanger wildlife, and winter snowmobiling scares them away from key sources of water and food in parks like Yellowstone.

The increased presence of more damaging activities was exacerbated by reduced budgets and staffs; there weren't enough rangers to monitor activity. The proposal failed in

Congress and was then pushed administratively in individual parks. Gaining the most attention nationally was the mandate to increase snowmobile use in Yellowstone National Park. National Park historian Richard West Sellars writes, it "had become a long-running struggle mainly involving the National Park Service, the Secretary of the Interior's office, off-road motor vehicle industry trade groups, gateway-community tourism interests, and environmental organizations." He notes the ongoing writing and rewriting of winter-use plans and environmental impact statements as well as the hundreds of thousands of comments provided by the public.

> In spite of Park Service scientific data showing impacts of snowmobile use on wildlife and air and water, upper-level politicians repeatedly interceded on behalf of the snowmobile industry and overrode Park Service decision making. In a 2007 interview, the recently resigned National Park Service director, Fran P. Mainella stated that decisions on Yellowstone winter use had been "decided at a level beyond our office. A pay grade higher than mine."[5]

The Bush administration's determination to open the parks to more destructive recreation and serving its political constituency led them to ignore the science and attempt to subvert the increased ecological focus of the park system.

There is a more existential and typical American conflict within this debate about snowmobiles in Yellowstone or jet skis in Olympic National Park. This is a fight over rights. As a nation we have never really faced and lived with the idea of limits, except economic. So when a national park sets rules protecting rivers, grasslands, forests, and various species of fish, bird, and animal this strikes many Americans who remain ignorant of ecology as arbitrary and unjustified, as a violation of their rights. On the other hand, environmental activists and many state and federal agency employees responsible for environmental health have embraced, to varying degrees, a land ethic. This ethic is informed by a greater interest in and understanding of ecology, ecosystem function, and the role and value of specific species; they sustain the idea that we bear a responsibility to protect and restore nature.

The Century of Restoration

This is best exemplified in the efforts to restore nature that gained momentum and widespread support through the 1990s. Even with the ideological barriers thrown up by the Bush administration, the 1990s and early 21st century have been a period of creative and exciting work in the area of environmental restoration. The nature of the work being done, the complexity of the projects, and the varied measures speaks to the persistent power of environmental thought and a commitment by a large section of the American population to restore and protect as much nature as possible. The reintroduction of wolves to Yellowstone National Park and the Northern Rockies, the Buffalo Commons idea, and the building of natural habitat, greenways, and community gardens in urban areas, along with a powerful dam and river restoration movement that has led to 900 dams being torn down by 2013, are all efforts consistent with the idea of creating a better balance between humans and nature. This also reflects a new approach to environmentalism that emphasizes pragmatism and community coalition building, and is generally less reliant on strong rhetoric and national legislation.

Bison and Grasslands

Frank and Deborah Popper proposed the Buffalo Commons in 1987. With the ongoing depopulation of much of the arid interior West, large stretches of the Great Plains lay empty. The rural plains alone lost half a million people between 1920 and 2000, and hundreds of western counties were populated by less than six people per square mile. The Poppers argued that it made more ecological and economic sense to return a vast portion of the long, undulating plains to the shortgrass prairie of blue grama and buffalograss it once was and restore the bison. What was required was a combination of efforts: federal acquisition of lands as well as private and tribal efforts to build greenways between healthy ecosystems and the restoration of overgrazed, underused land to bison ecosystems. It is more sustainable than cattle ranching and agriculture and could reduce the draining of the Oglala Aquifer. Supporters of this idea believe it will generate economic activity through tourism and niche businesses.

Some of this proposed plan has been achieved. Federal and state land acquisition has created healthier ecosystems for bison and other animals such as antelope, prairie dogs, and black-footed ferrets. The American Prairie Reserve, envisioned by the Nature Conservancy and other groups, is a 3-million-acre ecosystem composed mostly of public land made contiguous through strategic purchases of private acreage. Ninety-five percent of this commons in northeastern Montana has never been plowed, so there is a strong persistence of native grasses. The first bison were introduced into the reserve in 2005, and the population is approaching 500. The managers of this site are moving carefully, selecting for bison that are as genetically close to original plains bison as possible and seeking to avoid those that are descendants of bison crossbred with cattle. The *Kansas City Star* issued an editorial in 2009 supporting a western Kansas Buffalo Commons. Also, in 1992 native governments formed an Intertribal Bison Cooperative and currently manage more than 15,000 head of bison collectively. These ideas were not readily embraced by many western communities as the commons proposal appeared during the wise-use and sagebrush rebellion days; many rural westerners viewed this effort as a strategy for moving them off the land. They understand that private and environmental group acquisition of ranch lands for environmental reasons raises property prices and taxes, making it harder to pass the property down to their children. Creating a Buffalo Commons is a bold effort to restore a broad ecosystem or biome in order to restore the bison but also to develop a more sustainable method of using and living on the Great Plains than has been practiced for the last century and a half.

Tearing Down Dams

Restoration was not limited to bison. Various species of fish were also aided by environmental efforts in this era. In 1976 the Lower Elwha Klallam Indians of Washington State launched an effort that would culminate in one of the great acts of environmental restoration in

American history. The Elwha River was dammed in 1913, almost completely destroying the river's magnificent runs of approximately 500,000 salmon and steelhead in some years to a diminutive 3,000 to 5,000 by the 1970s. In an effort to force the Washington State Department of Fisheries to produce more fish, the tribe challenged the federal relicensing of the upstream dam that had been built in 1976. Within a few years local and national environmental groups joined the effort and pushed for dam removal and restoration of the salmon. Through a combination of regional activism, finding replacement hydroelectricity for the pulp mill that would be effected by removal of the two dams, and building relationships with the mill owners, the owners of the dam, federal agencies, and the local community, activists were able to craft legislation calling for removal of the dams if necessary for river and salmon restoration. The Elwha Restoration Act garnered the unanimous bipartisan support of the Washington State Congressional delegation and was signed in 1992 by Republican President George H. W. Bush. While this federal legislation was unprecedented in American history and therefore marked a high-water mark for river restoration efforts, conservative opposition stalled the appropriations needed for the project. Eventually, in September 2011, the process of dam removal on the Elwha River began. The largest dam removal in American history to this point, this has been a carefully planned and measured process. Specialists in fisheries, river hydrology, mammal biology, nearshore environments, and plant life conducted extensive studies prior to the removal of the two dams in order to create a baseline for understanding the recovery as comprehensively as possible. By fall 2013 large numbers of chum, pink, and even the vaunted king salmon were already returning up the river and spawning in habitat unused since 1913. The success of these high-head dam removals may provide scientific evidence and rationales for the removal of larger dams.

While the passage of the Elwha Restoration Act initiated the dam removal era in 1992, the destruction of the Edwards Dam on the Kennebec River of Maine in 1999 was the biggest restoration project to that point. The attendant media coverage and positive consequences provided the impetus for even more dam removals regionally and nationally. A relatively major removal and a costly one, restoration necessitated industry cooperation as well as agreements between state and local government. Also, the Federal Energy Regulatory Commission made an unprecedented decision to require dismantling of the dam for environmental reasons. Predictions of failure by the owners of the dam and local opponents were swept away by the surge of shad, striped bass, and other fish immediately migrating 17 miles upstream to the Fort Halifax dam, using river and tributary spawning sites unavailable since 1836. The Fort Halifax Dam was removed in 2008, and a town north of the dam site experienced its first shad harvest in 160 years, with residents collecting 350,000 of the fish. Activists on the Penobscot River, including native peoples, fought a long hard battle that culminated in free-flowing waters there as well. The removal of two dams and modification of another, paid for by a mix of federal and private dollars, has opened approximately 1,000 miles of habitat.

Further south, the Quaker Neck Dam was removed from the Neuse River of North Carolina in 1997–1998. Built in 1952, it had reduced a 700,000-pound American shad harvest to a mere 25,000 pounds by 1996. There was a fast, notable increase in striped bass and shad populations following removal. Sturgeon and herring have thrived as well. An unexpected benefit was the growth in numbers and size of the popular flat-head catfish that prey on newly available spawning fish. Wisconsinites have been active dam removers as well with at least 130 dams torn down in recent years (65 in 2013 alone) to include those on rivers such as the Baraboo,

Figure 13.2 Congress passed and President George H.W. Bush signed the Elwha River Restoration Act in 1992. Due to debates over dam removal and difficulties securing appropriations, dam removal did not actually begin until September 2011. It is the largest dam removal project in American history to this point. This is a photograph of the upper dam on the Elwha, Glines Canyon Dam, being removed. Because of a massive mountain of dirt, rocks, and old logs piled behind the dam and a desire to limit downstream damage, the dam was taken apart systematically over a long period of time. Salmon and steelhead are already returning to and spawning in the restored Elwha River. Courtesy of John Gussman.

Milwaukee, Manitowoc, Flambeau, and many others. Salmon, steelhead, and trout have thrived with dams removed from Bear Creek and Little Sandy River in Oregon, the Clearwater River in Idaho, and the Mad River of California. These are merely a few of the dozens of dams torn down in the West. Besides simple access to more habitat, the increased aeration that results from dam removal improves water quality, generating a rapid and diverse growth of aquatic insects and species dependent on them. The fish benefit from the richer and more complex food chain that grows with dam removal. Numerous species such as ravens, eagles, raccoons, bears, orcas, and humans also benefit from bringing fish back into the watershed ecosystem. These dam removal and river restoration programs reflect our increased understanding of the functioning of intact ecosystems and a determination to bring balance to a number of river ecosystems across the country. Many of these removal and restorations are common sense measures because of aging dams that provide little economic benefit. Some of them involve more conflict over water rights and access, but these efforts have the benefit of not requiring national legislation, so local coalitions can build the case by appealing to local issues and knowledge.

The Return of the Wolf

Salmon suffered many insults and injuries over the course of the 20th century but were never targeted for extermination the way wolves were. The federal government's program of restoring this once-despised species has to be one of the greatest turn-arounds in the government's relationship with a species in American history. Thirty-one Canadian gray wolves were reintroduced to Yellowstone National Park in 1995 and 1996 as part of a

Figure 13.3 One of the reasons wildlife biologists and park managers wanted to bring wolves back to Yellowstone National Park was to reduce the excessive elk population and restore ecological balance. Images such as these, of elk lolling about the grounds at Mammoth Hot Springs, easily and eagerly observed by excited tourists mere yards away, indicate the imbalance of nature created by predator removal. Courtesy of Bill Bouton.

broader plan by the U.S. Fish and Wildlife Service to reestablish the wolf in the northern Rockies. The population reached approximately 10 by 2011 with almost another 20 ranging the territory on the park borders. As anticipated, they immediately began hunting and eating elk, with that large ungulate representing about 80 percent of their diet. This has created a more balanced ecosystem as overgrazing has been brought under control with benefits for numerous other animals, birds, and fish. For example, elk used to congregate in huge herds in meadows and along riparian zones. Now their numbers are reduced, and they have broken into smaller herds that are mobile and use the forest more in order to avoid wolves. Consequently, riparian zones and aspen groves have recovered from overgrazing. Beaver populations are increasing, and the increasing number of dams that create pools and wetlands and reduce the carrying of sediments in streams has benefited cutthroat trout, moose, waterfowl, and numerous bugs, amphibians, and songbirds. Also, scavengers such as ravens, eagles, and bears have benefited from the elk carcasses left by wolves. The most visible way the Yellowstone ecosystem has changed is in the dramatic reduction in elk population and the practical disappearance of coyotes.[6]

Figure 13.4 As Lord Alfred Tennyson said, "nature red in tooth and claw." The return of wolves to Yellowstone National Park and other parts of that species' historical habitat has meant a restoration of a more realistic and healthier ecosystem with predators culling the herds of bison and elk, keeping those animals and their habitats healthier. National Park Service.

The wolf population of the northern Rockies into the interior Pacific Northwest has skyrocketed from a few individuals in the early 1990s to close to 1,800 by 2011. Wolves were reintroduced into Idaho's Salmon River Valley in 1995 and 1996. But some of this growth is also due to natural recolonization by wolves themselves as they migrate south from large, healthy populations north of the border in Canada into recovering, healthier national forests and parks, and even private lands, where ecosystems and prey populations have improved since the decades when wolves were aggressively exterminated. The number of wolves increased so substantially that after years of debate and court challenges the northern Rocky Mountain wolf was removed from the endangered list in 2009. That same year Idaho announced a wolf hunting season with a harvest limit of 220 wolves; 188 wolves were killed in the hunt.

Natural recolonization by species across the continent speaks to the success of game and predator protection laws, habitat restoration, and a desire by Americans to see more varied wildlife, including what are often referred to as charismatic mega fauna—elk, wolves, bears, mountain lions, moose, and so forth. It is true also, however, that structural changes in the economy and in agricultural production, along with regular creation of parks, wildlife refuges, and deliberate ecological restoration, have allowed land to return to forest and grass in many places, increasing habitat and food for these species. Not only have wolves repopulated a great deal of range in the northern Rockies and the Pacific Northwest, but cougars and panthers[7] have extended their range as well, moving into areas they had been exterminated from more than a century earlier. Black bears have also rebounded in western states, the South, and the Northeast. There is a downside to this. Runners killed by mountain lions and campers attacked by bears all serve as a reminder that coexistence with predators carries risks.

Building Ecological Resilience

Restoration efforts extend beyond individual species to programs for rebuilding important ecosystem infrastructure. An excellent example of this is the current oyster reef construction project in Matagorda Bay, Texas. There was once a 500-acre oyster reef in the bay, but overharvest, sedimentation, and hurricane damage destroyed it years ago. The Nature Conservancy along with the U.S. Army Corps of Engineers, and with a U.S. Fish and Wildlife Service Grant of $3.8 million as well as private donations, is currently building an 80-acre reef. Oyster reefs provide habitat and food for a number of fish and other species. Oysters also function as filters cleaning water, while reefs help slow storm surge. This can mitigate the impact of hurricanes. The cornerstone of a flourishing bay ecosystem, the reef would also contribute to the local oyster and fishing economy.

In Washington State federal and state agencies as well as municipal government have launched efforts to improve the nearshore habitat in Puget Sound. Many of the beaches in Puget Sound are armored; this means they are reinforced with stone and cement rip rap, a collection of rocks, boulders, and pieces of cement placed on shorelines to reduce erosion. Seawalls sit between the water and the shoreline and railroad lines, docks, boat launches, and other structures interfere in nearshore natural processes. With scientific research showing the importance of these ecosystems for sustaining crab, juvenile salmon, smelt, herring, and

Figure 13.5 This photograph shows how simple ecological restoration can be. By digging a hole through a decades-old road and building a bridge, Puget Sound ocean waters in Washington were once again reconnected to an old estuary. Estuaries are crucial elements of a healthy nearshore ecosystem providing food and protection for numerous birds, juvenile salmon, other fish, seals, oysters, shrimp, clams, and other species. Courtesy of Bryan Crane.

other important saltwater species and the birds, fish, and mammals dependent on them, great efforts are being made to remove structures interfering with nearshore functions. On Whidbey Island, Washington, the U.S. Navy has opened access to a wetland that was a coastal estuary before farmers diked it off and drained it for farmland in the 19th century.

This will provide important habitat for juvenile salmon and other species. South of Tacoma, Washington, a bay is being completely restored with armoring removed, natural grasses and plants reintroduced, and removal of a barrier between the bay and Puget Sound. These are but a few of numerous attempts to restore nearshore habitats, and while often carrying a large price tag, they provide numerous ecological and economic benefits and help create a more resilient ecosystem overall.

Across the nation land trusts and local land conservation groups use a combination of conservation easements and acquisition of critical properties to protect habitat, create green spaces in urban areas, restore and protect riparian zones, and secure properties that are essential to watershed health and aquifer recharge. Because these agreements can be taxpayer funded or provide tax benefits to landowners, they provide locally based, non-legislative solutions to pressing problems created by constant growth and sprawl. While these efforts are sometimes piecemeal and limited in scope, they also help generate coalition building across political lines that increase buy-in to environmental programs and help bridge the environmental schism in America.

Green Cities

Environmentalism has grown in complexity in recent years and embraced numerous strategies as a more diverse constituency grows. There have been a number of efforts to bring nature back into the city, to break down an urban-nature divide and create more healthy options for city dwellers; community garden and urban farming programs deal with all of these issues. Across the nation, food activists have squatted on vacant lots and have planted in school yards and on church property in order to produce food for the local community. While the majority of Americans still eschew growing their own food, communities have turned to local food production for a number of reasons.

Urban farming has several goals. One is to bring organic, healthy foods back into city diets. Another is to introduce healthy fruits and vegetables into urban food deserts where grocery stores with produce do not exist. Those communities typically suffer high rates of childhood obesity, malnutrition, and Type 2 Diabetes; urban farming can redress that problem. While the production of healthy food is important in and of itself, urban farming brings other benefits. Notably, urban farmers bring in not only gardens but flowers, insects, songbirds, and many other species, creating oases of nature in city landscapes. These gardens, and increasingly orchards and native plant gardens, may be the only point of contact with multiple species of plants, bugs, songbirds, and animals for urban children of limited means. The creativity of these efforts is stunning and offers hope for future endeavors. Rooftop gardens on apartment complexes and businesses, the top floor of a parking garage in Seattle converted to food gardens, and creative water collection systems and median planting all suggest the energy and innovative ideas flowing into reconstructing a healthier relationship with nature. Organizations such as Green Spaces Alliance in San Antonio focus on community gardens as a way to provide healthy food for a variety of communities and to create patches of nature and ecosystem within the city.

With 38 gardens in a variety of neighborhoods, from the impoverished east and west sides to senior citizens housing, these San Antonio gardens provide fresh, organic food, exercise, and a contact with nature that is difficult in urban areas. Moreover, they are sites that are used for permaculture training, master naturalist classes, and activities for children. Other cities, such as Houston, Philadelphia, Detroit, and Seattle, have created extensive urban gardening and farming operations that not only produce healthy organic produce in food deserts, but also create commercial opportunities for urban farmers and jobs for unemployed youth.

Sustainable urban farming does what writer and activist Wendell Berry calls solving for pattern.[8] With the reduced use of machinery and fossil fuels they contribute much less to climate change than industrial agriculture. The use of organic fertilizers rather than nitrates reduces eutrophication and does not add to environmental crises such as the Gulf of Mexico dead zone. Pollinators such as bees, moths, and butterflies do better in these farms with reduced chemical pesticides. Finally, these farms help to reduce urban heating, help capture carbon, and provide patches of ecosystem that can benefit a number of species and provide city kids opportunities for play and learning about nature.

Figure 13.6 Community gardening and school gardening is not a new phenomenon. Settlement houses did the same thing in the early 20th century. Community gardens in food deserts, such as this one in southeastern San Antonio, can provide numerous benefits. In this photo, preschool children are learning about plants, soil, and gardening. These gardens are classrooms about nature and food and also provide healthy produce for communities with a strong need. Finally, such gardens can help create a love of nature that will provide benefits in the future. Courtesy of the author.

Environmental Success and Abundance

The environmental consensus that fostered effective and comprehensive environmental reform at the national level in the 1950s through the 1970s is long gone, but a majority of Americans still attest to their belief that a healthy environment is important. Successes like the rebuilding of once endangered species, such as the bald eagle, kestrel, alligator, sea otter, and others, along with the rebounding of mammals such as deer, moose, elk, bears, and wolves, testify to the success of environmental efforts. On the one hand, there are clear signs of a healthy, recovering natural world. On the other, there is continuing fragmentation of large ecosystems, declining water quality from streams to coastal estuaries and the ocean itself, not to mention a quickly worsening climate change crisis. These issues are more complex than older problems such as endangered species protection, banning of above-ground nuclear weapons testing, or ending the use of DDT. Responding effectively to agriculturally-based environmental crises requires spending more money, paying for the real cost of food, and questioning the assumptions of abundance and consumption.

But the ideology of abundance impedes meaningful environmental reform and creates new sets of problems. This appears in many forms and places, but in no place is it so obvious and damaging as on our plates and in our bellies. An article on restaurants and food in Omaha, Nebraska, from *Saveur* (a culinary magazine) raved about a restaurant with 2,200 seats and 70 New York strips being cooked at one time. San Antonio hosts a "meatopia" festival, and the rise in the popularity of barbecue in recent years illuminates the degree to which Americans love their meat. But the obsession of living a life of food and meat plentitude exacts a heavy impact on human and environmental health. Livestock producers began mixing antibiotics and hormones in with feed in the 1950s in order to fight infections that are inevitable in dirty, muddy feedlots where animals are packed together by the thousands, and to increase the speed of animal growth. Also, cattle do not naturally eat grain and inevitably become sick. This necessitates increased antibiotics use. One estimate in 2004 put the amount of antibiotics fed to American livestock at approximately 50 million pounds a year. By the 1980s efficiency was taken to such extremes that, in many industrial feedlots in America and Europe, business owners began feeding cattle an evil stew of antibiotics, feed, grain, animal feces, and ground-up remains of slaughtered and diseased cattle. Actions such as these increased yields and drove down prices, putting small, traditional grazing operations out of business while reinforcing the belief that all Americans are entitled to a regular diet of meat. The outbreak of Creutzfeldt-Jacob disease, also known by the easier to remember moniker "mad cow disease," in 1986 in Great Britain led to 43 deaths and temporarily gained Americans' attention. The British government ordered the extermination of their cattle herds, slaughtering 4.4 million livestock and imposing restrictions on feeding cattle to cattle. In 1997 the FDA banned the use of meat products from livestock for feed. But since then livestock producers have shown great creativity in finding new sources of food in the face of rising corn prices. Heat-treated and processed, euthanized cats and dogs (with flea collars and tags ground up also) and road killed animals such as raccoons and opossums are regularly used in cattle feed. So is stale candy, chewing gum, heat-treated garbage, and processed chickens and chicken feathers. The industrialization of meat production has kept prices for meat low but in a way that many people view as unethical and that creates new health threats.

The continued intensification of industrial meat production in the late 20th century, while providing "cheap" beef, pork, and poultry, generates numerous environmental problems. The overuse of antibiotics has resulted in strains of bacteria that are antibiotic resistant. While it is true that doctors overprescribe and Americans overuse or improperly use antibiotics in a manner that increases the bacteria's evolutionary adaptation to the medicines, the widespread excessive use of antibiotics in feedlots and poultry sheds has worsened the situation. The feedlot system and the rapid speed of carcass disassembly in meat packing plants have also led to regular outbreaks of salmonella and campylobacter jejuni in poultry. E. coli bacteria are persistent in ground beef and have resulted in numerous outbreaks and deaths over the last several decades. One way E. coli has entered diets and sickened and killed Americans, is through pollution from ranches and feedlots into produce agriculture, as well as from eating undercooked beef. It is difficult to contain the waste generated by these industrial meat production facilities. Nationally, animals produced for food create 89,000 pounds of excrement per second, outstripping that produced by humans. Thirty-five hundred cattle produce 40,000 tons of manure annually, so feedlots with thousands of cattle or pigs have to build complex lagoon systems for storing, drying, and disposing of the waste. Inevitably feces and urine escape, and 35,000 miles of rivers in 22 states have been contaminated by run-off waste. The problem of waste management is so extreme that the Department of Agriculture provides large grants of money, sometimes paying the majority of the cost of a system improvement, to try to keep the waste from entering the environment and affecting the health of humans and the natural world. This is but one example of how Americans don't pay the real cost when buying "cheap" beef or pork. Even with these efforts, pollution remains a regular problem. In a recent Kansas incident a feedlot owner discharged 3.7 million gallons of wastewater into a public creek. Another feedlot had so much runoff discharge that the ammonia measurement was 550 times normal two miles downstream from the source. Cattle waste runoff can contain 100 times more nitrogen and phosphorus than from grazing land. Thus, eutrophication is also a result of intentional and accidental discharge of waste from these sites. This also creates costs that are not absorbed in the price of meat but rather in nature or in damage that has to be mitigated by government programs.

The production and consumption of large quantities of meat is deeply wasteful in a world of shrinking resources. It requires 11 times more gas or oil to produce one calorie of meat versus one plant calorie. It is difficult to slake the thirst of the meat industry with 25,000 gallons of water necessary for one pound of meat versus 25 gallons for a pound of wheat. To magnify the scale of waste, more than 70 percent of the grains and cereals grown in this country are fed to cattle, pigs, and chicken. Meat-heavy diets also accelerate global warming. Global meat production constitutes 14–22 percent of all greenhouse gas emissions, whereas bean, grain, and vegetable farming create significantly lower amounts. Also, cattle produce 2.5–4.7 ounces of methane for every pound of meat. Methane, while only lasting 12 years, is 23 times as powerful as carbon dioxide as a greenhouse gas.

The American ideology of abundance, predicated on externalizing many of the costs of resource extraction and food production, now drives some of our greatest environmental crises. Yet the topics of consumption, the assumptions of constant GDP growth, and the American dream of prosperity and a rich diet remain at the margins of debate in our society.

Current Energy Use and Pollution Crises

Americans' continued reliance on automobiles and oil has required a continued political and military presence in the Middle East to ensure access to petroleum. Even though Canada provides the majority of oil imported into the nation, Persian Gulf states still are integral to the functioning of the United States and world economy. While mileage standards for cars have improved modestly over the last two decades, the continued popularity of Humvees, large pickup trucks, and SUVs offsets the benefits accrued from those who drive smaller sedans and hybrids. The refrain of oil independence gained popularity in recent years when prices temporarily rose over $4 a gallon in 2008 and almost $4 a gallon nationally again in 2011. The result has been the aggressive pursuit of offshore oil, increased extraction of tar sand oils in Canada, and the dramatic growth of the hydraulic fracturing (fracking) industry in the United States. These resource extraction economies ensure continued reliance on fossil fuels and foreclose discussions of alternative energy use and development as "too expensive." Of course, the fuel industries create a wide array of external costs not captured in the cost of oil but absorbed by nature and humans and paid for, to some degree, by municipal, state, and the federal government.

The nation was stunned by the explosion on the deck of the Gulf of Mexico off-shore drilling rig on April 20th, 2010. Eleven men were killed in the explosion. As if that wasn't enough, the nation watched in surprise and increasing disbelief as a plume of oil flowed from the well site 13,000 feet below the surface of the gulf day after day, with British Petroleum (BP) and the U.S. government completely unable to stop the spill. Disbelief was heightened by news media's 24-hour coverage of the oil gushing out, shown via a deep-water camera. Disbelief turned to stupefaction as numerous stopgap efforts to stop the flow, repeatedly failed. When BP finally capped the well 87 days later, more than 200 million gallons of oil had polluted gulf waters. There is some confusion over the impacts to the Gulf ecosystem and economy because much of the oil was not visible and seemed to have dissipated. In reality it mostly ended up on the floor of the Gulf, on coral reefs, and in other locations. Millions of gallons still reached

Figure 13.7 This satellite photo of the Deepwater Horizon oil spill shows the extent of the oil's spread. NASA.

on beaches and required extensive clean-up operation. The spill also damaged the regional tourism, fishing, and shellfish industries.

Approximately 900 dolphins were killed, and in the years following the spill they grounded themselves at three to four times the rate they did prior to the event. Grounding is when marine species swim into shallow water and strand themselves for reasons not clearly understood. Also, sea-turtles are grounding themselves and dying in higher numbers. Thousands of birds died from coatings of oil, and there has been damage to fisheries. But the mortality rate to various species has been harder to measure, and long-term studies currently being conducted will better reveal the impact of the spill.

While many Americans were appalled and expressed shock that such a thing could happen, undermining constant critical rhetoric about the incompetence of government and the competence of corporations, there was little in the way of protest. Environmental groups demanded reforms, and BP has been assessed high fines to pay for those suffering from the economic impact and to conduct cleanup and restoration; there was nothing like the national outcry following the far less damaging Santa Barbara oil spill. And although the Obama administration did institute a moratorium on new permits for off-shore drilling for a period after the spill, creating great consternation from the right side of the nation and cries of "drill baby drill!," he and the democratic party did not push through reforms that would prevent such a spill from happening again. The environmental schism and Americans' acceptance of a certain amount of environmental degradation as the cost of maintaining prosperity and a high standard of consumption, are certainly two reasons for the muted response.

The conflict between jobs and environment is clearly evident in the debate over the Keystone Pipeline expansion to allow oil from the tar sand fields in Alberta, Canada, to reach the refineries on the Gulf Coast in Texas. President Obama was pressured for several years by Republicans and oil industry leaders, the Chamber of Commerce, and others to approve the pipeline in order to achieve petroleum independence and create American jobs. The anger from these same groups at the moratorium on new off-shore drilling operations following the BP Deepwater Horizon oil spill fueled increased demands for the Keystone Pipeline expansion. Environmentalists, particularly those focused on climate change such as the group 350.org, have demanded that Obama not approve the pipeline expansion and have organized numerous protests. The reasons for their objection are primarily due to the destructive nature of tar sands operations, the likelihood of spills from the pipeline, and the fact that the pipeline expansion will subsidize continued reliance on fossil fuels, thus worsening climate change. Advocates for the pipeline emphasize jobs created, but opponents argue that the number of jobs would be limited; the rhetoric of energy independence was applied early in the debate, but the pipeline expansion is designed to export and sell the oil in foreign markets, not for domestic use. In a society and world economy where abundance and prosperity are predicated on a regular supply of petroleum the pipeline expansion is perceived as necessary. To those seeking to support the development of alternative energy and slow down environmental degradation from mining pollution, pipeline spills, and automotive pollution, and for those who are also trying to slow the pace of climate change, the Keystone Pipeline expansion must be stopped.

The constant quest for "cheap" fuel has engendered a boom in hydraulic fracturing mining, otherwise known as "fracking." In this process mining companies inject a mix of

Figure 13.8 While no national outcry developed as it did with the Santa Barbara oil spill in 1969, communities along the Gulf Coast did protest the recklessness of British Petroleum and the inability of BP or the federal government to stop the spill for such a long stretch of time. Courtesy of Infrogmation of New Orleans.

sand, water, and up to 600 chemicals, including known carcinogens and toxins such as lead, formaldehyde, and mercury, into the shaft at high pressure; it then breaks up shale and releases natural gas. One fracking operation on one site can use several million gallons of water and thousands of gallons of chemicals, the ingredients of which companies are not legally required to disclose. A site can be fracked multiple times, so the amount of chemicals and water used is extraordinary, and with close to half a million fracking sites in the nation and growing quickly in number, the amount of water and chemicals injected into the ground is stunning, particularly as water shortages are occurring more frequently. In Texas in 2011, 632 million barrels of water were used for fracking. Less than half of fracking fluid is recovered; the rest is left in the shaft to infiltrate into wells and aquifers. Many communities are losing their local water supplies due to contamination from fracking, and water is being diverted from farming and ranching operations. The fluid that is recovered

Figure 13.9 This photograph shows a hydraulic fracturing operation in action. Courtesy of Joshua Doubek.

is hauled, usually in uncovered trucks with some slopping over the sides, and dumped in open-air pits for the water to evaporate and many of the chemicals to simply blow away. This allows these chemicals to enter the local air supply and food chain.

The impacts of fracking that most Americans are familiar with are the outbreaks of small earthquakes in the 3 to 4 range on the Richter scale in areas that have historically been earthquake free and the phenomenon of people lighting natural gas flowing from their faucets. The national government and state governments have done little to provide protections for property owners and for communities where fracking has taken over. One reason the federal government is weak on this issue is because Vice-President Dick Cheney pushed through a measure referred to as the Halliburton loophole, which exempts fracking companies (such as Halliburton, for whom Dick Cheney was once CEO) from the Clean Air Act, Clean Water Act, and the Safe Drinking Water Act. Eastern Wyoming, North Dakota, Pennsylvania, Colorado, Texas, and other locations have seen their landscapes and local economies transformed and degraded by these boom economies and fracking operations. Fracking has certainly kept the price of natural gas lower than it would normally be, but it is in no way "cheap" gas. The environmental impacts that companies and consumers are not paying for are numerous and onerous. Road construction, mining sites, evaporation pits, and other changes to the land are destroying habitat across the nation and drastically reducing property values and rural communities' tax base. When the companies move on they make no effort to restore the habitat unless the property owner has specified it in their lease, which is rare. Recent studies indicate that air pollution from mining operations is responsible for cancers, immune system disorders, pneumonia, asthma, and other ailments leading to sickness and death.

The millions of gallons of water and chemicals hauled to sites by trucks destroy local roads and worsen air pollution and greenhouse gas emissions. Moreover, methane released during the fracking process speeds the rate of climate change. Finally, the persistent extraction of natural gas supports continued reliance on fossil fuels and prevents meaningful climate change policies or reforms in this nation. If the externalities of fuel acquired through fracking and regular petroleum operations were included in the price at the gas pump, then solar and wind power would start to look much more affordable.

America is at an odd juncture in the early part of the 21st century. Americans by and large appreciate nature, assume recreational time in national parks and wilderness areas to be an automatic part of the American dream, and also understand basic ecological principles. A majority of Americans now believe in climate change and also believe humans to be responsible for climate change. Yet, habits die hard. We are a nation founded on abundance. Americans have built a vision of themselves, their place in the world, their relationship with nature, and the expectations for their lives on an ideology of abundance that is deeply unrealistic and damaging. As the Antarctic ice sheet slips into the ocean and the world's waters inevitably rise, what changes will Americans make in their own economic practices, behaviors, and relationships with the natural world?

Document 13.1 An Evangelical Declaration on the Care of Creation, *Evangelical Environment Network*, 1993

Strong interest in environmentalism by evangelical Christians began to grow in the 1980s, a key moment being the creation of the Au Sable Institute for Environmental Studies in 1980. This institute has brought together scholars and church leaders in forums and classes, and ongoing discussion, while helping generate numerous research papers, essays, and books on the need for Christian environmentalism. After multiple meetings and publications over the years, the Evangelical Environment Network was created in 1993. Later that same year, the group issued the "Evangelical Declaration on the Care of Creation." The movement continued to grow with the creation of covenant congregations dedicated to environmental issues and the distribution of thousands of copies of *Let the Earth Be Glad: A Starter Kit for Evangelical Churches to Care for God's Creation* to evangelical congregations.

The issuing of this declaration marked the larger entry of some evangelical Christian leaders into the environmental movement. The popular association of environmentalism with left-wing politics and ideologies in the minds of many Americans has made many conservative and moderate Christians in America uncomfortable with environmental activism. Environmental evangelical activism might change that fear.

1. Compare the uses of scripture to endorse a healthy attitude about the environment to the way scripture was deployed for dominion in the colonial period of American history. How does the language differ, and what arguments are made for a Christian environmentalism?
2. How does this declaration describe humanity's uses of nature and what results? In what way does this approach differ with or correlate with other forms of environmentalism you have studied?

The Earth is the Lord's, and the fullness thereof—Psalm 24:1

As followers of Jesus Christ, committed to the full authority of the Scriptures, and aware of the ways we have degraded creation, we believe that biblical faith is essential to the solution of our ecological problems.

Because we worship and honor the Creator, we seek to cherish and care for the creation.

Because we have sinned, we have failed in our stewardship of creation. Therefore we repent of the way we have polluted, distorted, or destroyed so much of the Creator's work.

Because in Christ God has healed our alienation from God and extended to us the first fruits of the reconciliation of all things, we commit ourselves to working in the power of the Holy Spirit to share the Good News of Christ in word and deed, to work for the reconciliation of all people in Christ, and to extend Christ's healing to suffering creation.

Because we await the time when even the groaning creation will be restored to wholeness, we commit ourselves to work vigorously to protect and heal that creation for the honor and glory of the Creator—whom we know dimly through creation, but meet fully through Scripture and in Christ. We and our children face a growing crisis in the health of the creation in which we are embedded, and through which, by God's grace, we are sustained. Yet we continue to degrade that creation.

These degradations of creation can be summed up as 1) land degradation; 2) deforestation; 3) species extinction; 4) water degradation; 5) global toxification; 6) the alteration of atmosphere; 7) human and cultural degradation . . .

First, God calls us to confess and repent of attitudes which devalue creation, and which twist or ignore biblical revelation to support our misuse of it. Forgetting that "the earth is the Lord's," we have often simply used creation and forgotten our responsibility to care for it.

Second, our actions and attitudes toward the earth need to proceed from the center of our faith, and be rooted in the fullness of God's revelation in Christ and the Scriptures. We resist both ideologies which would presume the Gospel has nothing to do with the care of non-human creation and also ideologies which would reduce the Gospel to nothing more than the care of that creation.

Third, we seek carefully to learn all that the Bible tells us about the Creator, creation, and the human task. In our and words we declare that full good news for all creation which is still waiting "with eager longing for the revealing of the children of God" (Rom. 8:19).

Fourth, we seek to understand what creation reveals about God's divinity, sustaining presence, and everlasting power, and what creation teaches us of its God-given order and the principles by which it works . . .

The Creator's concern is for all creatures. God declares all creation "good" (Gen. 1:31); promises care in a covenant with all creatures (Gen. 9:9–17); delights in creatures which have no human apparent usefulness (Job 39–41); and wills, in Christ, "to reconcile all things to himself" (Col. 1:20).

Men, women, and children, have a unique responsibility to the Creator; at the same time we are creatures, shaped by the same processes and embedded in the same systems of physical, chemical, and biological interconnections which sustain other creatures.

Men, women, and children, created in God's image, also have a unique responsibility for creation. Our actions should both sustain creation's fruitfulness and preserve creation's powerful testimony to its Creator . . .

We have ignored our creaturely limits and have used the earth with greed, rather than care.

The earthly result of human sin has been a perverted stewardship, a patchwork of garden and wasteland in which the waste is increasing. "There is no faithfulness, no love, no acknowledgement of God in the land . . . Because of this the land mourns, and all who live in it waste away" (Hosea 4:1, 3). Thus, one consequence of our misuse of the earth is an unjust denial of God's created bounty to other human beings, both now and in the future . . .

We call on Christians to work for godly, just, and sustainable economies which reflect God's sovereign economy and enable men, women and children to flourish along with all the diversity of creation. We recognize that poverty forces people to degrade creation in order to survive; therefore we support the development of just, free economies which empower the poor and create abundance without diminishing creation's bounty.

We commit ourselves to work for responsible public policies which embody the principles of biblical stewardship of creation.

We invite Christians—individuals, congregations and organizations—to join with us in this evangelical declaration on the environment, becoming a covenant people in an ever-widening circle of biblical care for creation . . .

We make this declaration knowing that until Christ returns to reconcile all things, we are called to be faithful stewards of God's good garden, our earthly home.

Document 13.2 Copy of Email from American Petroleum Institute to Its Membership—Obtained by Greenpeace, *American Petroleum Institute—President and CEO Jack N. Gerard,* 2009

The scientific evidence for climate change has built steadily over the last three decades with approximately 100 percent of the scientific community acknowledging the reality of global warming and almost all scientists attributing this change to human actions. A review of 928 scientific paper abstracts on global warming found none that did not accept that humans are a major contributing factor to global warming.

Reflecting a hardening of positions since the moderate 1970s, when members of both the Republican and Democratic parties embraced or supported environmental issues, the Republican Party has made opposition to climate change and corresponding regulation or legislation a central plank in its platform. Leading Republicans such as Senator Inhofe of Oklahoma derided climate science as "junk science" and organized by liberals to expand government power and hamper business. Business interests used Astroturfing to try and prevent meaningful reforms. Astroturfing is the strategy used by industry and allies to create ambiguity in Americans' minds on this subject.

1. Evaluate the strategy described in the letter. How consistent is this with the astroturfing tactics created by the tobacco companies? According to this letter, what fears and concerns are being cultivated in Americans?
2. Are these legitimate arguments or merely tactics to prevent American support for climate change or legislation?

Dear API Member Company CEO/Executive,

As I have outlined in the past few editions of the weekly "Executive Update," API is coordinating a series of "Energy Citizen" rallies in about 20 states across the country during the last two weeks of Congress's August recess. Most of these will be held at noontime, though some may be at different times in order to piggyback on other events. Thanks to the leadership of API's Executive Committee, I am pleased to report that we have strong support for this first-ever effort moving ahead. Now we are asking all API members to get involved.

The objective of these rallies is to put a human face on the impacts of unsound energy policy and to aim a loud message at those states' U.S. senators to avoid the mistakes embodied in the House climate bill and the Obama Administration's tax increases to our industry. Senate Majority Leader Senator Harry Reid reportedly has pushed back consideration of climate legislation to late September to allow Senators time to get their constituents' views during the August recess. It's important that our views be heard.

At the rallies, we will focus our message on two points: the adverse impacts of unsound energy policy (e.g., Waxman-Markey-like legislation, tax increases, and access limitations) on jobs and on consumers' energy costs. And we will call on the Senate to oppose unsound energy policy and "get it right." Recent opinion research that Harris Interactive conducted for API demonstrates that messages on Waxman-Markey-like legislation work extremely well and are very persuasive with the general public and policy influential. After hearing that Waxman-Markey like legislation could increase the costs of gasoline to around $4 and lead to significant job losses, these audiences change their opinions on the bill significantly. Opposition to the bill within the policy influential cohort grew 23 points, from 40% to 63%; with a 19 point increase in those who now "strongly" oppose the legislation. The data clearly demonstrate the softness of support of the current approach and very strong opposition when people are educated about the potential job losses and increased energy costs to all unsound proposals (e.g. Waxman-Markey-like legislation, tax increases, and access limitations).

We have identified 11 states with a significant industry presence and 10 other states where we have assets on the ground. We also have attracted allies from a broad range of interests: the Chamber of Commerce and NAM, the trucking industry, the agricultural sector, small business, and many others, including a significant number of consumer groups, which have pledged to have their membership join in the events in states where they have a strong presence. We also are collaborating closely with the allied oil and natural gas industry associations on these events. While such efforts are never easy and the risk of failure is always present, we must move aggressively in preparation for the post-Labor Day debate on energy, climate and taxes. The measure of success for these events will be the diversity of the participants expressing the same message, as well as turnouts of several hundred attendees. In the 11 states with an industry core, our member company local leadership—including your facility manager's commitment to provide significant attendance—is essential to achieving the participation level that Senators cannot ignore. In addition, please include all vendors, suppliers, contractors, retirees and others who have an interest in our success.

To be clear, API will provide the up-front resources to ensure logistical issues do not become a problem. This includes contracting with a highly experienced management company that has produced successful rallies for presidential campaigns, corporations and interest groups. It also includes coordination with the other interests who share our views on the issues, providing a field coordinator in each state, conducting a comprehensive communications and advocacy activation plan for each state, and serving as central manager for all events. We are asking all API members to assist in these August activities. The size of the company does not matter, and every participant adds to the strength of our collective voice. We need two actions from each participating company.

Actions Needed

Please provide us with the name of one central coordinator for your company's involvement in the rallies. (We will look to this person as your representative to assist the overall effort.) If you will let me know ASAP, we can be in touch quickly and provide that person with additional details about the project. Please indicate to your company leadership your strong support for employee participation in the rallies. (Unfortunately, we are already experiencing some delay from your regional people since they are not yet aware that headquarters supports the effort.) I believe that expression of support to your company leadership is a fundamental predicate to organizing quickly and achieving success in this endeavor.

The list of tentative venues is attached. Please treat this information as sensitive and ask those in your company to do so as well, as some of these places may be subject to change, and we don't want critics to know our game plan. You can assume with confidence that the advocates for Waxman-Markey-like legislation and the critics of oil and gas are going to be very active, particularly during the August recess. Once the list of venues and exact rally dates are determined, we will contact your company's coordinator to distribute the information internally and to coordinate transportation to the venues, if required, for your employees. In the meantime, your company's coordinator could assist us by telling us in which of the venues listed below your company has facilities or employees who can participate.

I look forward to working with you to make the August rally project and the other advocacy steps we are undertaking to deliver the policy outcomes we support with measurable results. Don't hesitate to call me with questions.

All the best,
JACK
Jack N. Gerard
President & CEO
API

Document 13.3 Salmon People, *Umatilla, for the Columbia River Inter-Tribal Fish Commission, Jeremy FiveCrows*

Source: http://www.critfc.org/salmon-culture/we-are-all-salmon-people/, accessed on June 18, 2013. Reprinted by permission of the Columbia River Inter-Tribal Fish Commission.

Native peoples in the Pacific Northwest lost access to their fish resources over the course of the 19th century. This is part of a much broader story of Indians across the continent losing access to land, grass, game, resources, and sacred spaces and struggling to both survive and preserve their cultures. In recent decades, tribes have worked to regain access to lands and resources and pressure state and federal governments to restore access to treaty protected resources. Native peoples have generally taken a strong conservation approach, emphasizing the poverty of reservation lands (Indians were given poor lands, and the best farm and grazing land, as well as that rich in mineral resources, was distributed to non-Indian settlers and businesses) and the need for jobs and a sustainable economy. But they also argue for the importance of restoring cultural traditions. Hence, the Makah Indians of the Olympic Peninsula began hunting whales again in the 1990s, and plains tribes are seeking to restore the bison on the grasslands. Northwest and Alaskan tribal peoples have been key participants in efforts to restore fisheries for cultural, economic, and environmental reasons and in some cases, are far ahead of state and federal government responses to threats posed by climate change.

1. How does the writer of this document employ history, culture, and ecology as part of the definition of place? What role do salmon play in this?
2. What is the importance of fishing, and what are the Columbia River Indians doing to try and restore the salmon?

We are all Salmon People

In Sahaptin, the word for salmon used in sacred ceremonies is "*wy-kan-ush.*" Also in Sahaptin, the word "*pum*" means "people." The tribal cultures in the Columbia River Basin could rightly be called *Wy-Kan-Ush-Pum* or "Salmon People" for how completely these sacred fish shaped their culture, diets, societies, and religions.

Visit the Celilo Longhouse on the banks of the Columbia River during their annual First Salmon feast and you'll probably be struck by how much reverence is paid to the fish. There is the smell of cooking salmon hovering in the air, tantalizing visitors with its savory aroma. The longhouse throbs with ancient songs of thanksgiving for *wy-kan-ush*—songs that have been sung up and down the Columbia River for thousands of years. From the fishnets lying on the ground to the fish stories of tribal fishers, there is no question that salmon are at the center of this gathering. It is difficult if not impossible to come away from the First Salmon ceremony without seeing the salmon in a different light. Salmon are one of the most important aspects of tribal culture.

Cultures Based on Salmon

To call salmon a staple of the tribal diet would be an understatement. Historically, the typical tribal member ate almost a pound of salmon every day, but salmon represented much more than a source of nutrition—they shaped our societies and our religions.

From a tribal legend, we learn that when the Creator was preparing to bring forth people onto the earth, He called a grand council of all creation. From them, He asked for a gift for these new creatures—a gift to help the people survive, since they would be quite helpless and require much assistance from them all. The very first to come forward was Salmon, who offered his body to feed the people. The second to come forward was Water, who promised to be the home to the salmon. In turn, everyone else gathered at the council gave the coming humans a gift, but it is significant that the very first two were Salmon and Water. In accordance with their sacrifice, these two receive a place of honor at traditional feasts throughout the Columbia Basin. These ceremonies always begin with a blessing on and the drinking of water, followed by a prayer of thanksgiving on and the serving of *wy-kan-ush*, the salmon. This ceremony reinforces the central role that salmon and water play in the health of Indian people and their culture.

Fishing: A Fundamental Tribal Right

Fishing for salmon is just as integral an aspect of tribal culture as consuming it. For many, fishing trips shape their appreciation for the land, the waters, and the salmon. It is without question that salmon are worth our time, energy, and sometimes even risking our lives. That is why it was troubling when the number of salmon that returned up the Columbia River each year grew smaller and smaller. By the 1960s the numbers had dipped so low that tribal concern turned into alarm that we might lose our sacred fish. At that time, the tribes did not have the political voice or power to fight the decline.

Fortunately those days are past. Today the tribes are doing everything in their power to make sure that salmon return to as many of their traditional waters as they can. Enormous amounts of resources are being poured into this effort, and tribal youth are joining the fight to save salmon. Every year, more and more tribal members are becoming fish biologists, environmental engineers, and other scientists who are offering their minds as well as their hearts for the protection of the salmon, the water, and ultimately, their traditional way of life.

***Wy-Kan-Ush-Pum:* We Are All Salmon People**

We can't completely restore salmon alone, but the power of *wy-kan-ush* is reason for hope. The world over, salmon affect the cultures of the people in which they come in contact. The wildly different traditional cultures of Japanese Ainu, Pacific Northwest Coastal tribes, the Norwegian coastal areas, and the Russian Far East each have salmon returning to their lands and each share a reverence and gratitude for the bounty that salmon provide. The modern Pacific Northwest is no different. Salmon have shaped the culture of the newcomers to this region just as they shaped tribal cultures before them. Salmon are the icon of this place. They are valued as food, as a resource, and as a representation of the wildness and wilderness for which the Pacific Northwest is known. They shape our land-use policies and power grid. Whether they realize it or not, every single person in the Northwest is a *Wy-Kan-Ush-Pum*. We are all Salmon People. Let us all work together to protect and restore salmon—this fish that unites us.

Notes

1 Kevin Grandia, "Senate's Chief Climate Denier Makes Copenhagen Cameo," Grist.org, 17 December 2009. Accessed on 17 February 2014, grist.org/article/2009-12-17-senator-inhofe-climate-denier-copenhagen/
2 Ibid.
3 Ibid.
4 Diane Urbani de la Paz, "'Prayers Answered' National Figures Join Peninsula Leaders at Dam Ceremony," *Peninsula Daily News* (Port Angeles, Washington) 18 September 2011.
5 Richard West Sellars, *Preserving Nature in the National Parks: A History* (New Haven, CT: Yale University Press, 2009), 300, 301.
6 My father and I visited Yellowstone National Park in the Autumn of 1993, two years before the wolf releases. On a four-day trip we saw seven coyotes. At one point my father was taking a nap by the Madison River and a coyote woke him by pulling his hat off his face. They then engaged in a bizarre "dances with wolves" sort of scene where my father put his hat on his face then the coyote pulled it off. I have been to Yellowstone National Park approximately six times since the wolf reintroduction and have since seen only one coyote.
7 The various names refer to the same species but within a particular geographic and ecosystem context. For example, panthers are generally associated with the Southeast and wetland and swamp ecosystems. Pumas generally refer to the same species in a desert setting in the West and Southwest. Cougar and mountain lion seems to be interchangeable for this species in many regions, but in the Midwest *cougar* is more commonly used, while in the Northwest, *mountain lion* has more currency.
8 Wendell Berry, *The Gift of Good Land: Further Essays Cultural and Agricultural* (Berkeley, CA: Counterpoint Press, 2009).

Further Reading

Crane, Jeff. *Finding the River: An Environmental History of the Elwha.* Corvallis: Oregon State University Press, 2011.

Conkin, Paul K. *A Revolution Down on the Farm: The Transformation of American Agriculture since 1929.* Lexington: The University Press of Kentucky, 2008.

Fagin, Dan. *Toms River: A Story of Science and Salvation*. New York: Bantam, 2013.
Fiege, Mark. *The Republic of Nature: An Environmental History of the United States*. Seattle: University of Washington Press, 2012.
Greene, Jeffrey. *Water from Stone: The Story of Selah, Bamberger Ranch Preserve*. College Station: Texas A&M University Press, 2007.
Lowry, William R. *Dam Politics: Restoring America's Rivers*. Washington, DC: Georgetown University Press, 2003.
Oreskes, Naomi and Erik M. Conway. *Merchants of Doubt: How a Handful of Scientists Obscured the Truth on Issues from Tobacco Smoke to Global Warming*. New York: Bloomsbury Press, 2011.
Orr, David. *Down to the Wire: Confronting Climate Collapse*. New York: Oxford University Press, 2012.
———. *Earth in Mind: On Education, Environment, and the Human Prospect*. Washington, DC: Island Press, 2004.
Pilkey, Orrin H. and Keith C. Pilkey. *Global Climate Change: A Primer*. Durham, NC: Duke University Press, 2011.
Pollan, Michael. *The Omnivore's Dilemma: A Natural History of Four Meals*. New York: Penguin Books, 2006.
Sellars, Richard West. *Preserving Nature in the National Parks: A History*. New Haven, CT: Yale University Press, 2009.
Sellers, Christopher C. *Crabgrass Crucible: Suburban Nature & the Rise of Environmentalism in Twentieth Century America*. Chapel Hill: The University of North Carolina Press, 2012.
Summitt, April R. *Contested Waters: An Environmental History of the Colorado River*. Boulder: University Press of Colorado, 2013.
Weart, Spencer R. *The Discovery of Global Warming*. Cambridge, MA: Harvard University Press, 2008.
Wilber, Tom. *Under the Surface: Fracking, Fortunes, and the Fate of the Marcellus Shale*. Ithaca, NY: Cornell University Press, 2012.

Epilogue

The Greatest Peril of Abundance

Climate change emerged as the great environmental crisis of the late 20th and early 21st century. While the science of climate change is improving quickly, there is some risk to predicting global temperature increases and corresponding changes in climate patterns and environment. However, many of these changes are already clearly underway and support many of the predictions of the last two decades. Global temperatures are predicted to increase by 2 to 4 degrees Celsius or 3.6 to 7.2 degrees Fahrenheit by the end of the century. Even with immediate, drastic cuts to greenhouse gas emissions (which isn't happening), it is almost certain that the world's oceans will rise by at least 3 feet within the next century with some predictions of sea rise of 6 or 7 feet. Downtown Miami floods on a regular basis now, and globally other coastal areas are already under water. What does this sea rise mean for society and nature? Besides the potential loss of coastal areas and the prohibitive costs involved in trying to preserve cities in the face of invading waters is the loss of key coastal habitat. Coastal estuaries are critical nurseries for any number of species on every coastline of the country and world. They are vulnerable to even small rises in ocean levels. The other consequences of climate change are manifold: the expansion of the range of tropical diseases such as malaria and dengue fever; radical changes to the hydrological cycles of rivers globally, affecting fisheries and water supplies; and transformations to growing seasons and weather, influencing crop production around the world, are merely a few of the problems. Also, there is the possibility of shifts in ocean currents as well as atmospheric changes in the jet stream. Worsening droughts and large episodic rain events with extensive flooding followed by long periods with no rain, winter freezes in subtropical regions, and superstorms are just a few of the problems that are already occurring and will surely worsen.

Climate change is consistently referred to as a "wicked problem." This helps explain the slow response to the crisis. The warming of air and water temperatures with resulting climatic and ecosystem change is not easily seen by many observers, particularly in the United States. There is no clear disaster (so far) that scares people and catches their attention in the way the Santa Barbara Oil Spill did in 1969. At the same time that there has been some limited, legitimate scientific debate about the speed and impacts of climate change, business interests and wealthy elites opposed to government regulation have sown confusion in the minds of the American people. They have intentionally done this so that acknowledgement of, one, the reality of climate change and, two, the role of humans in causing it has been rendered much more difficult. That has obviously prevented meaningful reforms and regulations. A wicked problem is also one that is a symptom of a whole bevy of other deeply rooted issues that are difficult to confront and fix, such as consumption and reliance on "cheap" energy. Finally, competing interests view the problem and its resolution in different ways based on self-interest; this adds to stasis and prevents action. What makes the climate change issue such a wicked problem is that the longer we wait to act, the worse the crisis will be. There is a threshold beyond which it will be impossible to reverse global climate and ecosystem change on a devastating scale. While there is debate over where the threshold is, if the permafrost melts, the Greenland ice sheet melts, and the Arctic melts, the amount of fresh water flowing into the oceans and the vast quantities of carbon dioxide and methane added to the atmosphere will bring irrevocable, ruinous change. One way of tackling this wicked problem is to gain a better comprehension of what climate change actually is and attempt to understand the impacts.

The major greenhouse gases are carbon dioxide (CO_2), methane (CH_4), water vapor (H_2O), and nitrous oxide (N_2O). While water vapor constitutes the largest portion of the greenhouse gases, its affect has been relatively constant through history. However, CO_2 has increased dramatically since the dawn of the industrial revolution, measuring approximately 280 parts per million (ppm) at that time, 315 ppm in 1958, and increasing to approximately 390 ppm by 2010. It is still growing at an approximate rate of 2 ppm per year. The general agreement by climatologists is that 350 ppm is the tipping point beyond which global warming worsens and increases its pace. Warming is accelerated by the presence and increase of another greenhouse gas, methane. While constituting a small portion of the mix of greenhouse gases, it is 23 times as strong as carbon dioxide. Fortunately methane's average lifespan is approximately 12 years versus the minimum of a millennium for carbon dioxide. Increased global warming gases create both warming and positive feedback, an increase of water vapor in the atmosphere leading to increased heat retention and further acceleration of atmospheric heating.

The impacts of climate change are myriad, complex, sometimes confusing, and difficult to forecast. Generally speaking, it is already causing widespread global glacial retreat and disappearance. Many glaciers in the North Cascades of Washington State have disappeared already, while glaciers in Glacier National Park shrunk by approximately 60 percent from 1966 to 2011. Several of these glaciers are expected to disappear by 2030. Himalayan Glaciers are turning into water faster than any others in the world with some retreating an average of 50 meters a year. Almost all of the Himalayan glaciers are expected to be gone by 2050. Melting in the Arctic is endangering polar bears and other species, while the Antarctic ecosystems are destabilized and changing by a mixed pattern of melting, increased ice, and contact between distinct penguin species normally physically isolated from each other. Another devastating problem is the loss of permafrost under the soil in Alaska, northern Canada, and Siberia. This layer of ice that has been in place for millennia is now melting, causing numerous problems such as houses collapsing, road failure, and so forth. More important, and illustrative of the problem of climate change building on itself, as the permafrost layer melts, it exposes trapped plant material to decomposition. This frees even more methane into the atmosphere even as carbon dioxide molecules trapped in the ice are also released, warming the atmosphere ever more rapidly.

The world's oceans conceal another huge, potentially catastrophic change as temperatures increase. Methane ices are stored at the bottom of the sea, typically below 1,000 feet. In colder bodies of water it can be found as shallow as 300 feet. These deposits originate in decaying organic material that is essentially frozen in place by both cold water temperatures and high water pressure. Much of this methane ice is located on continental shelves, increasing their vulnerability to disturbance from oil drilling. Also, warmer overall waters and shifts in warm water currents threaten to release vast amounts of methane gas. For example, the Deepwater Horizon oil spill in the Gulf of Mexico in 2010 released a large quantity of gas, and plumes of methane gas bubbles have been observed on the Arctic continental shelf. An estimated 8 million tons of methane are currently discharged annually off the coast of Siberia, roughly equivalent to the assumed amount of total methane released by all the world's oceans just a few years ago.

Ocean acidification is yet another unspooling crisis whose pace and dimensions may be beyond the will and ability of humans to solve or mitigate. Ocean acidity has increased 30 percent since the start of the industrial revolution with some predictions of increased acidification of 100 percent or more by 2100. This shift has ramifications across the ocean's foods chains, on species such as oysters, lobsters, shrimp, types of plankton, and coral. The historical chemical balance in the world's oceans, 8.0 ph. before the industrial revolution, allows these species to extract calcium carbonate out of the water and use it to build their shells. Increased acidity is interfering with the ability of many of these species to do this. Many organisms' shells are growing thinner, rendering them more fragile and vulnerable; recent collapses in some west coast oyster fisheries are believed to be related to increased acidification. Important bacteria that serve a critical function in the ocean system by breaking down chemical compounds are vulnerable to even slight increases in acidity. Many species of plankton, a cornerstone of the ocean food chain, have already been adversely affected by changes in the chemical composition of the oceans. Diminishment or loss of integral species such as these ripple across the food chain from the smallest of fish to whales and humans.

A tiny shellfish little known by most people, and suffering from climate change and acidification, holds the potential to cause global environmental and food production disruption. Pterepods are tiny snails found throughout the world's oceans and are an important food source for cod, whale, and salmon as well as zooplankton and other species. Their thin shells of aragonite are being dissolved by acidic waters. A mere 10 percent reduction of pteropods in Alaska's salmon rich waters would lead to a 20 percent weight reduction of some species of salmon. This not only means a reduced harvest for human use and less nutrition for predator species such as bears and orca, it also means that salmon might not have the strength or energy to reach their spawning sites, leaving large areas of spawning habitat underutilized or unused. Extrapolate that 10 percent reduction across the numerous species dependent on this species, or imagine a likely much larger decline, and the consequences are grim indeed. It is even more difficult to conceptualize the consequences if this species went extinct. They would be ecologically and economically devastating.

Another damaging blow to ecosystems and the food chain arises from the impact of ocean acidification and the bleaching of the oceans' reefs. The basis of ecosystems that constitute a third of marine life and an even larger proportion of shallow water organisms, reefs occupy less than 1 percent of ocean habitat. More than 4,000 fish species inhabit coral reefs, and a healthy one generates approximately 35 tons of fish per year per square kilometer. In addition to supporting a wide diversity and abundance of species, these reefs are the cornerstones of fishing and tourism economies around the world and produce large amounts of food necessary for human survival and for other species. There are a number of factors in the death of coral reefs, but the key problem is warming water temperatures. Corals survive within a narrow temperature range. When sea water warms by a few degrees, the coral reject the algae with which they exist symbiotically and that give them their brilliant colors. Since these algae produce carbohydrates through photosynthesis and feed the coral, their rejection by the coral means death for the reefs. The death of the coral reefs in the world's oceans would be cataclysmic for natural diversity, biological abundance, and humans.

Back on land, forest fires in the United States have already become much worse in recent years. The season lasts two months longer and destroys twice as much land as four decades ago. The fires themselves rage hotter, travel faster, and are much more dangerous. One conflagration in the summer of 2013 was reported to have crossed a snowfield, and the Yarnell Fire of June 2013 killed 19 hotshot firefighters, the worst firefighting loss of life since 1933. The reasons for the wildfire problem are multiple and tied to climate change in numerous ways. Understanding these myriad impacts helps illuminate the overlapping and complex impacts of climate change in other areas as well. The American West is in perpetual drought at this point, and a shift in the Gulf jet stream is keeping moist air from moving from the Gulf of Mexico into the region. Periodic heavy moisture makes the situation more flammable because of the rapid growth of trees and underbrush that then transforms into highly flammable tinder as it quickly dries. Increased air temperatures, in addition to extended droughts throughout the West, render the forests drier. This increases the rate of ignition from lightning strikes. Moreover, the burning of forests accelerates climate change because trees capture carbon; when they burn, the carbon is released, adding to the load in the atmosphere and increasing the speed of climate change.

Adding to these problems are the infestations and expansion of pine and spruce bark beetles, which have destroyed 42 million acres of forest since 1996. These insects have become even more destructive because winters at higher elevations are no longer cold enough to kill them and stop expansion of the population. Their range and elevation expand as winters grow warmer and shorter, and dead trees catch fire more easily and burn faster and hotter, increasing the speed and intensity of fires. Destruction of the trees increases the absorption of heat into the soil and leaves snowpack exposed to sunlight, increasing the rate of snowmelt in winter and thereby reducing water for spring and summer. This not only means a drier forest but also an interruption and skewing of river hydrological cycles that is catastrophic for a variety of fish. At the same time, streams are degraded or lost to erosion and overheating. This results in the further loss of the numerous species dependent on those fish, ranging from insects in the river that consume the carcasses to predators such as eagles and bears and so many more.

The United States and many other industrial nations have been lackadaisical in their response to the threat, but in the United States the topic has been tainted with a hysterical edge. This arises from several issues but most likely from the implicitly understood fact that appropriately responding to climate change will require large expenditures and dramatic changes in lifestyle. British novelist Ian McEwan captures the conundrum quite cogently through a character in his novel *Solar*:

> She approved of his mission and loyally read climate-change stories in the press. But she told him once that to take the matter seriously would be to think about it all the

> time. Everything else shrank before it. And so, like everyone she knew, she could not take it seriously, not entirely. Daily life would not permit it.[1]

Of those most concerned with changes in lifestyle, business practices, and the economic impacts therein are the petroleum companies. Moreover, they understand that if Americans get behind the effort to slow climate change, it will result in increased government power, higher taxes, and a series of regulations that will interfere with their economic interests. Many of these companies, particularly Exxon Mobil, Chevron, and Amoco, have spent hundreds of millions of dollars in efforts to lobby against cap-and-trade regulation as well as finance research and publicity that reject the reality of climate change. These companies have funded astroturf operations to try and sow as much chaos, confusion, and ambiguity around climate change as possible. Their ability to coarsen the debate makes this a truly wicked problem and prevents meaningful reform.

The art of astroturfing was created in the fight by big tobacco against the scientific evidence of smoking's negative health impacts. A fake grassroots group given the name the Advancement for Sound Science Coalition was created by the public relations firm APCO worldwide and funded by tobacco giant Philip Morris. In order to avoid obvious criticism that they functioned simply as hired guns for tobacco, the group also targeted government regulations and climate change science. One tactic developed by this group was to refer to peer-reviewed science that damaged their clients as "junk science" while calling the results produced by tobacco funded labs "sound science." The central strategy of astroturfing is to look like a grassroots group and create grassroots activism while also sowing as much confusion and ambiguity as possible through terms such as junk science and targeted media campaigns. As noted in a memo by the Brown and Williamson tobacco company, "doubt is our product since it is the means of competing with the 'body of fact' that exists in the mind of the general public. It is also the means of establishing a controversy."[2]

Exxon Mobil embraced this strategy of creating maximum uncertainty to prevent legislation or regulation. The company ponied up $16 million between 1998 and 2005 to denier organizations. Although stating its intention to stop funding such organizations in 2008, money continued to flow from the corporation to select groups. Astroturfing organizations also receive support from Koch Energy. The Koch brothers have funded numerous efforts to limit government regulation and power and have attracted the media spotlight in recent years because of their support of the Tea Party. While that may seem unrelated, the disturbances created by Tea Party politicking, the stasis of the political status quo, and the coarsening of rhetoric, serves industry interests by decreasing the odds of meaningful climate change reform occurring.

> A report in March 2010 revealed that Koch Industries, one of the largest and wealthiest private corporations in America, is a leading contributor to global warming deniers and groups opposing clean energy reform . . . between 2005 and 2008 Exxon Mobil spent $8.9 million, while foundations controlled by Koch Industries doled out $24.9 million to the climate denial lobby. Recipients of the Koch money include the Heritage Foundation ($1,620,000), the Cato Institute ($1,028,400), and the Atlas Economic Research Foundation ($113,800).[3]

Business interests learned a lesson from the era of environmental consensus. Fearing regulation and higher taxes, they have created enough confusion and ambiguity among the American people to prevent unity on the most pressing environmental problem of this era. Moreover, when business interests do acknowledge the possibility of climate change, they ask the question, are prevention measures worth the economic cost? Knowing Americans' interest in comfort and leisure pursuits, opponents often emphasize threats to resource and monetary abundance. The potential economic costs, impacts on lifestyle, a stalling or decline in the gross domestic product, inflation, and so forth have become the shield against meaningful climate change action. The ideology of abundance centrally located in the American identity makes change difficult because for the first time in American history, the nation's citizens will have to intentionally reject abundance and consumption, something no political leader is willing to ask them to do.

Climate change does not mean the end of the world as we know it, nor does it mean the end of the Earth. Some activists have viewed and described the threat in this apocalyptic manner, and climate deniers have used that extreme view as a straw man, an end of the world harum-scarum, in order to denounce the science and activism of those seeking to stop or slow climate change. At the same time many activists and organizations attempting to inform and mobilize support for initiatives to slow warming of the planet have emphasized mitigation, small changes in consumption patterns, and techno-optimistic solutions. Baby steps might seem like the only way to get buy-in from citizens so used to life in a land of plenty, but Americans are left struggling to understand the real impact and the best response. What the warming of the earth does mean is widespread, hard to predict, transformative changes to climate, ecosystems, and human communities.

If the oceans rise 6 feet by the end of the century and the coral reefs die and the tropical ocean species move north while cold-water species die or are displaced, humanity can adjust to this. Drier deserts, wetter forests and jungle, less snow, and stronger hurricanes are all changes that will result in dislocations of populations, the spread of disease, famines, economic crises, and the transformation of the natural landscape. This too can be addressed or lived with. Scientists, industry, and government will create new hybrid crops, means for moving water to deserts, massive desalination plants, heat absorbing satellite panels, and expanded fish farming operations and in the process bring nature further into service of humanity's needs. Climate change at a minimum simply accelerates the homogenization of world environments and man's dominion over a severely reduced natural world. While the earth will not end, much of what we love and need from nature will disappear or change dramatically.

Americans have benefited from an abundance of natural resource wealth and land over their history. In the narrative sweep of the American story, expansion and growth are consistent tropes told again and again to emphasize the nation's greatness and divine ordination. Much of that economic growth and the corresponding military and political power came at the cost of displacing other peoples and great ecological destruction. To not acknowledge this truth or to simply refer to it as transformation is to dodge the question of American culpability for environmental degradation and destruction. To do this is to be deeply irresponsible in a time when we and the rest of the world face onerous environmental and climatic tests and challenges. The long 20th century includes a successful history of responses to this damage and repairs to the environment nationwide that were frankly inconceivable in an earlier time. Some of these reforms and repairs have even moved the nation closer to a land ethic, to reconstructing a healthier relationship between humanity and nature. However, it seems the limit to far-reaching reform is reached when the economic cost appears too high. This is where we sit today. The intersection of economy and nature is still a busy and dangerous place. The scale of this crisis begs bigger questions: If nature can no longer subsidize American capitalism and culture, what happens next? If consumption and the assumption of never-ending economic growth exceeding population increase are the root cause of this crisis, then what are the possible solutions? As always, the future of America will rely on nature and our ability to strike the right balance between ecological and economic health.

Notes

1 Ian McEwan, *Solar* (New York: Anchor Books, 2010), 191.
2 Quoted in Orrin H. Pilkey and Keith C. Pilkey, *Global Climate Change: A Primer* (Durham, NC: Duke University Press, 2011), 43.
3 Ibid., 49.

Index